# Applied
# Nonlinear
# Control

# Applied Nonlinear Control

JEAN-JACQUES E. SLOTINE

*Massachusetts Institute of Technology*

WEIPING LI

*Massachusetts Institute of Technology*

Prentice Hall
Englewood Cliffs, New Jersey 07632

*Library of Congress Cataloging-in-Publication Data*

Slotine, J.-J. E. (Jean-Jacques E.)
   Applied nonlinear control / Jean-Jacques E. Slotine, Weiping Li

      p.   cm.
   Includes bibliographical references.
   ISBN 0-13-040890-5
   1, Nonlinear control theory.   I. Li, Weiping.   II. Title.
QA402.35.S56 **1991**                                    90-33365
629.8′312—dc20                                           CIP

Editorial/production supervision and
      interior design: Jennifer Wenzel
Cover design: Karen Stephens
Manufacturing Buyer: Lori Bulwin

Printed in the United States of America

10   9

ISBN   0-13-040890-5

Prentice-Hall International (UK) Limited, *London*
Prentice-Hall of Australia Pty. Limited, *Sydney*
Prentice-Hall Canada Inc., *Toronto*
Prentice-Hall Hispanoamericana, S.A., *Mexico*
Prentice-Hall of India Private Limited, *New Delhi*
Prentice-Hall of Japan, Inc., *Tokyo*
Simon & Schuster Asia Pte. Ltd., *Singapore*
Editora Prentice-Hall do Brasil, Ltda., *Rio de Janeiro*

*To Our Parents*

# Contents

## 9.   Control of Multi-Input Physical Systems      392

# Preface

In recent years, the availability of powerful low-cost microprocessors has spurred great advances in the theory and applications of nonlinear control. In terms of theory, major strides have been made in the areas of feedback linearization, sliding control, and nonlinear adaptation techniques. In terms of applications, many practical nonlinear control systems have been developed, ranging from digital "fly-by-wire" flight control systems for aircraft, to "drive-by-wire" automobiles, to advanced robotic and space systems. As a result, the subject of nonlinear control is occupying an increasingly important place in automatic control engineering, and has become a necessary part of the fundamental background of control engineers.

This book, based on a course developed at MIT, is intended as a textbook for senior and graduate students, and as a self-study book for practicing engineers. Its objective is to present the fundamental results of modern nonlinear control while keeping the mathematical complexity to a minimum, and to demonstrate their use and implications in the design of practical nonlinear control systems. Although a major motivation of this book is to detail the many recent developments in nonlinear control, classical techniques such as phase plane analysis and the describing function method are also treated, because of their continued practical importance.

In order to achieve our fundamental objective, we have tried to bring the following features to this book:

- **Readability:** Particular attention is paid to the readability of the book by carefully organizing the concepts, intuitively interpreting the major results, and selectively using the mathematical tools. The readers are only assumed to have had one introductory control course. No mathematical background beyond ordinary differential equations and elementary matrix algebra is required. For each new result, interpretation is emphasized rather than mathematics. For each major result, we try to ask and answer the following key questions: What does the result intuitively and physically mean? How can it be applied to practical problems? What is its relationship to other theorems? All major concepts and results are demonstrated by examples. We believe that learning and generalization from examples are crucial for proficiency in applying any theoretical result.

- **Practicality:** The choice and emphasis of materials is guided by the basic objective of making an engineer or student capable of dealing with practical control problems in industry. Some results of mostly theoretical interest are not included. The selected materials, in one way or another, are intended to allow readers to gain insights into the solution of real problems.

- **Comprehensiveness:** The book contains both classical materials, such as Lyapunov analysis and describing function techniques, and more modern topics such as feedback linearization, adaptive control, and sliding control. To facilitate digestion, asterisks are used to indicate sections which, given their relative complexity, can be safely skipped in a first reading.

- **Currentness:** In the past few years, a number of major results have been obtained in nonlinear control, particularly in nonlinear control system design and in robotics. It is one of the objectives of this book to present these new and important developments, and their implications, in a clear, easily understandable fashion. The book can thus be used as a reference and a guide to the active literature in these fields.

The book is divided into two major parts. Chapters 2-5 present the major *analytical* tools that can be used to study a nonlinear system, while chapters 6-9 treat the major nonlinear controller *design* techniques. Each chapter is supplied with exercises, allowing the reader to further explore specific aspects of the material discussed. A detailed index and a bibliography are provided at the end of the book.

The material included exceeds what can be taught in one semester or self-learned in a short period. The book can be studied in many ways, according to the particular interests of the reader or the instructor. We recommend that a first reading include a detailed study of chapter 3 (basic Lyapunov theory), sections 4.5-4.7 (Barbalat's lemma and passivity tools), section 6.1 and parts of sections 6.2-6.4 (feedback linearization), chapter 7 (sliding control), sections 8.1-8.3 and 8.5 (adaptive control of linear and nonlinear systems), and chapter 9 (control of multi-input physical systems). Conversely, sections denoted with an asterisk can be skipped in a first reading.

Many colleagues, students, and friends greatly contributed to this book through stimulating discussions and judicious suggestions. Karl Hedrick provided us with continued enthusiasm and encouragement, and with many valuable comments and suggestions. Discussions with Karl Aström and Semyon Meerkov helped us better define the tone of the book and its mathematical level. Harry Asada, Jo Bentsman, Marika DiBenedetto, Olav Egeland, Neville Hogan, Marija Ilic, Lars Nielsen, Ken Salisbury, Sajhendra Singh, Mark Spong, David Wormley, and Dana Yoerger provided many useful suggestions and much moral support. Barbara Hove created most of the nicer drawings in the book; Günter Niemeyer's expertise and energy was invaluable in setting up the computing and word processing environments; Hyun Yang greatly helped with the computer simulations; all three provided us with extensive technical and editorial comments. The book also greatly benefited from the interest and enthusiasm of many students who took the course at MIT.

Partial summer support for the first author towards the development of the book was provided by Gordon Funds. Finally, the energy and professionalism of Tim Bozik and Jennifer Wenzel at Prentice-Hall were very effective and highly appreciated.

**Jean-Jacques E. Slotine**
**Weiping Li**

# Applied
# Nonlinear
# Control

# Chapter 1
# Introduction

The subject of nonlinear control deals with the analysis and the design of nonlinear control systems, *i.e.*, of control systems containing at least one nonlinear component. In the analysis, a nonlinear closed-loop system is assumed to have been designed, and we wish to determine the characteristics of the system's behavior. In the design, we are given a nonlinear plant to be controlled and some specifications of closed-loop system behavior, and our task is to construct a controller so that the closed loop system meets the desired characteristics. In practice, of course, the issues of design and analysis are intertwined, because the design of a nonlinear control system usually involves an iterative process of analysis and design.

This introductory chapter provides the background for the specific analysis and design methods to be discussed in the later chapters. Section 1.1 explains the motivations for embarking on a study of nonlinear control. The unique and rich behaviors exhibited by nonlinear systems are discussed in section 1.2. Finally, section 1.3 gives an overview of the organization of the book.

## 1.1  Why Nonlinear Control ?

Linear control is a mature subject with a variety of powerful methods and a long history of successful industrial applications. Thus, it is natural for one to wonder why so many researchers and designers, from such broad areas as aircraft and spacecraft control, robotics, process control, and biomedical engineering, have recently showed an active interest in the development and applications of nonlinear control methodologies. Many reasons can be cited for this interest:

• **Improvement of existing control systems**: Linear control methods rely on the key assumption of small range operation for the linear model to be valid. When the required operation range is large, a linear controller is likely to perform very poorly or to be unstable, because the nonlinearities in the system cannot be properly compensated for. Nonlinear controllers, on the other hand, may handle the nonlinearities in large range operation directly. This point is easily demonstrated in robot motion control problems. When a linear controller is used to control robot motion, it neglects the nonlinear forces associated with the motion of the robot links. The controller's accuracy thus quickly degrades as the speed of motion increases, because many of the dynamic forces involved, such as Coriolis and centripetal forces, vary as the square of the speed. Therefore, in order to achieve a pre-specified accuracy in robot tasks such as pick-and-place, arc welding and laser cutting, the speed of robot motion, and thus productivity, has to be kept low. On the other hand, a conceptually simple nonlinear controller, commonly called computed torque controller, can fully compensate the nonlinear forces in the robot motion and lead to high accuracy control for a very large range of robot speeds and a large workspace.

• **Analysis of hard nonlinearities:** Another assumption of linear control is that the system model is indeed linearizable. However, in control systems there are many nonlinearities whose discontinuous nature does not allow linear approximation. These so-called "hard nonlinearities" include Coulomb friction, saturation, dead-zones, backlash, and hysteresis, and are often found in control engineering. Their effects cannot be derived from linear methods, and nonlinear analysis techniques must be developed to predict a system's performance in the presence of these inherent nonlinearities. Because such nonlinearities frequently cause undesirable behavior of the control systems, such as instabilities or spurious limit cycles, their effects must be predicted and properly compensated for.

• **Dealing with model uncertainties:** In designing linear controllers, it is usually necessary to assume that the parameters of the system model are reasonably well known. However, many control problems involve uncertainties in the model parameters. This may be due to a slow time variation of the parameters (*e.g.*, of ambient air pressure during an aircraft flight), or to an abrupt change in parameters (*e.g.,* in the inertial parameters of a robot when a new object is grasped). A linear controller based on inaccurate or obsolete values of the model parameters may exhibit significant performance degradation or even instability. Nonlinearities can be intentionally introduced into the controller part of a control system so that model uncertainties can be tolerated.  Two classes of nonlinear controllers for this purpose are robust controllers and adaptive controllers.

• **Design Simplicity:** Good nonlinear control designs may be simpler and more intuitive than their linear counterparts. This *a priori* paradoxical result comes from the fact that nonlinear controller designs are often deeply rooted in the physics of the plants. To take a very simple example, consider a swinging pendulum attached to a hinge, in the vertical plane.  Starting from some arbitrary initial angle, the pendulum will oscillate and progressively stop along the vertical.  Although the pendulum's behavior could be analyzed close to equilibrium by linearizing the system, physically its stability has very little to do with the eigenvalues of some linearized system matrix: it comes from the fact that the total mechanical energy of the system is progressively dissipated by various friction forces (*e.g.*, at the hinge), so that the pendulum comes to rest at a position of minimal energy.

There may be other related or unrelated reasons to use nonlinear control techniques, such as cost and performance optimality.  In industrial settings, ad-hoc extensions of linear techniques to control advanced machines with significant nonlinearities may result in unduly costly and lengthy development periods, where the control code comes with little stability or performance guarantees and is extremely hard to transport to similar but different applications. Linear control may require high quality actuators and sensors to produce linear behavior in the specified operation range, while nonlinear control may permit the use of less expensive components with nonlinear characteristics.  As for performance optimality, we can cite bang-bang type controllers, which can produce fast response, but are inherently nonlinear.

Thus, the subject of nonlinear control is an important area of automatic control. Learning basic techniques of nonlinear control analysis and design can significantly enhance the ability of a control engineer to deal with practical control problems effectively.  It also provides a sharper understanding of the real world, which is inherently nonlinear.  In the past, the application of nonlinear control methods had been limited by the computational difficulty associated with nonlinear control design and analysis. In recent years, however, advances in computer technology have greatly relieved this problem.  Therefore, there is currently considerable enthusiasm for the research and application of nonlinear control methods.  The topic of nonlinear control design for large range operation has attracted particular attention because, on the one hand, the advent of powerful microprocessors has made the implementation of nonlinear controllers a relatively simple matter, and, on the other hand, modern technology, such as high-speed high-accuracy robots or high-performance aircrafts, is demanding control systems with much more stringent design specifications. Nonlinear control occupies an increasingly conspicuous position in control engineering, as reflected by the ever-increasing number of papers and reports on nonlinear control research and applications.

## 1.2   Nonlinear System Behavior

Physical systems are inherently nonlinear. Thus, all control systems are nonlinear to a certain extent. Nonlinear control systems can be described by nonlinear differential equations. However, if the operating range of a control system is small, and if the involved nonlinearities are smooth, then the control system may be reasonably approximated by a linearized system, whose dynamics is described by a set of linear differential equations.

### NONLINEARITIES

Nonlinearities can be classified as *inherent (natural)* and *intentional (artificial)*. Inherent nonlinearities are those which naturally come with the system's hardware and motion. Examples of inherent nonlinearities include centripetal forces in rotational motion, and Coulomb friction between contacting surfaces. Usually, such nonlinearities have undesirable effects, and control systems have to properly compensate for them. Intentional nonlinearities, on the other hand, are artificially introduced by the designer. Nonlinear control laws, such as adaptive control laws and bang-bang optimal control laws, are typical examples of intentional nonlinearities.

Nonlinearities can also be classified in terms of their mathematical properties, as *continuous* and *discontinuous*. Because discontinuous nonlinearities cannot be locally approximated by linear functions, they are also called "hard" nonlinearities. Hard nonlinearities (such as, *e.g.*, backlash, hysteresis, or stiction) are commonly found in control systems, both in small range operation and large range operation. Whether a system in small range operation should be regarded as nonlinear or linear depends on the magnitude of the hard nonlinearities and on the extent of their effects on the system performance. A detailed discussion of hard nonlinearities is provided in section 5.2.

### LINEAR SYSTEMS

Linear control theory has been predominantly concerned with the study of linear time-invariant (LTI) control systems, of the form

$$\dot{\mathbf{x}} = \mathbf{A}\,\mathbf{x} \qquad\qquad (1.1)$$

with $\mathbf{x}$ being a vector of states and $\mathbf{A}$ being the system matrix. LTI systems have quite simple properties, such as

- a linear system has a *unique equilibrium point* if **A** is nonsingular;

- the equilibrium point is stable if all eigenvalues of **A** have negative real parts, *regardless of initial conditions*;

- the transient response of a linear system is composed of the natural modes of the system, and the general solution can be solved analytically;

- in the presence of an external input **u**(*t*), *i.e.*, with

$$\dot{\mathbf{x}} = \mathbf{A}\mathbf{x} + \mathbf{B}\mathbf{u} \tag{1.2}$$

the system response has a number of interesting properties. First, it satisfies the *principle of superposition*. Second, the asymptotic stability of the system (1.1) implies bounded-input bounded-output stability in the presence of **u**. Third, a sinusoidal input leads to a sinusoidal output of the same frequency.

## AN EXAMPLE OF NONLINEAR SYSTEM BEHAVIOR

The behavior of nonlinear systems, however, is much more complex. Due to the lack of linearity and of the associated superposition property, nonlinear systems respond to external inputs quite differently from linear systems, as the following example illustrates.

**Example 1.1:** A simplified model of the motion of an underwater vehicle can be written

$$\dot{v} + |v| \, v = u \tag{1.3}$$

where $v$ is the vehicle velocity and $u$ is the control input (the thrust provided by a propeller). The nonlinearity $|v| \, v$ corresponds to a typical "square-law" drag.

Assume that we apply a unit step input in thrust $u$, followed 5 seconds later by a negative unit step input. The system response is plotted in Figure 1.1. We see that the system settles much faster in response to the positive unit step than it does in response to the subsequent negative unit step. Intuitively, this can be interpreted as reflecting the fact that the "apparent damping" coefficient $|v|$ is larger at high speeds than at low speeds.

Assume now that we repeat the same experiment but with larger steps, of amplitude 10. Predictably, the difference between the settling times in response to the positive and negative steps is even more marked (Figure 1.2). Furthermore, the settling speed $v_s$ in response to the first step is *not* 10 times that obtained in response to the first unit step in the first experiment, as it would be in a linear system. This can again be understood intuitively, by writing that

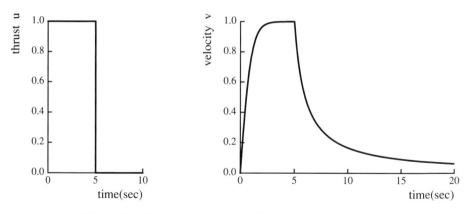

**Figure 1.1 :** Response of system (1.3) to unit steps

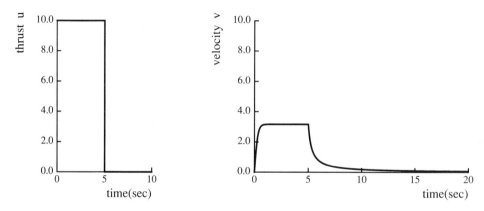

**Figure 1.2 :** Response of system (1.3) to steps of amplitude 10

$$u = 1 \quad \Rightarrow \quad 0 + |v_s| v_s = 1 \quad \Rightarrow \quad v_s = 1$$

$$u = 10 \quad \Rightarrow \quad 0 + |v_s| v_s = 10 \quad \Rightarrow \quad v_s = \sqrt{10} \approx 3.2$$

Carefully understanding and effectively controlling this nonlinear behavior is particularly important if the vehicle is to move in a large dynamic range and change speeds continually, as is typical of industrial remotely-operated underwater vehicles (R.O.V.'s).                    ☐

## SOME COMMON NONLINEAR SYSTEM BEHAVIORS

Let us now discuss some common nonlinear system properties, so as to familiarize ourselves with the complex behavior of nonlinear systems and provide a useful background for our study in the rest of the book.

## Multiple Equilibrium Points

Nonlinear systems frequently have more than one equilibrium point (an equilibrium point is a point where the system can stay forever without moving, as we shall formalize later). This can be seen by the following simple example.

**Example 1.2: A first-order system**

Consider the first order system

$$\dot{x} = -x + x^2 \tag{1.4}$$

with initial condition $x(0) = x_o$. Its linearization is

$$\dot{x} = -x \tag{1.5}$$

The solution of this linear equation is $x(t) = x_o e^{-t}$. It is plotted in Figure 1.3(a) for various initial conditions. The linearized system clearly has a unique equilibrium point at $x = 0$.

By contrast, integrating equation $dx/(-x + x^2) = dt$, the actual response of the nonlinear dynamics (1.4) can be found to be

$$x(t) = \frac{x_o e^{-t}}{1 - x_o + x_o e^{-t}}$$

This response is plotted in Figure 1.3(b) for various initial conditions. The system has two equilibrium points, $x = 0$ and $x = 1$ , and its qualitative behavior strongly depends on its initial condition.                                                   □

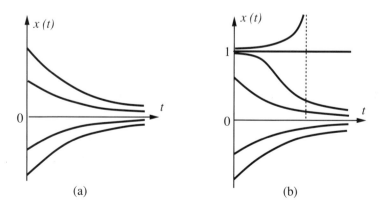

**Figure 1.3** : Responses of the linearized system (a) and the nonlinear system (b)

The issue of motion stability can also be discussed with the aid of the above example. For the linearized system, stability is seen by noting that for *any* initial condition, the motion always converges to the equilibrium point $x = 0$. However, consider now the actual nonlinear system. While motions starting with $x_o < 1$ will indeed converge to the equilibrium point $x = 0$, those starting with $x_o > 1$ will go to infinity (actually in finite time, a phenomenon known as finite escape time). This means that the stability of nonlinear systems may depend on initial conditions.

In the presence of a bounded external input, stability may also be dependent on the input value. This input dependence is highlighted by the so-called bilinear system

$$\dot{x} = xu$$

If the input $u$ is chosen to be $-1$, then the state $x$ converges to 0. If $u = 1$, then $|x|$ tends to infinity.

## Limit Cycles

Nonlinear systems can display oscillations of fixed amplitude and fixed period without external excitation. These oscillations are called limit cycles, or self-excited oscillations. This important phenomenon can be simply illustrated by a famous oscillator dynamics, first studied in the 1920's by the Dutch electrical engineer Balthasar Van der Pol.

### Example 1.3: Van der Pol Equation

The second-order nonlinear differential equation

$$m\ddot{x} + 2c(x^2 - 1)\dot{x} + kx = 0 \tag{1.6}$$

where $m$, $c$ and $k$ are positive constants, is the famous Van der Pol equation. It can be regarded as describing a mass-spring-damper system with a position-dependent damping coefficient $2c(x^2 - 1)$ (or, equivalently, an RLC electrical circuit with a nonlinear resistor). For large values of $x$, the damping coefficient is positive and the damper removes energy from the system. This implies that the system motion has a convergent tendency. However, for small values of $x$, the damping coefficient is negative and the damper adds energy into the system. This suggests that the system motion has a divergent tendency. Therefore, because the nonlinear damping varies with $x$, the system motion can neither grow unboundedly nor decay to zero. Instead, it displays a sustained oscillation independent of initial conditions, as illustrated in Figure 1.4. This so-called limit cycle is sustained by periodically releasing energy into and absorbing energy from the environment, through the damping term. This is in contrast with the case of a conservative mass-spring system, which does not exchange energy with its environment during its vibration.    □

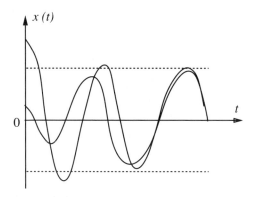

**Figure 1.4 :** Responses of the Van der Pol oscillator

Of course, sustained oscillations can also be found in linear systems, in the case of marginally stable linear systems (such as a mass-spring system without damping) or in the response to sinusoidal inputs. However, limit cycles in nonlinear systems are different from linear oscillations in a number of fundamental aspects. First, the amplitude of the self-sustained excitation is independent of the initial condition, as seen in Figure 1.2, while the oscillation of a marginally stable linear system has its amplitude determined by its initial conditions. Second, marginally stable linear systems are very sensitive to changes in system parameters (with a slight change capable of leading either to stable convergence or to instability), while limit cycles are not easily affected by parameter changes.

Limit cycles represent an important phenomenon in nonlinear systems. They can be found in many areas of enginering and nature. Aircraft wing fluttering, a limit cycle caused by the interaction of aerodynamic forces and structural vibrations, is frequently encountered and is sometimes dangerous. The hopping motion of a legged robot is another instance of a limit cycle. Limit cycles also occur in electrical circuits, *e.g.*, in laboratory electronic oscillators. As one can see from these examples, limit cycles can be undesirable in some cases, but desirable in other cases. An engineer has to know how to eliminate them when they are undesirable, and conversely how to generate or amplify them when they are desirable. To do this, however, requires an understanding of the properties of limit cycles and a familiarity with the tools for manipulating them.

**Bifurcations**

As the parameters of nonlinear dynamic systems are changed, the stability of the equilibrium point can change (as it does in linear systems) and so can the number of equilibrium points. Values of these parameters at which the qualitative nature of the system's motion changes are known as *critical* or *bifurcation* values. The phenomenon of bifurcation, *i.e.*, quantitative change of parameters leading to qualitative change of system properties, is the topic of bifurcation theory.

For instance, the smoke rising from an incense stick (smokestacks and cigarettes are old-fashioned) first accelerates upwards (because it is lighter than the ambient air), but beyond some critical velocity breaks into swirls. More prosaically, let us consider the system described by the so-called undamped Duffing equation

$$\ddot{x} + \alpha x + x^3 = 0$$

(the damped Duffing equation is $\ddot{x} + c\dot{x} + \alpha x + \beta x^3 = 0$, which may represent a mass-damper-spring system with a hardening spring). We can plot the equilibrium points as a function of the parameter $\alpha$. As $\alpha$ varies from positive to negative, one equilibrium point splits into *three* points ($x_e = 0, \sqrt{\alpha}, -\sqrt{\alpha}$), as shown in Figure 1.5(a). This represents a qualitative change in the dynamics and thus $\alpha = 0$ is a critical bifurcation value. This kind for bifurcation is known as a *pitchfork*, due to the shape of the equilibrium point plot in Figure 1.5(a).

Another kind of bifurcation involves the emergence of limit cycles as parameters are changed. In this case, a pair of complex conjugate eigenvalues $p_1 = \gamma + j\omega$, $p_2 = \gamma - j\omega$ cross from the left-half plane into the right-half plane, and the response of the unstable system diverges to a limit cycle. Figure 1.5(b) depicts the change of typical system state trajectories (states are $x$ and $\dot{x}$) as the parameter $\alpha$ is varied. This type of bifurcation is called a Hopf bifurcation.

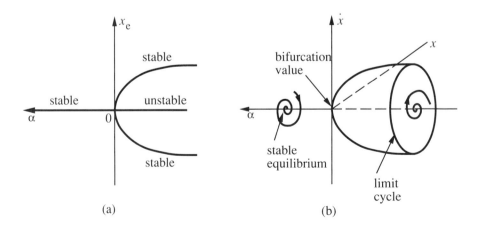

(a)                                                      (b)

**Figure 1.5** : (a) a pitchfork bifurcation; (b) a Hopf bifurcation

**Chaos**

For stable linear systems, small differences in initial conditions can only cause small differences in output. Nonlinear systems, however, can display a phenomenon called *chaos*, by which we mean that the system output is extremely sensitive to initial conditions. The essential feature of chaos is the unpredictability of the system output. Even if we have an exact model of a nonlinear system and an extremely accurate computer, the system's response in the long-run still cannot be well predicted.

Chaos must be distinguished from random motion. In random motion, the system model or input contain uncertainty and, as a result, the time variation of the output cannot be predicted exactly (only statistical measures are available). In chaotic motion, on the other hand, the involved problem is deterministic, and there is little uncertainty in system model, input, or initial conditions.

As an example of chaotic behavior, let us consider the simple nonlinear system

$$\ddot{x} + 0.1\dot{x} + x^5 = 6\sin t$$

which may represent a lightly-damped, sinusoidally forced mechanical structure undergoing large elastic deflections. Figure 1.6 shows the responses of the system corresponding to two almost identical initial conditions, namely $x(0) = 2, \dot{x}(0) = 3$ (thick line) and $x(0) = 2.01, \dot{x}(0) = 3.01$ (thin line). Due to the presence of the strong nonlinearity in $x^5$, the two responses are radically different after some time.

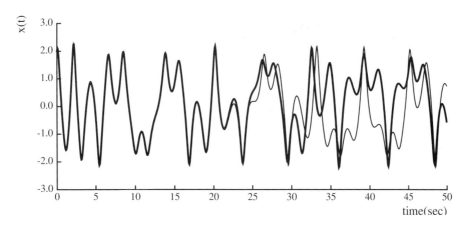

**Figure 1.6 :** Chaotic behavior of a nonlinear system

Chaotic phenomena can be observed in many physical systems. The most commonly seen physical problem is turbulence in fluid mechanics (such as the swirls of our incense stick). Atmospheric dynamics also display clear chaotic behavior, thus making long-term weather prediction impossible. Some mechanical and electrical systems known to exhibit chaotic vibrations include buckled elastic structures, mechanical systems with play or backlash, systems with aeroelastic dynamics, wheel-rail dynamics in railway systems, and, of course, feedback control devices.

Chaos occurs mostly in *strongly* nonlinear systems. This implies that, for a given system, if the initial condition or the external input cause the system to operate in a highly nonlinear region, it increases the possibility of generating chaos. Chaos cannot occur in linear systems. Corresponding to a sinusoidal input of arbitrary magnitude, the linear system response is always a sinusoid of the same frequency. By contrast, the output of a given nonlinear system may display sinusoidal, periodic, or chaotic behaviors, depending on the initial condition and the input magnitude.

In the context of feedback control, it is of course of interest to know when a nonlinear system will get into a chaotic mode (so as to avoid it) and, in case it does, how to recover from it. Such problems are the object of active research.

**Other behaviors**

Other interesting types of behavior, such as jump resonance, subharmonic generation, asynchronous quenching, and frequency-amplitude dependence of free vibrations, can also occur and become important in some system studies. However, the above description should provide ample evidence that nonlinear systems can have considerably richer and more complex behavior than linear systems.

# 1.3  An Overview of the Book

Because nonlinear systems can have much richer and more complex behaviors than linear systems, their analysis is much more difficult. Mathematically, this is reflected in two aspects. First, nonlinear equations, unlike linear ones, cannot in general be solved analytically, and therefore a complete understanding of the behavior of a nonlinear system is very difficult. Second, powerful mathematical tools like Laplace and Fourier transforms do not apply to nonlinear systems.

As a result, there are no systematic tools for predicting the behavior of nonlinear systems, nor are there systematic procedures for designing nonlinear control systems. Instead, there is a rich inventory of powerful analysis and design tools, each best applicable to particular classes of nonlinear control problems. It is the objective of this book to present these various tools, with particular emphasis on their powers and limitations, and on how they can be effectively combined.

This book is divided into two major parts. Part I (chapters 2-5) presents the major *analytical* tools that can be used to study a nonlinear system. Part II (chapters 6-9) discusses the major nonlinear controller *design* techniques. Each part starts with a short introduction providing the background for the main issues and techniques to be discussed.

In chapter 2, we further familiarize ourselves with some basic nonlinear system behaviors, by studying second-order systems using the simple graphical tools provided by so-called phase plane analysis. Chapter 3 introduces the most fundamental analysis tool to be used in this book, namely the concept of a Lyapunov function and its use in nonlinear stability analysis. Chapter 4 studies selected advanced topics in stability analysis. Chapter 5 discusses an approximate nonlinear system analysis method, the describing function method, which aims at extending to nonlinear systems some of the desirable and intuitive properties of linear frequency response analysis.

The basic idea of chapter 6 is to study under what conditions the dynamics of a nonlinear system can be algebraically transformed in that of a linear system, on which linear control design techniques can in turn be applied. Chapters 7 and 8 then study how to reduce or practically eliminate the effects of model uncertainties on the stability and performance of feedback controllers for linear or nonlinear systems, using so-called robust and adaptive approaches. Finally, chapter 9 extensively discusses the use of known physical properties to simplify and enhance the design of controllers for complex multi-input nonlinear systems.

The book concentrates on nonlinear systems represented in continuous-time form. Even though most control systems are implemented digitally, nonlinear physical systems are continuous in nature and are hard to meaningfully discretize, while digital control systems may be treated as continuous-time systems in analysis and design if high sampling rates are used. Given the availability of cheap computation, the most common practical case when it may be advantageous to consider sampling explicitly is when *measurements* are sparse, as *e.g.*, in the case of underwater vehicles using acoustic navigation. Some practical issues involved in the digital implementation of controllers designed from continuous-time formulations are discussed in the introduction to Part II.

## 1.4  Notes and References

Detailed discussions of bifurcations and chaos can be found, *e.g.*, in [Guckenheimer and Holmes, 1983] and in [Thompson and Stewart, 1986], from which the example of Figure 1.6 is adapted.

# Part I
# Nonlinear Systems Analysis

The objective of this part is to present various tools available for analyzing nonlinear control systems. The study of these nonlinear analysis techniques is important for a number of reasons. First, theoretical analysis is usually the least expensive way of exploring a system's characteristics. Second, simulation, though very important in nonlinear control, has to be guided by theory. Blind simulation of nonlinear systems is likely to produce few results or misleading results. This is especially true given the great richness of behavior that nonlinear systems can exhibit, depending on initial conditions and inputs. Third, the design of nonlinear controllers is always based on analysis techniques. Since design methods are usually based on analysis methods, it is almost impossible to master the design methods without first studying the analysis tools. Furthermore, analysis tools also allow us to assess control designs after they have been made, and, in case of inadequate performance, they may also suggest directions of modifying the control designs.

It should not come as a surprise that no universal technique has been devised for the analysis of all nonlinear control systems. In linear control, one can analyze a system in the time domain or in the frequency domain. However, for nonlinear control systems, none of these standard approaches can be used, since direct solution of nonlinear differential equations is generally impossible, and frequency domain transformations do not apply.

While the analysis of nonlinear control systems is difficult, serious efforts have been made to develop appropriate theoretical tools for it. Many methods of nonlinear control system analysis have been proposed. Let us briefly describe some of these methods before discussing their details in the following chapters.

## Phase plane analysis

Phase plane analysis, discussed in chapter 2, is a graphical method of studying second-order nonlinear systems. Its basic idea is to solve a second order differential equation graphically, instead of seeking an analytical solution. The result is a family of system motion trajectories on a two-dimensional plane, called the phase plane, which allow us to visually observe the motion patterns of the system. While phase plane analysis has a number of important advantages, it has the fundamental disadvantage of being applicable only to systems which can be well approximated by a second-order dynamics. Because of its graphical nature, it is frequently used to provide intuitive insights about nonlinear effects.

## Lyapunov theory

Basic Lyapunov theory comprises two methods introduced by Lyapunov, the indirect method and the direct method. The indirect method, or linearization method, states that the stability properties of a nonlinear system in the close vicinity of an equilibrium point are essentially the same as those of its linearized approximation. The method serves as the theoretical justification for using linear control for physical systems, which are always inherently nonlinear. The direct method is a powerful tool for nonlinear system analysis, and therefore the so-called Lyapunov analysis often actually refers to the direct method. The direct method is a generalization of the energy concepts associated with a mechanical system: the motion of a mechanical system is stable if its total mechanical energy decreases all the time. In using the direct method to analyze the stability of a nonlinear system, the idea is to construct a scalar energy-like function (a Lyapunov function) for the system, and to see whether it decreases. The power of this method comes from its generality: it is applicable to all kinds of control systems, be they time-varying or time-invariant, finite dimensional or infinite dimensional. Conversely, the limitation of the method lies in the fact that it is often difficult to find a Lyapunov function for a given system.

Although Lyapunov's direct method is originally a method of stability analysis, it can be used for other problems in nonlinear control. One important application is the design of nonlinear controllers. The idea is to somehow formulate a scalar positive function of the system states, and then choose a control law to make this function decrease. A nonlinear control system thus designed will be guaranteed to be stable. Such a design approach has been used to solve many complex design problems, *e.g.*, in robotics and adaptive control. The direct method can also be used to estimate the performance of a control system and study its robustness. The important subject of Lyapunov analysis is studied in chapters 3 and 4, with chapter 3 presenting the main concepts and results in Lyapunov theory, and chapter 4 discussing some advanced topics.

### Describing functions

The describing function method is an approximate technique for studying nonlinear systems. The basic idea of the method is to approximate the nonlinear components in nonlinear control systems by linear "equivalents", and then use frequency domain techniques to analyze the resulting systems. Unlike the phase plane method, it is not restricted to second-order systems. Unlike Lyapunov methods, whose applicability to a specific system hinges on the success of a trial-and-error search for a Lyapunov function, its application is straightforward for nonlinear systems satisfying some easy-to-check conditions.

The method is mainly used to predict limit cycles in nonlinear systems. Other applications include the prediction of subharmonic generation and the determination of system response to sinusoidal excitation. The method has a number of advantages. First, it can deal with low order and high order systems with the same straightforward procedure. Second, because of its similarity to frequency-domain analysis of linear systems, it is conceptually simple and physically appealing, allowing users to exercise their physical and engineering insights about the control system. Third, it can deal with the "hard nonlinearities" frequently found in control systems without any difficulty. As a result, it is an important tool for practical problems of nonlinear control analysis and design. The disadvantages of the method are linked to its approximate nature, and include the possibility of inaccurate predictions (false predictions may be made if certain conditions are not satisfied) and restrictions on the systems to which it applies (for example, it has difficulties in dealing with systems with multiple nonlinearities).

# Chapter 2
# Phase Plane Analysis

Phase plane analysis is a graphical method for studying second-order systems, which was introduced well before the turn of the century by mathematicians such as Henri Poincare. The basic idea of the method is to generate, in the state space of a second-order dynamic system (a two-dimensional plane called the phase plane), motion trajectories corresponding to various initial conditions, and then to examine the qualitative features of the trajectories. In such a way, information concerning stability and other motion patterns of the system can be obtained. In this chapter, our objective is to gain familiarity with nonlinear systems through this simple graphical method.

Phase plane analysis has a number of useful properties. First, as a graphical method, it allows us to visualize what goes on in a nonlinear system starting from various initial conditions, without having to solve the nonlinear equations analytically. Second, it is not restricted to small or smooth nonlinearities, but applies equally well to strong nonlinearities and to "hard" nonlinearities. Finally, some practical control systems can indeed be adequately approximated as second-order systems, and the phase plane method can be used easily for their analysis. Conversely, of course, the fundamental disadvantage of the method is that it is restricted to second-order (or first-order) systems, because the graphical study of higher-order systems is computationally and geometrically complex.

# 2.1  Concepts of Phase Plane Analysis

## 2.1.1  Phase Portraits

The phase plane method is concerned with the graphical study of second-order autonomous systems described by

$$\dot{x}_1 = f_1(x_1, x_2) \tag{2.1a}$$

$$\dot{x}_2 = f_2(x_1, x_2) \tag{2.1b}$$

where $x_1$ and $x_2$ are the states of the system, and $f_1$ and $f_2$ are nonlinear functions of the states. Geometrically, the state space of this system is a plane having $x_1$ and $x_2$ as coordinates. We will call this plane the *phase plane*.

Given a set of initial conditions $\mathbf{x}(0) = \mathbf{x}_o$, Equation (2.1) defines a solution $\mathbf{x}(t)$. With time $t$ varied from zero to infinity, the solution $\mathbf{x}(t)$ can be represented geometrically as a curve in the phase plane. Such a curve is called a phase plane *trajectory*. A family of phase plane trajectories corresponding to various initial conditions is called a *phase portrait* of a system.

To illustrate the concept of phase portrait, let us consider the following simple system.

**Example 2.1: Phase portrait of a mass-spring system**

The governing equation of the mass-spring system in Figure 2.1(a) is the familiar linear second-order differential equation

$$\ddot{x} + x = 0 \tag{2.2}$$

Assume that the mass is initially at rest, at length $x_o$. Then the solution of the equation is

$$x(t) = x_o \cos t$$

$$\dot{x}(t) = -x_o \sin t$$

Eliminating time $t$ from the above equations, we obtain the equation of the trajectories

$$x^2 + \dot{x}^2 = x_o^2$$

This represents a circle in the phase plane. Corresponding to different initial conditions, circles of different radii can be obtained. Plotting these circles on the phase plane, we obtain a phase portrait for the mass-spring system (Figure 2.1.b).    □

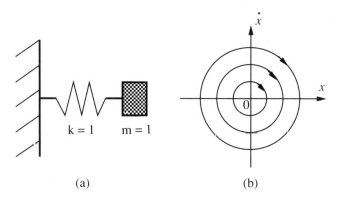

**Figure 2.1** : A mass-spring system and its phase portrait

The power of the phase portrait lies in the fact that once the phase portrait of a system is obtained, the nature of the system response corresponding to various initial conditions is directly displayed on the phase plane. In the above example, we easily see that the system trajectories neither converge to the origin nor diverge to infinity. They simply circle around the origin, indicating the marginal nature of the system's stability.

A major class of second-order systems can be described by differential equations of the form

$$\ddot{x} + f(x, \dot{x}) = 0 \tag{2.3}$$

In state space form, this dynamics can be represented as

$$\dot{x}_1 = x_2$$

$$\dot{x}_2 = -f(x_1, x_2)$$

with $x_1 = x$ and $x_2 = \dot{x}$. Most second-order systems in practice, such as mass-damper-spring systems in mechanics, or resistor-coil-capacitor systems in electrical engineering, can be represented in or transformed into this form. For these systems, the states are $x$ and its derivative $\dot{x}$. Traditionally, the phase plane method is developed for the dynamics (2.3), and the phase plane is defined as the plane having $x$ and $\dot{x}$ as coordinates. But it causes no difficulty to extend the method to more general dynamics of the form (2.1), with the $(x_1, x_2)$ plane as the phase plane, as we do in this chapter.

## 2.1.2  Singular Points

An important concept in phase plane analysis is that of a singular point. A singular point is an equilibrium point in the phase plane. Since an equilibrium point is defined as a point where the system states can stay forever, this implies that $\dot{\mathbf{x}} = \mathbf{0}$, and using (2.1),

$$f_1(x_1, x_2) = 0 \qquad f_2(x_1, x_2) = 0 \tag{2.4}$$

The values of the equilibrium states can be solved from (2.4).

For a linear system, there is usually only one singular point (although in some cases there can be a *continuous* set of singular points, as in the system $\ddot{x} + \dot{x} = 0$, for which all points on the real axis are singular points). However, a nonlinear system often has more than one isolated singular point, as the following example shows.

**Example 2.2: A nonlinear second-order system**

Consider the system

$$\ddot{x} + 0.6\,\dot{x} + 3\,x + x^2 = 0$$

whose phase portrait is plotted in Figure 2.2. The system has two singular points, one at $(0, 0)$ and the other at $(-3, 0)$. The motion patterns of the system trajectories in the vicinity of the two singular points have different natures. The trajectories move towards the point $x = 0$ while moving away from the point $x = -3$.  ∎

One may wonder why an equilibrium point of a second-order system is called a *singular* point. To answer this, let us examine the slope of the phase trajectories. From (2.1), the slope of the phase trajectory passing through a point $(x_1, x_2)$ is determined by

$$\frac{dx_2}{dx_1} = \frac{f_2(x_1, x_2)}{f_1(x_1, x_2)} \tag{2.5}$$

With the functions $f_1$ and $f_2$ assumed to be single valued, there is usually a definite value for this slope at any given point in phase plane. This implies that the phase trajectories will not intersect. At singular points, however, the value of the slope is $0/0$, *i.e.*, the slope is indeterminate. Many trajectories may intersect at such points, as seen from Figure 2.2. This indeterminacy of the slope accounts for the adjective "singular".

Singular points are very important features in the phase plane. Examination of the singular points can reveal a great deal of information about the properties of a

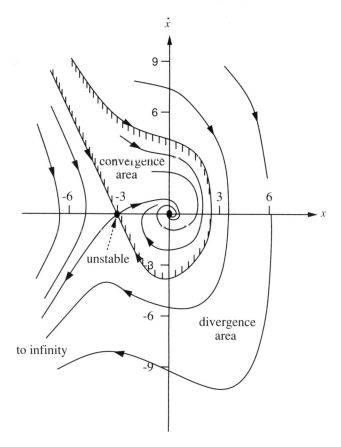

**Figure 2.2 :** The phase portrait of a nonlinear system

system. In fact, the stability of linear systems is uniquely characterized by the nature of their singular points. For nonlinear systems, besides singular points, there may be more complex features, such as limit cycles. These issues will be discussed in detail in sections 2.3 and 2.4.

Note that, although the phase plane method is developed primarily for second-order systems, it can also be applied to the analysis of first-order systems of the form

$$\dot{x} + f(x) = 0$$

The idea is still to plot $\dot{x}$ with respect to $x$ in the phase plane. The difference now is that the phase portrait is composed of a single trajectory.

**Example 2.3: A first-order system**

Consider the system

$$\dot{x} = -4x + x^3$$

There are three singular points, defined by $-4x + x^3 = 0$, namely, $x = 0, -2,$ and 2. The phase-portrait of the system consists of a single trajectory, and is shown in Figure 2.3. The arrows in the figure denote the direction of motion, and whether they point toward the left or the right at a particular point is determined by the sign of $\dot{x}$ at that point. It is seen from the phase portrait of this system that the equilibrium point $x = 0$ is stable, while the other two are unstable. $\qquad\square$

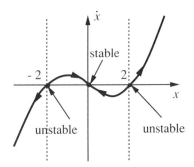

**Figure 2.3 :** Phase trajectory of a first-order system

## 2.1.3 Symmetry in Phase Plane Portraits

A phase portrait may have *a priori* known symmetry properties, which can simplify its generation and study. If a phase portrait is symmetric with respect to the $x_1$ or the $x_2$ axis, one only needs in practice to study half of it. If a phase portrait is symmetric with respect to both the $x_1$ and $x_2$ axes, only one quarter of it has to be explicitly considered.

Before generating a phase portrait itself, we can determine its symmetry properties by examining the system equations. Let us consider the second-order dynamics (2.3). The slope of trajectories in the phase plane is of the form

$$\frac{dx_2}{dx_1} = -\frac{f(x_1, x_2)}{\dot{x}}$$

Since symmetry of the phase portraits also implies symmetry of the slopes (equal in absolute value but opposite in sign), we can identify the following situations:

**Symmetry about the $x_1$ axis:** The condition is

$$f(x_1, x_2) = f(x_1, -x_2)$$

This implies that the function $f$ should be even in $x_2$. The mass-spring system in Example 2.1 satisfies this condition. Its phase portrait is seen to be symmetric about the $x_1$ axis.

**Symmetry about the $x_2$ axis**: Similarly,

$$f(x_1, x_2) = -f(-x_1, x_2)$$

implies symmetry with respect to the $x_2$ axis. The mass-spring system also satisfies this condition.

**Symmetry about the origin**: When

$$f(x_1, x_2) = -f(-x_1, -x_2)$$

the phase portrait of the system is symmetric about the origin.

## 2.2   Constructing Phase Portraits

Today, phase portraits are routinely computer-generated. In fact, it is largely the advent of the computer in the early 1960's, and the associated ease of quickly generating phase portraits, which spurred many advances in the study of complex nonlinear dynamic behaviors such as chaos. However, of course (as *e.g.*, in the case of root locus for linear systems), it is still practically useful to learn how to roughly sketch phase portraits or quickly verify the plausibility of computer outputs.

There are a number of methods for constructing phase plane trajectories for linear or nonlinear systems, such as the so-called analytical method, the method of isoclines, the delta method, Lienard's method, and Pell's method. We shall discuss two of them in this section, namely, the analytical method and the method of isoclines. These methods are chosen primarily because of their relative simplicity. The analytical method involves the analytical solution of the differential equations describing the systems. It is useful for some special nonlinear systems, particularly piece-wise linear systems, whose phase portraits can be constructed by piecing together the phase portraits of the related linear systems. The method of isoclines is a graphical method which can conveniently be applied to construct phase portraits for systems which cannot be solved analytically, which represent by far the most common case.

## ANALYTICAL METHOD

There are two techniques for generating phase plane portraits analytically. Both techniques lead to a functional relation between the two phase variables $x_1$ and $x_2$ in the form

$$g(x_1, x_2, c) = 0 \qquad (2.6)$$

where the constant $c$ represents the effects of initial conditions (and, possibly, of external input signals). Plotting this relation in the phase plane for different initial conditions yields a phase portrait.

The first technique involves solving equations (2.1) for $x_1$ and $x_2$ as functions of time $t$, *i.e.*,

$$x_1(t) = g_1(t) \qquad x_2(t) = g_2(t)$$

and then eliminating time $t$ from these equations, leading to a functional relation in the form of (2.6). This technique was already illustrated in Example 2.1.

The second technique, on the other hand, involves directly eliminating the time variable, by noting that

$$\frac{dx_2}{dx_1} = \frac{f_2(x_1, x_2)}{f_1(x_1, x_2)}$$

and then solving this equation for a functional relation between $x_1$ and $x_2$. Let us use this technique to solve the mass-spring equation again.

**Example 2.4: Mass-spring system**

By noting that $\ddot{x} = (d\dot{x}/dx)(dx/dt)$, we can rewrite (2.2) as

$$\dot{x}\frac{d\dot{x}}{dx} + x = 0$$

Integration of this equation yields

$$\dot{x}^2 + x^2 = x_o^2 \qquad \qquad \square$$

One sees that the second technique is more straightforward in generating the equations for the phase plane trajectories.

Most nonlinear systems cannot be easily solved by either of the above two techniques. However, for piece-wise linear systems, an important class of nonlinear systems, this method can be conveniently used, as the following example shows.

**Example 2.5: A satellite control system**

Figure 2.4 shows the control system for a simple satellite model. The satellite, depicted in Figure 2.5(a), is simply a rotational unit inertia controlled by a pair of thrusters, which can provide either a positive constant torque $U$ (positive firing) or a negative torque $-U$ (negative firing). The purpose of the control system is to maintain the satellite antenna at a zero angle by appropriately firing the thrusters. The mathematical model of the satellite is

$$\ddot{\theta} = u$$

where $u$ is the torque provided by the thrusters and $\theta$ is the satellite angle.

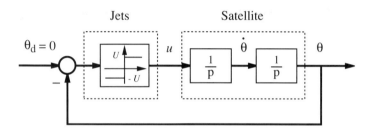

**Figure 2.4 :** Satellite control system

Let us examine on the phase plane the behavior of the control system when the thrusters are fired according to the control law

$$u(t) = \begin{cases} -U & \text{if } \theta > 0 \\ U & \text{if } \theta < 0 \end{cases} \tag{2.7}$$

which means that the thrusters push in the counterclockwise direction if $\theta$ is positive, and vice versa.

As the first step of the phase portrait generation, let us consider the phase portrait when the thrusters provide a positive torque $U$. The dynamics of the system is

$$\ddot{\theta} = U$$

which implies that $\dot{\theta} \, d\dot{\theta} = U \, d\theta$. Therefore, the phase trajectories are a family of parabolas defined by

$$\dot{\theta}^2 = 2U\theta + c_1$$

where $c_1$ is a constant. The corresponding phase portrait of the system is shown in Figure 2.5(b).

When the thrusters provide a negative torque $-U$, the phase trajectories are similarly found to be

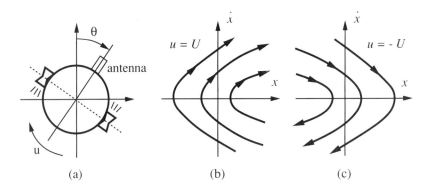

**Figure 2.5 :** Satellite control using on-off thrusters

$$\dot{\theta}^2 = -2Ux + c_1$$

with the corresponding phase portrait shown in Figure 2.5(c).

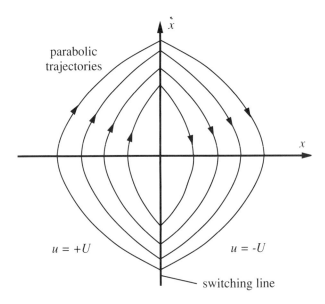

**Figure 2.6 :** Complete phase portrait of the control system

The complete phase portrait of the closed-loop control system can be obtained simply by connecting the trajectories on the left half of the phase plane in 2.5(b) with those on the right half of the phase plane in 2.5(c), as shown in Figure 2.6. The vertical axis represents a switching line, because the control input and thus the phase trajectories are switched on that line.  It is interesting to see that, starting from a nonzero initial angle, the satellite will oscillate in periodic motions under the action of the jets. One concludes from this phase portrait that the system is marginally stable, similarly to the mass-spring system in Example 2.1.  Convergence of the system to the zero angle can be obtained by adding rate feedback (Exercise 2.4).                    □

## THE METHOD OF ISOCLINES

The basic idea in this method is that of isoclines.  Consider the dynamics in (2.1). At a point $(x_1, x_2)$ in the phase plane, the slope of the tangent to the trajectory can be determined by (2.5).  An isocline is defined to be the locus of the points with a given tangent slope.  An isocline with slope $\alpha$ is thus defined to be

$$\frac{dx_2}{dx_1} = \frac{f_2(x_1, x_2)}{f_1(x_1, x_2)} = \alpha$$

This is to say that points on the curve

$$f_2(x_1, x_2) = \alpha f_1(x_1, x_2)$$

all have the same tangent slope $\alpha$.

In the method of isoclines, the phase portrait of a system is generated in two steps. In the first step, a field of directions of tangents to the trajectories is obtained. In the second step, phase plane trajectories are formed from the field of directions .

Let us explain the isocline method on the mass-spring system in (2.2).  The slope of the trajectories is easily seen to be

$$\frac{dx_2}{dx_1} = -\frac{x_1}{x_2}$$

Therefore, the isocline equation for a slope $\alpha$ is

$$x_1 + \alpha x_2 = 0$$

*i.e.*, a straight line. Along the line, we can draw a lot of short line segments with slope $\alpha$.  By taking $\alpha$ to be different values, a set of isoclines can be drawn, and a field of directions of tangents to trajectories are generated, as shown in Figure 2.7.  To obtain trajectories from the field of directions, we assume that the the tangent slopes are locally constant. Therefore, a trajectory starting from any point in the plane can be found by connecting a sequence of line segments.

Let us use the method of isoclines to study the Van der Pol equation, a nonlinear equation.

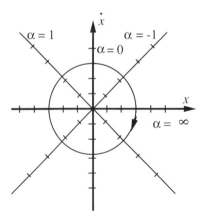

**Figure 2.7 :** Isoclines for the mass-spring system

**Example 2.6: The Van der Pol equation**

For the Van der Pol equation

$$\ddot{x} + 0.2(x^2 - 1)\dot{x} + x = 0$$

an isocline of slope $\alpha$ is defined by

$$\frac{d\dot{x}}{dx} = -\frac{0.2(x^2 - 1)\dot{x} + x}{\dot{x}} = \alpha$$

Therefore, the points on the curve

$$0.2(x^2 - 1)\dot{x} + x + \alpha\dot{x} = 0$$

all have the same slope $\alpha$.

By taking $\alpha$ of different values, different isoclines can be obtained, as plotted in Figure 2.8. Short line segments are drawn on the isoclines to generate a field of tangent directions. The phase portraits can then be obtained, as shown in the plot. It is interesting to note that there exists a closed curve in the portrait, and the trajectories starting from both outside and inside converge to this curve. This closed curve corresponds to a limit cycle, as will be discussed further in section 2.5. □

Note that the same scales should be used for the $x_1$ axis and $x_2$ axis of the phase plane, so that the derivative $dx_2/dx_1$ equals the geometric slope of the trajectories. Also note that, since in the second step of phase portrait construction we essentially assume that the slope of the phase plane trajectories is locally constant, more isoclines should be plotted in regions where the slope varies quickly, to improve accuracy.

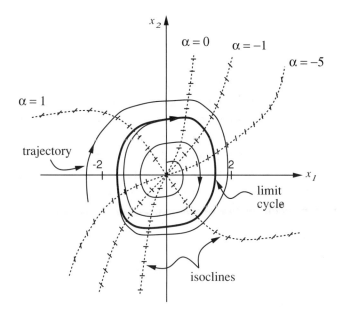

**Figure 2.8 :** Phase portrait of the Van der Pol equation

## 2.3   Determining Time from Phase Portraits

Note that time $t$ does not explicitly appear in the phase plane having $x_1$ and $x_2$ as coordinates. However, in some cases, we might be interested in the time information. For example, one might want to know the time history of the system states starting from a specific initial point. Another relevant situation is when one wants to know how long it takes for the system to move from a point to another point in a phase plane trajectory.  We now describe two techniques for computing time history from phase portraits.  Both techniques involve a step-by step procedure for recovering time.

### Obtaining time from  $\Delta t \approx \Delta x / \dot{x}$

In a short time $\Delta t$, the change of $x$ is approximately

$$\Delta x \approx \dot{x} \Delta t \tag{2.8}$$

where $\dot{x}$ is the velocity corresponding to the increment $\Delta x$. Note that for a $\Delta x$ of finite magnitude, the average value of velocity during a time increment should be used to improve accuracy.  From (2.8), the length of time corresponding to the increment $\Delta x$

is

$$\Delta t \approx \frac{\Delta x}{\dot{x}}$$

The above reasoning implies that, in order to obtain the time corresponding to the motion from one point to another point along a trajectory, one should divide the corresponding part of the trajectory into a number of small segments (not necessarily equally spaced), find the time associated with each segment, and then add up the results. To obtain the time history of states corresponding to a certain initial condition, one simply computes the time $t$ for each point on the phase trajectory, and then plots $x$ with respect to $t$ and $\dot{x}$ with respect to $t$.

**Obtaining time from** $t = \int (1/\dot{x}) \, dx$

Since $\dot{x} = dx/dt$, we can write $dt = dx/\dot{x}$. Therefore,

$$t - t_o = \int_{x_o}^{x} (1/\dot{x}) \, dx$$

where $x$ corresponds to time $t$ and $x_o$ corresponds to time $t_o$ . This equation implies that, if we plot a phase plane portrait with new coordinates $x$ and $(1/\dot{x})$, then the area under the resulting curve is the corresponding time interval.

## 2.4  Phase Plane Analysis of Linear Systems

In this section, we describe the phase plane analysis of linear systems. Besides allowing us to visually observe the motion patterns of linear systems, this will also help the development of nonlinear system analysis in the next section, because a nonlinear systems behaves similarly to a linear system around each equilibrium point.

The general form of a linear second-order system is

$$\dot{x}_1 = a x_1 + b x_2 \tag{2.9a}$$

$$\dot{x}_2 = c x_1 + d x_2 \tag{2.9b}$$

To facilitate later discussions, let us transform this equation into a scalar second-order differential equation. Note from (2.9a) and (2.9b) that

$$b \dot{x}_2 = b c x_1 + d(\dot{x}_1 - a x_1)$$

Consequently, differentiation of (2.9a) and then substitution of (2.9b) leads to

$$\ddot{x}_1 = (a+d)\dot{x}_1 + (cb-ad)x_1$$

Therefore, we will simply consider the second-order linear system described by

$$\ddot{x} + a\dot{x} + bx = 0 \tag{2.10}$$

To obtain the phase portrait of this linear system, we first solve for the time history

$$x(t) = k_1 e^{\lambda_1 t} + k_2 e^{\lambda_2 t} \qquad \text{for } \lambda_1 \neq \lambda_2 \tag{2.11a}$$

$$x(t) = k_1 e^{\lambda_1 t} + k_2 t\, e^{\lambda_1 t} \qquad \text{for } \lambda_1 = \lambda_2 \tag{2.11b}$$

where the constants $\lambda_1$ and $\lambda_2$ are the solutions of the characteristic equation

$$s^2 + as + b = (s - \lambda_1)(s - \lambda_2) = 0$$

The roots $\lambda_1$ and $\lambda_2$ can be explicitly represented as

$$\lambda_1 = (-a + \sqrt{a^2 - 4b})/2 \qquad \lambda_2 = (-a - \sqrt{a^2 - 4b})/2$$

For linear systems described by (2.10), there is only one singular point (assuming $b \neq 0$), namely the origin. However, the trajectories in the vicinity of this singularity point can display quite different characteristics, depending on the values of $a$ and $b$. The following cases can occur

1. $\lambda_1$ and $\lambda_2$ are both real and have the same sign (positive or negative)
2. $\lambda_1$ and $\lambda_2$ are both real and have opposite signs
3. $\lambda_1$ and $\lambda_2$ are complex conjugate with non-zero real parts
4. $\lambda_1$ and $\lambda_2$ are complex conjugates with real parts equal to zero

We now briefly discuss each of the above four cases.

## STABLE OR UNSTABLE NODE

The first case corresponds to a *node*. A node can be stable or unstable. If the eigenvalues are negative, the singularity point is called a *stable node* because both $x(t)$ and $\dot{x}(t)$ converge to zero exponentially, as shown in Figure 2.9(a). If both eigenvalues are positive, the point is called an *unstable node*, because both $x(t)$ and $\dot{x}(t)$ diverge from zero exponentially, as shown in Figure 2.9(b). Since the eigenvalues are real, there is no oscillation in the trajectories.

### SADDLE POINT

The second case (say $\lambda_1 < 0$ and $\lambda_2 > 0$) corresponds to a *saddle point* (Figure 2.9(c)). The phase portrait of the system has the interesting "saddle" shape shown in Figure 2.9(c). Because of the unstable pole $\lambda_2$, almost all of the system trajectories diverge to infinity. In this figure, one also observes two straight lines passing through the origin. The diverging line (with arrows pointing to infinity) corresponds to initial conditions which make $k_2$ (*i.e.*, the unstable component) equal zero. The converging straight line corresponds to initial conditions which make $k_1$ equal zero.

### STABLE OR UNSTABLE FOCUS

The third case corresponds to a focus. A *stable focus* occurs when the real part of the eigenvalues is negative, which implies that $x(t)$ and $\dot{x}(t)$ both converge to zero. The system trajectories in the vicinity of a stable focus are depicted in Figure 2.9(d). Note that the trajectories encircle the origin one or more times before converging to it, unlike the situation for a stable node. If the real part of the eigenvalues is positive, then $x(t)$ and $\dot{x}(t)$ both diverge to infinity, and the singularity point is called an *unstable focus*. The trajectories corresponding to an unstable focus are sketched in Figure 2.9(e).

### CENTER POINT

The last case corresponds to a center point, as shown in Figure 2.9(f). The name comes from the fact that all trajectories are ellipses and the singularity point is the center of these ellipses. The phase portrait of the undamped mass-spring system belongs to this category.

Note that the stability characteristics of linear systems are uniquely determined by the nature of their singularity points. This, however, is not true for nonlinear systems.

## 2.5  Phase Plane Analysis of Nonlinear Systems

In discussing the phase plane analysis of nonlinear systems, two points should be kept in mind. Phase plane analysis of nonlinear systems is related to that of linear systems, because the local behavior of a nonlinear system can be approximated by the behavior of a linear system. Yet, nonlinear systems can display much more complicated patterns in the phase plane, such as multiple equilibrium points and limit cycles. We now discuss these points in more detail.

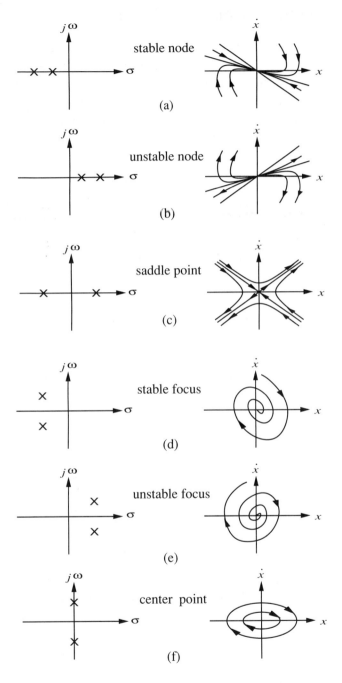

**Figure 2.9 :** Phase-portraits of linear systems

## LOCAL BEHAVIOR OF NONLINEAR SYSTEMS

In the phase portrait of Figure 2.2, one notes that, in contrast to linear systems, there are two singular points, $(0, 0)$ and $(-3, 0)$. However, we also note that the features of the phase trajectories in the neighborhood of the two singular points look very much like those of linear systems, with the first point corresponding to a stable focus and the second to a saddle point. This similarity to a linear system in the local region of each singular point can be formalized by linearizing the nonlinear system, as we now discuss.

If the singular point of interest is not at the origin, by defining the difference between the original state and the singular point as a new set of state variables, one can always shift the singular point to the origin. Therefore, without loss of generality, we may simply consider Equation (2.1) with a singular point at 0. Using Taylor expansion, Equations (2.1a) and (2.1b) can be rewritten as

$$\dot{x}_1 = a x_1 + b x_2 + g_1(x_1, x_2)$$

$$\dot{x}_2 = c x_1 + d x_2 + g_2(x_1, x_2)$$

where $g_1$ and $g_2$ contain higher order terms.

In the vicinity of the origin, the higher order terms can be neglected, and therefore, the nonlinear system trajectories essentially satisfy the linearized equation

$$\dot{x}_1 = a x_1 + b x_2$$

$$\dot{x}_2 = c x_1 + d x_2$$

As a result, the local behavior of the nonlinear system can be approximated by the patterns shown in Figure 2.9.

## LIMIT CYCLES

In the phase portrait of the nonlinear Van der Pol equation, shown in Figure 2.8, one observes that the system has an unstable node at the origin. Furthermore, there is a closed curve in the phase portrait. Trajectories inside the curve and those outside the curve all tend to this curve, while a motion started on this curve will stay on it forever, circling periodically around the origin. This curve is an instance of the so-called "limit cycle" phenomenon. Limit cycles are unique features of nonlinear systems.

In the phase plane, *a limit cycle is defined as an isolated closed curve*. The trajectory has to be both closed, indicating the periodic nature of the motion, and isolated, indicating the limiting nature of the cycle (with nearby trajectories converging or diverging from it). Thus, while there are many closed curves in the phase portraits of the mass-spring-damper system in Example 2.1 or the satellite system in Example 2.5, these are not considered limit cycles in this definition, because they are not isolated.

Depending on the motion patterns of the trajectories in the vicinity of the limit cycle, one can distinguish three kinds of limit cycles

1. **Stable Limit Cycles**: all trajectories in the vicinity of the limit cycle converge to it as $t \to \infty$ (Figure 2.10(a));

2. **Unstable Limit Cycles**: all trajectories in the vicinity of the limit cycle diverge from it as $t \to \infty$ (Figure 2.10(b));

3. **Semi-Stable Limit Cycles**: some of the trajectories in the vicinity converge to it, while the others diverge from it as $t \to \infty$ (Figure 2.10(c));

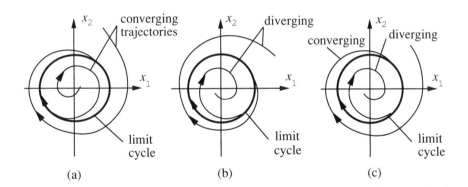

**Figure 2.10 :** Stable, unstable, and semi-stable limit cycles

As seen from the phase portrait of Figure 2.8, the limit cycle of the Van der Pol equation is clearly stable. Let us consider some additional examples of stable, unstable, and semi-stable limit cycles.

**Example 2.7: stable, unstable, and semi-stable limit cycles**

Consider the following nonlinear systems

$$\text{(a)} \quad \dot{x}_1 = x_2 - x_1(x_1{}^2 + x_2{}^2 - 1) \qquad \dot{x}_2 = -x_1 - x_2(x_1{}^2 + x_2{}^2 - 1) \qquad (2.12)$$

$$\text{(b)} \quad \dot{x}_1 = x_2 + x_1(x_1{}^2 + x_2{}^2 - 1) \qquad \dot{x}_2 = -x_1 + x_2(x_1{}^2 + x_2{}^2 - 1) \qquad (2.13)$$

$$\text{(c)} \quad \dot{x}_1 = x_2 - x_1(x_1{}^2 + x_2{}^2 - 1)^2 \qquad \dot{x}_2 = -x_1 - x_2(x_1{}^2 + x_2{}^2 - 1)^2 \qquad (2.14)$$

Let us study system (a) first. By introducing polar coordinates

$$r = ( x_1^2 + x_2^2)^{1/2} \qquad \theta = \tan^{-1}(x_2/x_1)$$

the dynamic equations (2.12) are transformed as

$$\frac{dr}{dt} = - r\,(r^2 - 1) \qquad \frac{d\theta}{dt} = -1$$

When the state starts on the unit circle, the above equation shows that $\dot{r}(t) = 0$. Therefore, the state will circle around the origin with a period $1/2\pi$. When $r < 1$, then $\dot{r} > 0$. This implies that the state tends to the circle from inside. When $r > 1$, then $\dot{r} < 0$. This implies that the state tends toward the unit circle from outside. Therefore, the unit circle is a stable limit cycle. This can also be concluded by examining the analytical solution of (2.12)

$$r(t) = \frac{1}{(1 + c_o e^{-2t})^{1/2}} \qquad \theta(t) = \theta_o - t$$

where

$$c_o = \frac{1}{r_o^2} - 1$$

Similarly, one can find that the system (b) has an unstable limit cycle and system (c) has a semi-stable limit cycle.                                                      ☐

## 2.6  Existence of Limit Cycles

As mentioned in chapter 1, it is of great importance for control engineers to predict the existence of limit cycles in control systems. In this section, we state three simple classical theorems to that effect. These theorems are easy to understand and apply.

The first theorem to be presented reveals a simple relationship between the existence of a limit cycle and the number of singular points it encloses. In the statement of the theorem, we use $N$ to represent the number of nodes, centers, and foci enclosed by a limit cycle, and $S$ to represent the number of enclosed saddle points.

**Theorem 2.1 (Poincare)**  *If a limit cycle exists in the second-order autonomous system (2.1), then $N = S + 1$ .*

This theorem is sometimes called the *index theorem*. Its proof is mathematically involved (actually, a family of such proofs led to the development of algebraic topology) and shall be omitted here. One simple inference from this theorem is that a limit cycle must enclose at least one equilibrium point. The theorem's result can be verified easily on Figures 2.8 and 2.10.

The second theorem is concerned with the asymptotic properties of the trajectories of second-order systems.

**Theorem 2.2 (Poincare-Bendixson)** *If a trajectory of the second-order autonomous system remains in a finite region* $\Omega$, *then one of the following is true:*

*(a) the trajectory goes to an equilibrium point*

*(b) the trajectory tends to an asymptotically stable limit cycle*

*(c) the trajectory is itself a limit cycle*

While the proof of this theorem is also omitted here, its intuitive basis is easy to see, and can be verified on the previous phase portraits.

The third theorem provides a sufficient condition for the non-existence of limit cycles.

**Theorem 2.3 (Bendixson)**    *For the nonlinear system (2.1), no limit cycle can exist in a region* $\Omega$ *of the phase plane in which* $\partial f_1 / \partial x_1 + \partial f_2 / \partial x_2$ *does not vanish and does not change sign.*

**Proof**: Let us prove this theorem by contradiction. First note that, from (2.5), the equation

$$f_2\, dx_1 - f_1\, dx_2 = 0 \tag{2.15}$$

is satisfied for any system trajectories, including a limit cycle. Thus, along the closed curve $L$ of a limit cycle, we have

$$\int_L (f_1\, dx_2 - f_2\, dx_1) = 0 \tag{2.16}$$

Using Stokes' Theorem in calculus, we have

$$\int_L (f_1\, dx_2 - f_2\, dx_1) = \iint \left( \frac{\partial f_1}{\partial x_1} + \frac{\partial f_2}{\partial x_2} \right) dx_1\, dx_2 = 0$$

where the integration on the right-hand side is carried out on the area enclosed by the limit cycle.

By Equation (2.16), the left-hand side must equal zero. This, however, contradicts the fact that the right-hand side cannot equal zero because by hypothesis $\partial f_1 / \partial x_1 + \partial f_2 / \partial x_2$ does not vanish and does not change sign.  □

Let us illustrate the result on an example.

**Example 2.8**: Consider the nonlinear system

$$\dot{x}_1 = g(x_2) + 4x_1 x_2^2$$

$$\dot{x}_2 = h(x_1) + 4x_1^2 x_2$$

Since

$$\frac{\partial f_1}{\partial x_1} + \frac{\partial f_2}{\partial x_2} = 4(x_1^2 + x_2^2)$$

which is always strictly positive (except at the origin), the system does not have any limit cycles anywhere in the phase plane.                                                                              ☐

The above three theorems represent very powerful results. It is important to notice, however, that they have no equivalent in higher-order systems, where exotic asymptotic behaviors other than equilibrium points and limit cycles can occur.

# 2.7  Summary

Phase plane analysis is a graphical method used to study second-order dynamic systems. The major advantage of the method is that it allows visual examination of the global behavior of systems.  The major disadvantage is that it is mainly limited to second-order systems (although extensions to third-order systems are often achieved with the aid of computer graphics). The phenomena of multiple equilibrium points and of limit cycles are clearly seen in phase plane analysis. A number of useful classical theorems for the prediction of limit cycles in second-order systems are also presented.

# 2.8  Notes and References

Phase plane analysis is a very classical topic which has been addressed by numerous control texts. An extensive treatment can be found in [Graham and McRuer, 1961].  Examples 2.2 and 2.3 are adapted from [Ogata, 1970].  Examples 2.5 and 2.6 and section 2.6 are based on [Hsu and Meyer, 1968].

# 2.9  Exercises

**2.1**    Draw the phase portrait and discuss the properties of the linear, unity feedback control system of open-loop transfer function

$$G(p) = \frac{10}{p\,(1 + 0.1\,p)}$$

**2.2**   Draw the phase portraits of the following systems, using isoclines

(a)   $\ddot{\theta} + \dot{\theta} + 0.5\,\theta = 0$

(b)   $\ddot{\theta} + \dot{\theta} + 0.5\,\theta = 1$

(c)   $\ddot{\theta} + \dot{\theta}^2 + 0.5\,\theta = 0$

**2.3**   Consider the nonlinear system

$$\dot{x} = y + x\,(x^2 + y^2 - 1)\,\sin\frac{1}{x^2 + y^2 - 1}$$

$$\dot{y} = -x + y\,(x^2 + y^2 - 1)\,\sin\frac{1}{x^2 + y^2 - 1}$$

Without solving the above equations explicitly, show that the system has infinite number of limit cycles. Determine the stability of these limit cycles. (*Hint*: Use polar coordinates.)

**2.4**   The system shown in Figure 2.10 represents a satellite control system with rate feedback provided by a gyroscope. Draw the phase portrait of the system, and determine the system's stability.

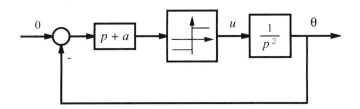

**Figure 2.10 :** Satellite control system with rate feedback

# Chapter 3
# Fundamentals of Lyapunov Theory

Given a control system, the first and most important question about its various properties is whether it is stable, because an unstable control system is typically useless and potentially dangerous. Qualitatively, a system is described as stable if starting the system somewhere near its desired operating point implies that it will stay around the point ever after. The motions of a pendulum starting near its two equilibrium points, namely, the vertical up and down positions, are frequently used to illustrate unstable and stable behavior of a dynamic system. For aircraft control systems, a typical stability problem is intuitively related to the following question: will a trajectory perturbation due to a gust cause a significant deviation in the later flight trajectory? Here, the desired operating point of the system is the flight trajectory in the absence of disturbance. Every control system, whether linear or nonlinear, involves a stability problem which should be carefully studied.

The most useful and general approach for studying the stability of nonlinear control systems is the theory introduced in the late 19th century by the Russian mathematician Alexandr Mikhailovich Lyapunov. Lyapunov's work, *The General Problem of Motion Stability*, includes two methods for stability analysis (the so-called linearization method and direct method) and was first published in 1892. The linearization method draws conclusions about a nonlinear system's local stability around an equilibrium point from the stability properties of its linear approximation. The direct method is not restricted to local motion, and determines the stability properties of a nonlinear system by constructing a scalar "energy-like" function for the system and examining the function's time variation. For over half a century, however,

Lyapunov's pioneering work on stability received little attention outside Russia, although it was translated into French in 1908 (at the instigation of Poincare), and reprinted by Princeton University Press in 1947. The publication of the work of Lur'e and a book by La Salle and Lefschetz brought Lyapunov's work to the attention of the larger control engineering community in the early 1960's. Many refinements of Lyapunov's methods have since been developed. Today, Lyapunov's linearization method has come to represent the theoretical justification of linear control, while Lyapunov's direct method has become the most important tool for nonlinear system analysis and design. Together, the linearization method and the direct method constitute the so-called Lyapunov stability theory.

The objective of this and the next chapter is to present Lyapunov stability theory and illustrate its use in the analysis and the design of nonlinear systems. To prevent mathematical complexity from obscuring the theoretical concepts, this chapter presents the most basic results of Lyapunov theory in terms of autonomous (*i.e.*, time-invariant) systems, leaving more advanced topics to chapter 4. This chapter is organized as follows. In section 3.1, we provide some background definitions concerning nonlinear systems and equilibrium points. In section 3.2, various concepts of stability are described to characterize different aspects of system behavior. Lyapunov's linearization method is presented in section 3.3. The most useful theorems in the direct method are studied in section 3.4. Section 3.5 is devoted to the question of how to use these theorems to study the stability of particular classes of nonlinear systems. Section 3.6 sketches how the direct method can be used as a powerful way of designing controllers for nonlinear systems.

# 3.1 Nonlinear Systems and Equilibrium Points

Before addressing the main problems of defining and determining stability in the next sections, let us discuss some relatively simple background issues.

### NONLINEAR SYSTEMS

A nonlinear dynamic system can usually be represented by a set of nonlinear differential equations in the form

$$\dot{\mathbf{x}} = \mathbf{f}(\mathbf{x}, t) \tag{3.1}$$

where $\mathbf{f}$ is a $n\times 1$ nonlinear vector function, and $\mathbf{x}$ is the $n\times 1$ state vector. A particular value of the state vector is also called a point because it corresponds to a point in the state-space. The number of states $n$ is called the *order* of the system. A solution $\mathbf{x}(t)$ of the equations (3.1) usually corresponds to a curve in state space as $t$ varies from zero to infinity, as already seen in phase plane analysis for the case $n = 2$. This curve is generally referred to as a *state trajectory* or a *system trajectory*.

It is important to note that although equation (3.1) does not explicitly contain the control input as a variable, it is directly applicable to feedback control systems. The reason is that equation (3.1) can represent the *closed-loop* dynamics of a feedback control system, with the control input being a function of state $\mathbf{x}$ and time $t$, and therefore disappearing in the closed-loop dynamics. Specifically, if the plant dynamics is

$$\dot{\mathbf{x}} = \mathbf{f}(\mathbf{x}, \mathbf{u}, t)$$

and some control law has been selected

$$\mathbf{u} = \mathbf{g}(\mathbf{x}, t)$$

then the closed-loop dynamics is

$$\dot{\mathbf{x}} = \mathbf{f}[\mathbf{x}, \mathbf{g}(\mathbf{x}, t), t]$$

which can be rewritten in the form (3.1). Of course, equation (3.1) can also represent dynamic systems where no control signals are involved, such as a freely swinging pendulum.

A special class of nonlinear systems are *linear systems*. The dynamics of linear systems are of the form

$$\dot{\mathbf{x}} = \mathbf{A}(t)\mathbf{x}$$

where $\mathbf{A}(t)$ is an $n \times n$ matrix.

## AUTONOMOUS AND NON-AUTONOMOUS SYSTEMS

Linear systems are classified as either time-varying or time-invariant, depending on whether the system matrix $\mathbf{A}$ varies with time or not. In the more general context of nonlinear systems, these adjectives are traditionally replaced by "autonomous" and "non-autonomous".

**Definition 3.1**     *The nonlinear system (3.1) is said to be <u>autonomous</u> if* $\mathbf{f}$ *does not depend explicitly on time*, i.e., if the system's state equation can be written

$$\dot{\mathbf{x}} = \mathbf{f}(\mathbf{x}) \tag{3.2}$$

*Otherwise, the system is called <u>non-autonomous</u> .*

Obviously, linear time-invariant (LTI) systems are autonomous and linear time-varying (LTV) systems are non-autonomous. The second-order systems studied in chapter 2 are all autonomous.

Strictly speaking, all physical systems are non-autonomous, because none of their dynamic characteristics is strictly time-invariant. The concept of an autonomous system is an idealized notion, like the concept of a linear system. In practice, however, system properties often change very slowly, and we can neglect their time variation without causing any practically meaningful error.

It is important to note that for control systems, the above definition is made on the *closed-loop dynamics*. Since a control system is composed of a controller and a plant (including sensor and actuator dynamics), the non-autonomous nature of a control system may be due to a time-variation either in the plant or in the control law. Specifically, a time-invariant plant with dynamics

$$\dot{\mathbf{x}} = \mathbf{f}(\mathbf{x}, \mathbf{u})$$

may lead to a non-autonomous closed-loop system if a controller dependent on time $t$ is chosen, *i.e.*, if $\mathbf{u} = \mathbf{g}(\mathbf{x}, t)$. For example, the closed-loop system of the simple plant $\dot{x} = -x + u$ can be nonlinear and non-autonomous by choosing $u$ to be nonlinear and time-varying (*e.g.*, $u = -x^2 \sin t$). In fact, adaptive controllers for linear time-invariant plants usually make the closed-loop control systems nonlinear and non-autonomous.

The fundamental difference between autonomous and non-autonomous systems lies in the fact that the state trajectory of an autonomous system is independent of the initial time, while that of a non-autonomous system generally is not. As we will see in the next chapter, this difference requires us to consider the initial time explicitly in defining stability concepts for non-autonomous systems, and makes the analysis more difficult than that of autonomous systems.

It is well known that the analysis of linear time-invariant systems is much easier than that of linear time-varying systems. The same is true with nonlinear systems. Generally speaking, autonomous systems have relatively simpler properties and their analysis is much easier. For this reason, *in the remainder of this chapter, we will concentrate on the analysis of autonomous systems*, represented by (3.2). Extensions of the concepts and results to non-autonomous systems will be studied in chapter 4.

## EQUILIBRIUM POINTS

It is possible for a system trajectory to correspond to only a single point. Such a point is called an equilibrium point. As we shall see later, many stability problems are naturally formulated with respect to equilibrium points.

**Definition 3.2** *A state* $\mathbf{x}^*$ *is an equilibrium state (or equilibrium point) of the system if once* $\mathbf{x}(t)$ *is equal to* $\mathbf{x}^*$, *it remains equal to* $\mathbf{x}^*$ *for all future time.*

Mathematically, this means that the constant vector $\mathbf{x}^*$ satisfies

$$\mathbf{0} = \mathbf{f}(\mathbf{x}^*) \tag{3.3}$$

Equilibrium points can be found by solving the nonlinear algebraic equations (3.3).

A linear time-invariant system

$$\dot{\mathbf{x}} = \mathbf{A}\,\mathbf{x} \tag{3.4}$$

has a single equilibrium point (the origin $\mathbf{0}$) if $\mathbf{A}$ is nonsingular. If $\mathbf{A}$ is singular, it has an infinity of equilibrium points, which are contained in the null-space of the matrix $\mathbf{A}$, *i.e.*, the subspace defined by $\mathbf{A}\mathbf{x} = \mathbf{0}$. This implies that the equilibrium points are not isolated, as reflected by the example $\ddot{x} + \dot{x} = 0$, for which all points on the $x$ axis of the phase plane are equilibrium points.

A nonlinear system can have several (or infinitely many) isolated equilibrium points, as seen in Example 1.1. The following example involves a familiar physical system.

### Example 3.1: The Pendulum

Consider the pendulum of Figure 3.1, whose dynamics is given by the following nonlinear autonomous equation

$$MR^2\,\ddot{\theta} + b\,\dot{\theta} + MgR\,\sin\theta = 0 \tag{3.5}$$

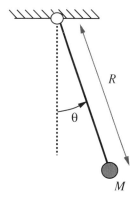

**Figure 3.1** : The pendulum

where $R$ is the pendulum's length, $M$ its mass, $b$ the friction coefficient at the hinge, and $g$ the gravity constant. Letting $x_1 = \theta$, $x_2 = \dot{\theta}$, the corresponding state-space equation is

$$\dot{x}_1 = x_2 \tag{3.6a}$$

$$\dot{x}_2 = -\frac{b}{MR^2}x_2 - \frac{g}{R}\sin x_1 \tag{3.6b}$$

Therefore, the equilibrium points are given by

$$x_2 = 0, \quad \sin x_1 = 0$$

which leads to the points $(0\ [2\pi],\ 0)$ and $(\pi\ [2\pi],\ 0)$. Physically, these points correspond to the pendulum resting exactly at the vertical up and down positions.     □

In linear system analysis and design, for notational and analytical simplicity, we often transform the linear system equations in such a way that the equilibrium point is the origin of the state-space. We can do the same thing for nonlinear systems (3.2), about a *specific* equilibrium point. Let us say that the equilibrium point of interest is $\mathbf{x}^*$. Then, by introducing a new variable

$$\mathbf{y} = \mathbf{x} - \mathbf{x}^*$$

and substituting $\mathbf{x} = \mathbf{y} + \mathbf{x}^*$ into equations (3.2), a new set of equations on the variable $\mathbf{y}$ are obtained

$$\dot{\mathbf{y}} = \mathbf{f}(\mathbf{y} + \mathbf{x}^*) \tag{3.7}$$

One can easily verify that there is a one-to-one correspondence between the solutions of (3.2) and those of (3.7), and that in addition, $\mathbf{y} = \mathbf{0}$, the solution corresponding to $\mathbf{x} = \mathbf{x}^*$, is an equilibrium point of (3.7). Therefore, instead of studying the behavior of the equation (3.2) in the neighborhood of $\mathbf{x}^*$, one can equivalently study the behavior of the equations (3.7) in the neighborhood of the origin.

## NOMINAL MOTION

In some practical problems, we are not concerned with stability around an equilibrium point, but rather with the stability of a *motion*, *i.e*, whether a system will remain close to its original motion trajectory if slightly perturbed away from it, as exemplified by the aircraft trajectory control problem mentioned at the beginning of this chapter. We can show that this kind of motion stability problem can be transformed into an equivalent stability problem around an equilibrium point, although the equivalent system is now non-autonomous.

Let $\mathbf{x}^*(t)$ be the solution of equation (3.2), *i.e.*, the nominal motion trajectory, corresponding to initial condition $\mathbf{x}^*(0) = \mathbf{x}_O$. Let us now perturb the initial condition to be $\mathbf{x}(0) = \mathbf{x}_O + \delta\mathbf{x}_O$ and study the associated variation of the motion error

$$\mathbf{e}(t) = \mathbf{x}(t) - \mathbf{x}^*(t)$$

as illustrated in Figure 3.2. Since both $\mathbf{x}^*(t)$ and $\mathbf{x}(t)$ are solutions of (3.2), we have

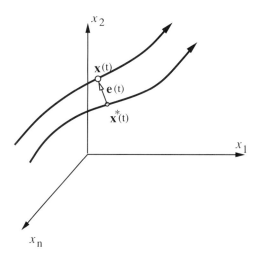

**Figure 3.2** : Nominal and Perturbed Motions

$$\dot{\mathbf{x}}^* = \mathbf{f}(\mathbf{x}^*) \qquad \mathbf{x}(0) = \mathbf{x}_O$$

$$\dot{\mathbf{x}} = \mathbf{f}(\mathbf{x}) \qquad \mathbf{x}(0) = \mathbf{x}_O + \delta\mathbf{x}_O$$

then $\mathbf{e}(t)$ satisfies the following non-autonomous differential equation

$$\dot{\mathbf{e}} = \mathbf{f}(\mathbf{x}^* + \mathbf{e}, t) - \mathbf{f}(\mathbf{x}^*, t) = \mathbf{g}(\mathbf{e}, t) \tag{3.8}$$

with initial condition $\mathbf{e}(0) = \delta\mathbf{x}_O$. Since $\mathbf{g}(\mathbf{0}, t) = \mathbf{0}$, the new dynamic system, with $\mathbf{e}$ as state and $\mathbf{g}$ in place of $\mathbf{f}$, has an equilibrium point at the origin of the state space. Therefore, instead of studying the deviation of $\mathbf{x}(t)$ from $\mathbf{x}^*(t)$ for the original system, we may simply study the stability of the perturbation dynamics (3.8) with respect to the equilibrium point $\mathbf{0}$. Note, however, that the perturbation dynamics is non-autonomous, due to the presence of the nominal trajectory $\mathbf{x}^*(t)$ on the right-hand side. *Each* particular nominal motion of an autonomous system corresponds to an equivalent non-autonomous system, whose study requires the non-autonomous system analysis techniques to be presented in chapter 4.

Let us now illustrate this important transformation on a specific system.

**Example 3.2**: Consider the autonomous mass-spring system

$$m\ddot{x} + k_1 x + k_2 x^3 = 0$$

which contains a nonlinear term reflecting the hardening effect of the spring. Let us study the stability of the motion $x^*(t)$ which starts from initial position $x_o$.

Assume that we slightly perturb the initial position to be $x(0) = x_o + \delta x_o$. The resulting system trajectory is denoted as $x(t)$. Proceeding as before, the equivalent differential equation governing the motion error $e$ is

$$m\ddot{e} + k_1 e + k_2 [ e^3 + 3 e^2 x^*(t) + 3 e x^{*2}(t) ] = 0$$

Clearly, this is a non-autonomous system.     □

Of course, one can also show that for non-autonomous nonlinear systems, the stability problem around a nominal motion can also be transformed as a stability problem around the origin for an equivalent non-autonomous system.

Finally, note that if the original system is autonomous and *linear*, in the form (3.4), then the equivalent system is still autonomous, since it can be written

$$\dot{\mathbf{e}} = \mathbf{A}\,\mathbf{e}$$

## 3.2  Concepts of Stability

In the beginning of this chapter, we introduced the intuitive notion of stability as a kind of well-behavedness around a desired operating point. However, since nonlinear systems may have much more complex and exotic behavior than linear systems, the mere notion of stability is not enough to describe the essential features of their motion. A number of more refined stability concepts, such as asymptotic stability, exponential stability and global asymptotic stability, are needed. In this section, we define these stability concepts formally, for autonomous systems, and explain their practical meanings.

A few simplifying notations are defined at this point. Let $\mathbf{B}_R$ denote the spherical region (or ball) defined by $\| \mathbf{x} \| < R$ in state-space, and $\mathbf{S}_R$ the sphere itself, defined by $\| \mathbf{x} \| = R$.

## STABILITY AND INSTABILITY

Let us first introduce the basic concepts of stability and instability.

**Definition 3.3**     *The equilibrium state* $\mathbf{x} = \mathbf{0}$ *is said to be* <u>stable</u> *if, for any* $R > 0$, *there exists* $r > 0$, *such that if* $\|\mathbf{x}(0)\| < r$, *then* $\|\mathbf{x}(t)\| < R$ *for all* $t \geq 0$ . *Otherwise, the equilibrium point is* <u>unstable</u>.

Essentially, stability (also called *stability in the sense of Lyapunov*, or *Lyapunov stability*) means that the system trajectory can be kept arbitrarily close to the origin by starting sufficiently close to it. More formally, the definition states that the origin is stable, if, given that we do not want the state trajectory $\mathbf{x}(t)$ to get out of a ball of arbitrarily specified radius $\mathbf{B}_R$ , a value $r(R)$ can be found such that starting the state from within the ball $\mathbf{B}_r$ at time 0 guarantees that the state will stay within the ball $\mathbf{B}_R$ thereafter. The geometrical implication of stability is indicated by curve 2 in Figure 3.3. Chapter 2 provides examples of stable equilibrium points in the case of second-order systems, such as the origin for the mass-spring system of Example 2.1, or stable nodes or foci in the local linearization of a nonlinear system.

Throughout the book, we shall use the standard mathematical abbreviation symbols:

$\forall$     to mean "for any"
$\exists$     for "there exists"
$\in$     for "in the set"
$=>$     for "implies that"

Of course, we shall say interchangeably that $A$ implies $B$, or that $A$ is a *sufficient condition* of $B$, or that $B$ is a *necessary condition* of $A$. If $A => B$ and $B => A$ , then $A$ and $B$ are *equivalent*, which we shall denote by $A <=> B$ .

Using these symbols, Definition 3.3 can be written

$$\forall R > 0 , \exists r > 0 , \quad \| \mathbf{x}(0) \| < r \quad => \quad \forall t \geq 0 , \| \mathbf{x}(t) \| < R$$

or, equivalently

$$\forall R > 0 , \exists r > 0 , \quad \mathbf{x}(0) \in \mathbf{B}_r \quad => \quad \forall t \geq 0 , \mathbf{x}(t) \in \mathbf{B}_R$$

Conversely, an equilibrium point is unstable if there exists at least *one* ball $\mathbf{B}_R$, such that for *every* $r > 0$, no matter how small, it is always possible for the system trajectory to start somewhere within the ball $\mathbf{B}_r$ and eventually leave the ball $\mathbf{B}_R$ (Figure 3.3). Unstable nodes or saddle points in second-order systems are examples of unstable equilibria. Instability of an equilibrium point is typically undesirable, because it often leads the system into limit cycles or results in damage to the involved mechanical or electrical components.

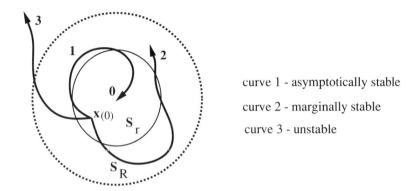

curve 1 - asymptotically stable

curve 2 - marginally stable

curve 3 - unstable

**Figure 3.3 :** Concepts of stability

It is important to point out the qualitative difference between instability and the intuitive notion of "blowing up" (all trajectories close to origin move further and further away to infinity). In linear systems, instability is equivalent to blowing up, because unstable poles always lead to exponential growth of the system states. However, for nonlinear systems, blowing up is only one way of instability. The following example illustrates this point.

**Example 3.3: Instability of the Van der Pol Oscillator**

The Van der Pol oscillator of Example 2.6 is described by

$$\dot{x}_1 = x_2$$

$$\dot{x}_2 = -x_1 + (1 - x_1^2)x_2$$

One easily shows that the system has an equilibrium point at the origin.

As pointed out in section 2.2 and seen in the phase portrait of Figure 2.8, system trajectories starting from any non-zero initial states all asymptotically approach a limit cycle. This implies that, if we choose $R$ in Definition 3.3 to be small enough for the circle of radius $R$ to fall completely within the closed-curve of the limit cycle, then system trajectories starting near the origin will eventually get out of this circle (Figure 3.4). This implies instability of the origin.

Thus, even though the state of the system does remain around the equilibrium point in a certain sense, it cannot stay *arbitrarily* close to it. This is the fundamental distinction between stability and instability. ◻

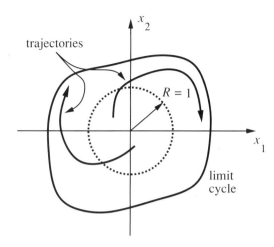

**Figure 3.4 :** Unstable origin of the Van der Pol Oscillator

## ASYMPTOTIC STABILITY AND EXPONENTIAL STABILITY

In many engineering applications, Lyapunov stability is not enough. For example, when a satellite's attitude is disturbed from its nominal position, we not only want the satellite to maintain its attitude in a range determined by the magnitude of the disturbance, *i.e.*, Lyapunov stability, but also require that the attitude gradually go back to its original value. This type of engineering requirement is captured by the concept of asymptotic stability.

**Definition 3.4**    *An equilibrium point* **0** *is* <u>*asymptotically stable*</u> *if it is stable, and if in addition there exists some $r > 0$ such that $\| \mathbf{x}(0) \| < r$ implies that $\mathbf{x}(t) \to \mathbf{0}$ as $t \to \infty$.*

Asymptotic stability means that the equilibrium is stable, and that in addition, states started close to **0** actually converge to **0** as time $t$ goes to infinity. Figure 3.3 shows that system trajectories starting from within the ball $\mathbf{B}_r$ converge to the origin. The ball $\mathbf{B}_r$ is called a *domain of attraction* of the equilibrium point (while *the* domain of attraction of the equilibrium point refers to the largest such region, *i.e.*, to the set of all points such that trajectories initiated at these points eventually converge to the origin). An equilibrium point which is Lyapunov stable but not asymptotically stable is called *marginally stable*.

One may question the need for the explicit stability requirement in the definition above, in view of the second condition of state convergence to the origin. However, it it easy to build counter-examples that show that state convergence does not necessarily imply stability. For instance, a simple system studied by Vinograd has trajectories of the form shown in Figure 3.5. All the trajectories starting from non-zero initial points within the unit disk first reach the curve **C** before converging to the origin.   Thus, the origin is *unstable* in the sense of Lyapunov, despite the state convergence.  Calling such a system unstable is quite reasonable, since a curve such as **C** may be outside the region where the model is valid – for instance, the subsonic and supersonic dynamics of a high-performance aircraft are radically different, while, with the problem under study using subsonic dynamic models, **C** could be in the supersonic range.

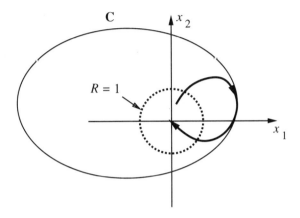

**Figure 3.5 :** State convergence does not imply stability

In many engineering applications, it is still not sufficient to know that a system will converge to the equilibrium point after infinite time. There is a need to estimate *how fast* the system trajectory approaches **0**.  The concept of *exponential stability* can be used for this purpose.

**Definition 3.5**   *An equilibrium point* **0** *is* <u>*exponentially stable*</u> *if there exist two strictly positive numbers* $\alpha$ *and* $\lambda$ *such that*

$$\forall \, t > 0, \quad \| \mathbf{x}(t) \| \leq \alpha \, \| \mathbf{x}(0) \| \, e^{-\lambda t} \tag{3.9}$$

*in some ball* $\mathbf{B}_r$ *around the origin.*

In words, (3.9) means that the state vector of an exponentially stable system converges to the origin faster than an exponential function.  The positive number $\lambda$ is often called the *rate* of exponential convergence. For instance, the system

$$\dot{x} = -(1 + \sin^2 x) \, x$$

is exponentially convergent to $x = 0$ with a rate $\lambda = 1$ . Indeed, its solution is

$$x(t) = x(0) \exp\left( \int_o^t [1 + \sin^2(x(\tau))] \, d\tau \right)$$

and therefore

$$|x(t)| \leq |x(0)| \, e^{-t}$$

Note that exponential stability implies asymptotic stability. But asymptotic stability does not guarantee exponential stability, as can be seen from the system

$$\dot{x} = -x^2, \quad x(0) = 1 \tag{3.10}$$

whose solution is $x = 1/(1 + t)$, a function slower than any exponential function $e^{-\lambda t}$ (with $\lambda > 0$).

The definition of exponential convergence provides an explicit bound on the state at any time, as seen in (3.9). By writing the positive constant $\alpha$ as $\alpha = e^{\lambda \tau_o}$, it is easy to see that, after a time of $\tau_o + (1/\lambda)$, the magnitude of the state vector decreases to less than 35% ($\approx e^{-1}$) of its original value, similarly to the notion of *time-constant* in a linear system. After $\tau_o + (3/\lambda)$, the state magnitude $\|\mathbf{x}(t)\|$ will be less than 5% ($\approx e^{-3}$) of $\|\mathbf{x}(0)\|$.

## LOCAL AND GLOBAL STABILITY

The above definitions are formulated to characterize the *local* behavior of systems, *i.e.*, how the state evolves after starting near the equilibrium point. Local properties tell little about how the system will behave when the initial state is some distance away from the equilibrium, as seen for the nonlinear system in Example 1.1. Global concepts are required for this purpose.

**Definition 3.6**   *If asymptotic (or exponential) stability holds for any initial states, the equilibrium point is said to be asymptotically (or exponentially) stable in the large. It is also called globally asymptotically (or exponentially) stable.*

For instance, in Example 1.2 the linearized system is globally asymptotically stable, but the original system is not. The simple system in (3.10) is also globally asymptotically stable, as can be seen from its solutions.

*Linear* time-invariant systems are either asymptotically stable, or marginally stable, or unstable, as can be be seen from the modal decomposition of linear system solutions; linear asymptotic stability is always global and exponential, and linear instability always implies exponential blow-up. *This explains why the refined notions of stability introduced here were not previously encountered in the study of linear systems. They are explicitly needed only for nonlinear systems.*

# 3.3  Linearization and Local Stability

Lyapunov's linearization method is concerned with the *local* stability of a nonlinear system. It is a formalization of the intuition that a nonlinear system should behave similarly to its linearized approximation for small range motions. Because all physical systems are inherently nonlinear, Lyapunov's linearization method serves as the fundamental *justification of using linear control techniques* in practice, *i.e.*, shows that stable design by linear control guarantees the stability of the original physical system locally.

Consider the autonomous system in (3.2), and assume that $\mathbf{f}(\mathbf{x})$ is continuously differentiable. Then the system dynamics can be written as

$$\dot{\mathbf{x}} = \left( \frac{\partial \mathbf{f}}{\partial \mathbf{x}} \right)_{\mathbf{x}=\mathbf{0}} \mathbf{x} + \mathbf{f}_{h.o.t.}(\mathbf{x}) \tag{3.11}$$

where $\mathbf{f}_{h.o.t.}$ stands for higher-order terms in $\mathbf{x}$. Note that the above Taylor expansion starts directly with the first-order term, due to the fact that $\mathbf{f}(\mathbf{0}) = \mathbf{0}$, since $\mathbf{0}$ is an equilibrium point. Let us use the constant matrix $\mathbf{A}$ to denote the Jacobian matrix of $\mathbf{f}$ with respect to $\mathbf{x}$ at $\mathbf{x} = \mathbf{0}$ (an $n \times n$ matrix of elements $\partial f_i / \partial x_j$)

$$\mathbf{A} = \left( \frac{\partial \mathbf{f}}{\partial \mathbf{x}} \right)_{\mathbf{x}=\mathbf{0}}$$

Then, the system

$$\dot{\mathbf{x}} = \mathbf{A}\,\mathbf{x} \tag{3.12}$$

is called the *linearization* (or *linear approximation*) of the original nonlinear system at the equilibrium point $\mathbf{0}$.

Note that, similarly, starting with a non-autonomous nonlinear system with a control input $\mathbf{u}$

$$\dot{\mathbf{x}} = \mathbf{f}(\mathbf{x}, \mathbf{u})$$

such that $\mathbf{f}(\mathbf{0}, \mathbf{0}) = \mathbf{0}$, we can write

$$\dot{\mathbf{x}} = \left( \frac{\partial \mathbf{f}}{\partial \mathbf{x}} \right)_{(\mathbf{x}=\mathbf{0},\,\mathbf{u}=\mathbf{0})} \mathbf{x} + \left( \frac{\partial \mathbf{f}}{\partial \mathbf{u}} \right)_{(\mathbf{x}=\mathbf{0},\,\mathbf{u}=\mathbf{0})} \mathbf{u} + \mathbf{f}_{h.o.t.}(\mathbf{x}, \mathbf{u})$$

where $\mathbf{f}_{h.o.t.}$ stands for higher-order terms in $\mathbf{x}$ and $\mathbf{u}$. Letting $\mathbf{A}$ denote the Jacobian matrix of $\mathbf{f}$ with respect to $\mathbf{x}$ at $(\mathbf{x} = \mathbf{0}, \mathbf{u} = \mathbf{0})$, and $\mathbf{B}$ denote the Jacobian matrix of $\mathbf{f}$ with respect to $\mathbf{u}$ at the same point (an $n \times m$ matrix of elements $\partial f_i / \partial u_j$, where $m$ is the number of inputs)

$$\Lambda = \left(\frac{\partial\,f}{\partial\,x}\right)_{(x=0,\,u=0)} \qquad\qquad B = \left(\frac{\partial\,f}{\partial\,u}\right)_{(x=0,\,u=0)}$$

the system

$$\dot{x} = A\,x + B\,u$$

is the linearization (or linear approximation) of the original nonlinear system at $(x = 0, u = 0)$.

Furthermore, the choice of a control law of the form $u = u(x)$ (with $u(0) = 0$) transforms the original non-autonomous system into an autonomous closed-loop system, having $x = 0$ as an equilibrium point. Linearly approximating the control law as

$$u \approx \left(\frac{d\,u}{d\,x}\right)_{x=0} x = G\,x$$

the closed-loop dynamics can be linearly approximated as

$$\dot{x} = f(x, u(x)) \approx (A + B\,G)\,x$$

Of course, the same linear approximation can be obtained by directly considering the autonomous closed-loop system

$$\dot{x} = f(x, u(x)) = f_1(x)$$

and linearizing the function $f_1$ with respect to $x$, at its equilibrium point $x = 0$.

In practice, finding a system's linearization is often most easily done simply by neglecting any term of order higher than 1 in the dynamics, as we now illustrate.

**Example 3.4:** Consider the system

$$\dot{x}_1 = x_2{}^2 + x_1 \cos x_2$$

$$\dot{x}_2 = x_2 + (x_1 + 1)\, x_1 + x_1 \sin x_2$$

Its linearized approximation about $x = 0$ is

$$\dot{x}_1 \approx 0 + x_1 \cdot 1 = x_1$$

$$\dot{x}_2 \approx x_2 + 0 + x_1 + x_1 x_2 \approx x_2 + x_1$$

The linearized system can thus be written

$$\dot{x} = \begin{bmatrix} 1 & 0 \\ 1 & 1 \end{bmatrix} x$$

A similar procedure can be applied for a controlled system. Consider the system

$$\ddot{x} + 4\,\dot{x}^5 + (x^2 + 1)\,u = 0$$

The system can be linearly approximated about $\mathbf{x} = \mathbf{0}$ as

$$\ddot{x} + 0 + (0 + 1)\,u \approx 0$$

*i.e.*, the linearized system can be written

$$\ddot{x} = -u$$

Assume that the control law for the original nonlinear system has been selected to be

$$u = \sin x + x^3 + \dot{x}\cos^2 x$$

then the linearized closed-loop dynamics is

$$\ddot{x} + \dot{x} + x = 0 \qquad\qquad\qquad \square$$

The following result makes precise the relationship between the stability of the linear system (3.12) and that of the original nonlinear system (3.2).

### Theorem 3.1 (Lyapunov's linearization method)

> • *If the linearized system is strictly stable (i.e, if all eigenvalues of* **A** *are strictly in the left-half complex plane), then the equilibrium point is asymptotically stable (for the actual nonlinear system).*

> • *If the linearized system is unstable (i.e, if at least one eigenvalue of* **A** *is strictly in the right-half complex plane), then the equilibrium point is unstable (for the nonlinear system).*

> • *If the linearized system is marginally stable (i.e, all eigenvalues of* **A** *are in the left-half complex plane, but at least one of them is on the jω axis), then one cannot conclude anything from the linear approximation (the equilibrium point may be stable, asymptotically stable, or unstable for the nonlinear system).*

While the proof of this theorem (which is actually based on Lyapunov's direct method, see Exercise 3.12) shall not be detailed, let us remark that its results are *intuitive*. A summary of the theorem is that it is true *by continuity*. If the linearized system is strictly stable, or strictly unstable, then, since the approximation is valid "not too far" from the equilibrium, the nonlinear system itself is locally stable, or locally unstable. However, if the linearized system is marginally stable, the higher-order terms in (3.11) can have a decisive effect on whether the nonlinear system is stable or unstable. As we shall see in the next section, simple nonlinear systems may be globally asymptotically stable while their linear approximations are only marginally stable: one simply cannot infer any stability property of a nonlinear system from its marginally stable linear approximation.

**Example 3.5:** As expected, it can be shown easily that the equilibrium points $(\theta = \pi \,[2\pi]\,,\,\dot\theta = 0)$ of the pendulum of Example 3.1 are unstable. Consider for instance the equilibrium point $(\theta = \pi\,,\,\dot\theta = 0)$. Since, in a neighborhood of $\theta = \pi$, we can write

$$\sin\theta \;=\; \sin\pi + \cos\pi\,(\theta - \pi) + h.o.t. \;=\; (\pi - \theta) + h.o.t.$$

thus, letting $\tilde\theta = \theta - \pi$, the system's linearization *about the equilibrium point* $(\theta = \pi\,,\,\dot\theta = 0)$ is

$$\ddot{\tilde\theta} + \frac{b}{MR^2}\,\dot{\tilde\theta} - \frac{g}{R}\,\tilde\theta \;=\; 0$$

Hence the linear approximation is unstable, and therefore so is the nonlinear system at this equilibrium point.                              □

**Example 3.6:** Consider the first order system

$$\dot{x} = ax + bx^5$$

The origin 0 is one of the two equilibrium points of this system. The linearization of this system around the origin is

$$\dot{x} = ax$$

The application of Lyapunov's linearization method indicates the following stability properties of the nonlinear system

- $a < 0$ : asymptotically stable;
- $a > 0$ : unstable;
- $a = 0$ : cannot tell from linearization.

In the third case, the nonlinear system is

$$\dot{x} = bx^5$$

The linearization method fails while, as we shall see, the direct method to be described can easily solve this problem.                              □

Lyapunov's linearization theorem shows that linear control design is a matter of *consistency*: one must design a controller such that the system remain in its "linear range". It also stresses major limitations of linear design: how large is the linear range? What is the *extent* of stability (how large is *r* in Definition 3.3) ? These questions motivate a deeper approach to the nonlinear control problem, Lyapunov's direct method.

# 3.4 Lyapunov's Direct Method

The basic philosophy of Lyapunov's direct method is the mathematical extension of a fundamental physical observation: if the total *energy* of a mechanical (or electrical) system is continuously dissipated, then the system, *whether linear or nonlinear*, must eventually settle down to an equilibrium point. Thus, we may conclude the stability of a system by examining the variation of a single *scalar* function.

Specifically, let us consider the nonlinear mass-damper-spring system in Figure 3.6, whose dynamic equation is

$$m\ddot{x} + b\dot{x}|\dot{x}| + k_o x + k_1 x^3 = 0 \tag{3.13}$$

with $b\dot{x}|\dot{x}|$ representing nonlinear dissipation or damping, and $(k_o x + k_1 x^3)$ representing a nonlinear spring term. Assume that the mass is pulled away from the natural length of the spring by a large distance, and then released. Will the resulting motion be stable? It is very difficult to answer this question using the definitions of stability, because the general solution of this nonlinear equation is unavailable. The linearization method cannot be used either because the motion starts outside the linear range (and in any case the system's linear approximation is only marginally stable). However, examination of the system energy can tell us a lot about the motion pattern.

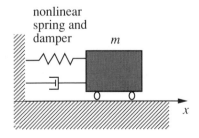

nonlinear
spring and
damper      *m*

*x*

**Figure 3.6 :** A nonlinear mass-damper-spring system

The total mechanical energy of the system is the sum of its kinetic energy and its potential energy

$$V(\mathbf{x}) = \frac{1}{2} m \dot{x}^2 + \int_o^x (k_o x + k_1 x^3) \, dx = \frac{1}{2} m \dot{x}^2 + \frac{1}{2} k_o x^2 + \frac{1}{4} k_1 x^4 \quad (3.14)$$

Comparing the definitions of stability and mechanical energy, one can easily see some relations between the mechanical energy and the stability concepts described earlier:

- zero energy corresponds to the equilibrium point ($\mathbf{x} = \mathbf{0}$, $\dot{\mathbf{x}} = \mathbf{0}$)
- asymptotic stability implies the convergence of mechanical energy to zero
- instability is related to the growth of mechanical energy

These relations indicate that the value of a scalar quantity, the mechanical energy, indirectly reflects the magnitude of the state vector; and furthermore, that the stability properties of the system can be characterized by the variation of the mechanical energy of the system.

The rate of energy variation during the system's motion is obtained easily by differentiating the first equality in (3.14) and using (3.13)

$$\dot{V}(\mathbf{x}) = m \dot{x} \ddot{x} + (k_o x + k_1 \, x^3) \, \dot{x} = \dot{x} \, (-b \dot{x} |\dot{x}|) = -b |\dot{x}|^3 \quad (3.15)$$

Equation (3.15) implies that the energy of the system, starting from some initial value, is continuously dissipated by the damper until the mass settles down, *i.e.*, until $\dot{x} = 0$. Physically, it is easy to see that the mass must finally settle down at the natural length of the spring, because it is subjected to a non-zero spring force at any position other than the natural length.

The direct method of Lyapunov is based on a generalization of the concepts in the above mass-spring-damper system to more complex systems. Faced with a set of nonlinear differential equations, the basic procedure of Lyapunov's direct method is to generate a scalar "energy-like" function for the dynamic system, and examine the time variation of that scalar function. In this way, conclusions may be drawn on the stability of the set of differential equations without using the difficult stability definitions or requiring explicit knowledge of solutions.

## 3.4.1  Positive Definite Functions and Lyapunov Functions

The energy function in (3.14) has two properties. The first is a property of the function itself: it is strictly positive unless both state variables $x$ and $\dot{x}$ are zero. The second property is a property associated with the dynamics (3.13): the function is monotonically decreasing when the variables $x$ and $\dot{x}$ vary according to (3.13). In Lyapunov's direct method, the first property is formalized by the notion of *positive definite functions*, and the second is formalized by the so-called Lyapunov functions.

Let us discuss positive definite functions first.

**Definition 3.7**  *A scalar continuous function $V(\mathbf{x})$ is said to be <u>locally positive definite</u> if $V(\mathbf{0}) = 0$ and, in a ball $\mathbf{B}_{R_o}$*

$$\mathbf{x} \neq \mathbf{0} \quad => \quad V(\mathbf{x}) > 0$$

*If $V(\mathbf{0}) = 0$ and the above property holds over the whole state space, then $V(\mathbf{x})$ is said to be <u>globally positive definite</u>.*

For instance, the function

$$V(\mathbf{x}) = \frac{1}{2} M R^2 x_2^2 + M R g (1 - \cos x_1)$$

which is the mechanical energy of the pendulum of Example 3.1, is locally positive definite. The mechanical energy (3.14) of the nonlinear mass-damper-spring system is globally positive definite. Note that, for that system, the kinetic energy $(1/2)\, m\, \dot{x}^2$ is *not* positive definite by itself, because it can equal zero for non-zero values of $x$.

The above definition implies that the function $V$ has a unique minimum at the origin $\mathbf{0}$. Actually, given any function having a *unique* minimum in a certain ball, we can construct a locally positive definite function simply by adding a constant to that function. For example, the function $V(\mathbf{x}) = x_1^2 + x_2^2 - 1$ is a lower bounded function with a unique minimum at the origin, and the addition of the constant 1 to it makes it a positive definite function. Of course, the function shifted by a constant has the same time-derivative as the original function.

Let us describe the geometrical meaning of locally positive definite functions. Consider a positive definite function $V(\mathbf{x})$ of two state variables $x_1$ and $x_2$. Plotted in a 3-dimensional space, $V(\mathbf{x})$ typically corresponds to a surface looking like an upward cup (Figure 3.7). The lowest point of the cup is located at the origin.

A second geometrical representation can be made as follows. Taking $x_1$ and $x_2$ as Cartesian coordinates, the level curves $V(x_1, x_2) = V_\alpha$ typically represent a set of ovals surrounding the origin, with each oval corresponding to a positive value of $V_\alpha$. These ovals, often called *contour curves*, may be thought as the sections of the cup by horizontal planes, projected on the $(x_1, x_2)$ plane (Figure 3.8). Note that the contour curves do not intersect, because $V(x_1, x_2)$ is uniquely defined given $(x_1, x_2)$.

A few related concepts can be defined similarly, in a local or global sense, *i.e.*, a function $V(\mathbf{x})$ is *negative definite* if $-V(\mathbf{x})$ is positive definite; $V(\mathbf{x})$ is *positive semi-definite* if $V(\mathbf{0}) = 0$ and $V(\mathbf{x}) \geq 0$ for $\mathbf{x} \neq \mathbf{0}$; $V(\mathbf{x})$ is *negative semi-definite* if $-V(\mathbf{x})$ is positive semi-definite. The prefix "semi" is used to reflect the possibility of $V$ being

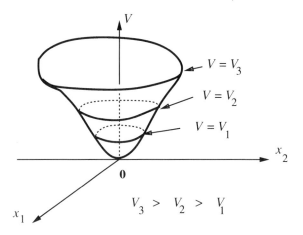

**Figure 3.7 :** Typical shape of a positive definite function $V(x_1, x_2)$

equal to zero for $\mathbf{x} \neq \mathbf{0}$. These concepts can be given geometrical meanings similar to the ones given for positive definite functions.

With $\mathbf{x}$ denoting the state of the system (3.2), a scalar function $V(\mathbf{x})$ actually represents an implicit function of time $t$. Assuming that $V(\mathbf{x})$ is differentiable, its derivative with respect to time can be found by the chain rule,

$$\dot{V} = \frac{dV(\mathbf{x})}{dt} = \frac{\partial V}{\partial \mathbf{x}} \dot{\mathbf{x}} = \frac{\partial V}{\partial \mathbf{x}} \mathbf{f}(\mathbf{x})$$

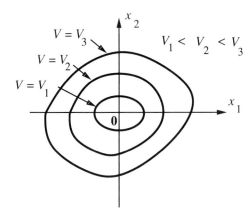

**Figure 3.8 :** Interpreting positive definite functions using contour curves

We see that, because **x** is required to satisfy the autonomous state equations (3.2), $\dot{V}$ only depends on **x**. It is often referred to as "the derivative of $V$ along the system trajectory". For the system (3.13), $\dot{V}(\mathbf{x})$ is computed in (3.15) and found to be negative. Functions such as $V$ in that example are given a special name because of their importance in Lyapunov's direct method.

**Definition 3.8** *If, in a ball* $\mathbf{B}_{R_o}$, *the function* $V(\mathbf{x})$ *is positive definite and has continuous partial derivatives, and if its time derivative along any state trajectory of system (3.2) is negative semi-definite, i.e.,*

$$\dot{V}(\mathbf{x}) \leq 0$$

*then* $V(\mathbf{x})$ *is said to be a <u>Lyapunov function</u> for the system (3.2).*

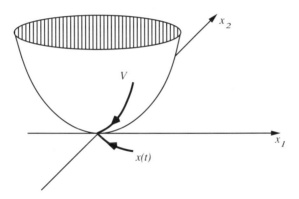

**Figure 3.9 :** Illustrating Definition 3.8 for $n = 2$

A Lyapunov function can be given simple geometrical interpretations. In Figure 3.9, the point denoting the value of $V(x_1, x_2)$ is seen to always point down a bowl. In Figure 3.10, the state point is seen to move across contour curves corresponding to lower and lower values of $V$.

## 3.4.2 Equilibrium Point Theorems

The relations between Lyapunov functions and the stability of systems are made precise in a number of theorems in Lyapunov's direct method. Such theorems usually have local and global versions. The local versions are concerned with stability properties in the neighborhood of equilibrium point and usually involve a locally positive definite function.

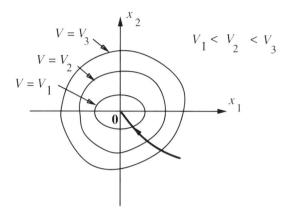

**Figure 3.10 :** Illustrating Definition 3.8 for $n = 2$ using contour curves

## LYAPUNOV THEOREM FOR LOCAL STABILITY

**Theorem 3.2 (Local Stability)** *If, in a ball* $\mathbf{B}_{R_o}$ *, there exists a scalar function* $V(\mathbf{x})$ *with continuous first partial derivatives such that*

- $V(\mathbf{x})$ *is positive definite (locally in* $\mathbf{B}_{R_o}$ *)*

- $\dot{V}(\mathbf{x})$ *is negative semi-definite (locally in* $\mathbf{B}_{R_o}$ *)*

*then the equilibrium point* $\mathbf{0}$ *is stable. If, actually, the derivative* $\dot{V}(\mathbf{x})$ *is locally negative definite in* $\mathbf{B}_{R_o}$ *, then the stability is asymptotic.*

The proof of this fundamental result is conceptually simple, and is typical of many proofs in Lyapunov theory.

**Proof**: Let us derive the result using the geometric interpretation of a Lyapunov function, as illustrated in Figure 3.9 in the case $n = 2$. To show stability, we must show that given any strictly positive number $R$, there exists a (smaller) strictly positive number $r$ such that any trajectory starting inside the ball $\mathbf{B}_r$ remains inside the ball $\mathbf{B}_R$ for all future time. Let $m$ be the *minimum of V on the sphere* $\mathbf{S}_R$ . Since $V$ is continuous and positive definite, $m$ exists and is strictly positive. Furthermore, since $V(\mathbf{0}) = 0$, there exists a ball $\mathbf{B}_r$ around the origin such that $V(\mathbf{x}) < m$ for any $\mathbf{x}$ inside the ball (Figure 3.11a). Consider now a trajectory whose initial point $\mathbf{x}(0)$ is within the ball $\mathbf{B}_r$ . Since $V$ is non-increasing along system trajectories, $V$ *remains strictly smaller than m*, and therefore the trajectory cannot possibly cross the outside sphere $\mathbf{S}_R$ . Thus, any trajectory starting inside the ball $\mathbf{B}_r$ remains inside the ball $\mathbf{B}_R$ , and therefore Lyapunov stability is guaranteed.

Let us now assume that $\dot{V}$ is negative definite, and show asymptotic stability, by contradiction. Consider a trajectory starting in some ball $\mathbf{B}_r$ as constructed above (*e.g.*, the ball $\mathbf{B}_r$

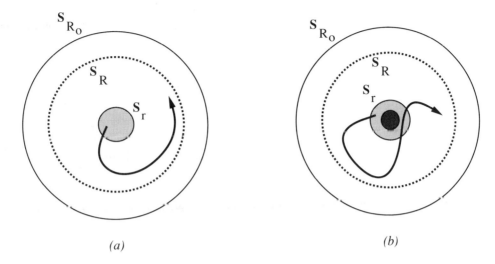

*(a)*                                              *(b)*

**Figure 3.11 :** Illustrating the proof of Theorem 3.2 for $n = 2$

corresponding to $R = R_o$). Then the trajectory will remain in the ball $\mathbf{B}_R$ for all future time. Since $V$ is lower bounded and decreases continually, $V$ tends towards a limit $L$, such that $\forall\, t \geq 0$, $V(\mathbf{x}(t)) \geq L$. Assume that this limit is not zero, *i.e.*, that $L > 0$. Then, since $V$ is continuous and $V(\mathbf{0}) = 0$, there exists a ball $\mathbf{B}_{r_o}$ that the system trajectory never enters (Figure 3.11b). But then, *since* $-\dot{V}$ *is also continuous and positive definite*, and since $\mathbf{B}_R$ is bounded, $-\dot{V}$ must remain larger than some strictly positive number $L_1$. This is a contradiction, because it would imply that $V(t)$ decreases from its initial value $V_o$ to a value strictly smaller than $L$, in a finite time smaller than $[V_o - L]/L_1$. Hence, all trajectories starting in $\mathbf{B}_r$ asymptotically converge to the origin.                          □

    In applying the above theorem for analysis of a nonlinear system, one goes through the two steps of choosing a positive definite function, and then determining its derivative along the path of the nonlinear systems. The following example illustrates this procedure.

### Example 3.7: Local Stability

A simple pendulum with viscous damping is described by

$$\ddot{\theta} + \dot{\theta} + \sin \theta = 0$$

Consider the following scalar function

$$V(\mathbf{x}) = (1 - \cos \theta) + \frac{\dot{\theta}^2}{2}$$

One easily verifies that this function is locally positive definite. As a matter of fact, this function represents the total energy of the pendulum, composed of the sum of the potential energy and the kinetic energy. Its time-derivative is easily found to be

$$\dot{V}(\mathbf{x}) = \dot{\theta} \sin \theta + \dot{\theta} \ddot{\theta} = -\dot{\theta}^2 \le 0$$

Therefore, by invoking the above theorem, one concludes that the origin is a stable equilibrium point. In fact, using physical insight, one easily sees the reason why $\dot{V}(\mathbf{x}) \le 0$, namely that the damping term absorbs energy. Actually, $\dot{V}$ is precisely the power dissipated in the pendulum. However, with this Lyapunov function, one cannot draw conclusions on the asymptotic stability of the system, because $\dot{V}(\mathbf{x})$ is only negative semi-definite.                                  ☐

The following example illustrates the asymptotic stability result.

**Example 3.8: Asymptotic stability**

Let us study the stability of the nonlinear system defined by

$$\dot{x}_1 = x_1 (x_1^2 + x_2^2 - 2) - 4 x_1 x_2^2$$

$$\dot{x}_2 = 4 x_1^2 x_2 + x_2 (x_1^2 + x_2^2 - 2)$$

around its equilibrium point at the origin. Given the positive definite function

$$V(x_1, x_2) = x_1^2 + x_2^2$$

its derivative $\dot{V}$ along any system trajectory is

$$\dot{V} = 2 (x_1^2 + x_2^2) (x_1^2 + x_2^2 - 2)$$

Thus, $\dot{V}$ is locally negative definite in the 2-dimensional ball $\mathbf{B}_2$, *i.e.*, in the region defined by $x_1^2 + x_2^2 < 2$. Therefore, the above theorem indicates that the origin is asymptotically stable.   ☐

## LYAPUNOV THEOREM FOR GLOBAL STABILITY

The above theorem applies to the local analysis of stability. In order to assert *global asymptotic stability* of a system, one might naturally expect that the ball $\mathbf{B}_{R_o}$ in the above local theorem has to be expanded to be the whole state-space. This is indeed necessary, but it is not enough. An additional condition on the function $V$ has to be satisfied: $V(\mathbf{x})$ must be *radially unbounded*, by which we mean that $V(\mathbf{x}) \to \infty$ as $\|\mathbf{x}\| \to \infty$ (in other words, as $\mathbf{x}$ tends to infinity in *any* direction). We then obtain the following powerful result:

**Theorem 3.3 (Global Stability)**   *Assume that there exists a scalar function V of the state* **x**, *with continuous first order derivatives such that*

- *$V(\mathbf{x})$ is positive definite*

- *$\dot{V}(\mathbf{x})$ is negative definite*

- *$V(\mathbf{x}) \to \infty$   as   $\|\mathbf{x}\| \to \infty$*

*then the equilibrium at the origin is globally asymptotically stable.*

> **Proof**: The proof is the same as in the local case, by noticing that the radial unboundedness of $V$, combined with the negative definiteness of $\dot{V}$, implies that, given any initial condition $\mathbf{x}_0$, the trajectories remain in the *bounded* region defined by $V(\mathbf{x}) \leq V(\mathbf{x}_0)$.   $\square$

The reason for the radial unboundedness condition is to assure that the contour curves (or contour surfaces in the case of higher order systems) $V(\mathbf{x}) = V_\alpha$ correspond to closed curves. If the curves are not closed, it is possible for the state trajectories to drift away from the equilibrium point, even though the state keeps going through contours corresponding to smaller and smaller $V_\alpha$'s. For example, for the positive definite function $V = [\, x_1^2/(1 + x_1^2)\,] + x_2^2$, the curves $V(\mathbf{x}) = V_\alpha$ for $V_\alpha > 1$ are open curves. Figure 3.12 shows the divergence of the state while moving toward lower and lower "energy" curves. Exercise 3.4 further illustrates this point on a specific system.

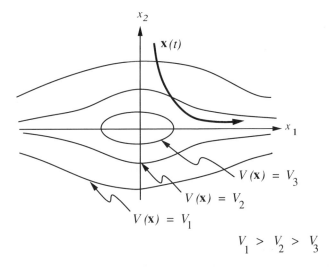

**Figure 3.12 :** Motivation of the radial unboundedness condition

**Example 3.9: A class of first-order systems**

Consider the nonlinear system

$$\dot{x} + c(x) = 0$$

where $c$ is any continuous function of the same sign as its scalar argument $x$, *i.e.*,

$$x\,c(x) > 0 \quad \text{for} \quad x \neq 0$$

Intuitively, this condition indicates that $-c(x)$ "pushes" the system back towards its rest position $x = 0$, but is otherwise arbitrary. Since $c$ is continuous, it also implies that $c(0) = 0$ (Figure 3.13).

Consider as the Lyapunov function candidate the square of the distance to the origin

$$V = x^2$$

The function $V$ is radially unbounded, since it tends to infinity as $|x| \to \infty$. Its derivative is

$$\dot{V} = 2\,x\,\dot{x} = = -\,2\,x\,c(x)$$

Thus $\dot{V} < 0$ as long as $x \neq 0$, so that $x = 0$ is a globally asymptotically stable equilibrium point.

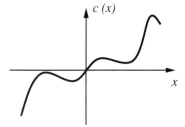

**Figure 3.13 :** The function $c(x)$

For instance, the system

$$\dot{x} = \sin^2 x - x$$

is globally asymptotically convergent to $x = 0$, since for $x \neq 0$, $\sin^2 x \leq |\sin x| < |x|$. Similarly, the system

$$\dot{x} = -\,x^3$$

is globally asymptotically convergent to $x = 0$. Notice that while this system's linear approximation ($\dot{x} \approx 0$) is inconclusive, even about local stability, the actual nonlinear system enjoys a strong stability property (*global asymptotic* stability).   □

**Example 3.10**: Consider the system

$$\dot{x}_1 = x_2 - x_1(x_1^2 + x_2^2)$$

$$\dot{x}_2 = -x_1 - x_2(x_1^2 + x_2^2)$$

The origin of the state-space is an equilibrium point for this system. Let $V$ be the positive definite function

$$V(\mathbf{x}) = x_1^2 + x_2^2$$

The derivative of $V$ along any system trajectory is

$$\dot{V}(\mathbf{x}) = 2x_1\dot{x}_1 + 2x_2\dot{x}_2 = -2(x_1^2 + x_2^2)^2$$

which is negative definite. Therefore, the origin is a globally asymptotically stable equilibrium point. Note that the globalness of this stability result also implies that the origin is the *only* equilibrium point of the system.                                                    □

## REMARKS

Many Lyapunov functions may exist for the same system. For instance, if $V$ is a Lyapunov function for a given system, so is

$$V_1 = \rho V^\alpha$$

where $\rho$ is any strictly positive constant and $\alpha$ is any scalar (not necessarily an integer) larger than 1. Indeed, the positive-definiteness of $V$ implies that of $V_1$, the positive-definiteness (or positive semi-definiteness) of $-\dot{V}$ implies that of $-\dot{V}_1$, and (the radial unboundedness of $V$ (if applicable) implies that of $V_1$.

  More importantly, for a given system, specific choices of Lyapunov functions may yield more precise results than others. Consider again the pendulum of Example 3.7. The function

$$V(\mathbf{x}) = \frac{1}{2}\dot{\theta}^2 + \frac{1}{2}(\dot{\theta} + \theta)^2 + 2(1 - \cos\theta)$$

is also a Lyapunov function for the system, because locally

$$\dot{V}(\mathbf{x}) = -(\dot{\theta}^2 + \theta\sin\theta) \leq 0$$

However, it is interesting to note that $\dot{V}$ is actually locally *negative definite*, and therefore, this modified choice of $V$, without obvious physical meaning, allows the asymptotic stability of the pendulum to be shown.

Along the same lines, it is important to realize that the theorems in Lyapunov analysis are all *sufficiency* theorems. If for a particular choice of Lyapunov function candidate $V$, the conditions on $\dot{V}$ are not met, one cannot draw any conclusions on the stability or instability of the system – the only conclusion one should draw is that a different Lyapunov function candidate should be tried.

## 3.4.3  Invariant Set Theorems

Asymptotic stability of a control system is usually a very important property to be determined. However, the equilibrium point theorems just described are often difficult to apply in order to assert this property. The reason is that it often happens that $\dot{V}$, the derivative of the Lyapunov function candidate, is only negative *semi*-definite, as seen in (3.15). In this kind of situation, fortunately, it is still possible to draw conclusions on asymptotic stability, with the help of the powerful *invariant set theorems*, attributed to La Salle. This section presents the local and global versions of the invariant set theorems.

The central concept in these theorems is that of invariant set, a generalization of the concept of equilibrium point.

**Definition 3.9**    *A set* **G** *is an* <u>*invariant set*</u> *for a dynamic system if every system trajectory which starts from a point in* **G** *remains in* **G** *for all future time.*

For instance, any equilibrium point is an invariant set. The domain of attraction of an equilibrium point is also an invariant set. A trivial invariant set is the whole state-space. For an autonomous system, any of the trajectories in state-space is an invariant set. Since limit cycles are special cases of system trajectories (closed curves in the phase plane), they are also invariant sets.

Besides often yielding conclusions on asymptotic stability when $\dot{V}$, the derivative of the Lyapunov function candidate, is only negative semi-definite, the invariant set theorems also allow us to *extend* the concept of Lyapunov function so as to describe convergence to dynamic behaviors more general than equilibrium, *e.g.*, convergence to a limit cycle.

Similarly to our earlier discussion of Lyapunov's direct method, we first discuss the local version of the invariant set theorems, and then the global version.

**LOCAL INVARIANT SET THEOREM**

The invariant set theorems reflect the intuition that the decrease of a Lyapunov function $V$ has to gradually vanish (*i.e.*, $\dot{V}$ has to converge to zero) because $V$ is lower bounded. A precise statement of this result is as follows.

**Theorem 3.4 (Local Invariant Set Theorem)** *Consider an autonomous system of the form (3.2), with* **f** *continuous, and let* $V(\mathbf{x})$ *be a scalar function with continuous first partial derivatives. Assume that*

- *for some* $l > 0$, *the region* $\Omega_l$ *defined by* $V(\mathbf{x}) < l$ *is bounded*

- $\dot{V}(\mathbf{x}) \leq 0$ *for all* **x** *in* $\Omega_l$

*Let* **R** *be the set of all points within* $\Omega_l$ *where* $\dot{V}(\mathbf{x}) = 0$ , *and* **M** *be the largest invariant set in* **R**. *Then, every solution* $\mathbf{x}(t)$ *originating in* $\Omega_l$ *tends to* **M** *as* $t \rightarrow \infty$.

Note that in the above theorem, the word "largest" is understood in the sense of set theory, *i.e.*, **M** is the *union* of all invariant sets (*e.g.*, equilibrium points or limit cycles) within **R**. In particular, if the set **R** is itself invariant (*i.e.*, if once $\dot{V} = 0$, then $\dot{V} \equiv 0$ for all future time), then **M** = **R**.

The geometrical meaning of the theorem is illustrated in Figure 3.14, where a trajectory starting from within the bounded region $\Omega_l$ is seen to converge to the largest invariant set **M**. Note that the set **R** is not necessarily connected, nor is the set **M**.

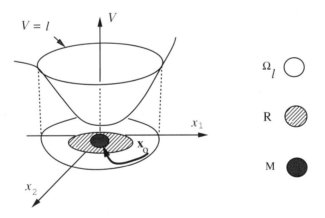

**Figure 3.14 :** Convergence to the largest invariant set **M**

The theorem can be proven in two steps, by first showing that $\dot{V}$ goes to zero, and then showing that the state converges to the largest invariant set within the set defined by $\dot{V} = 0$. We shall simply give a sketch of the proof, since the detailed proof of the second part involves a number of concepts in topology and real analysis which are not prerequisites of this text.

**Proof**: The first part of the proof involves showing that $\dot{V} \to 0$ for any trajectory starting from a point in $\Omega_l$, using a result in functional analysis known as Barbalat's lemma, which we shall detail in section 4.3.

Specifically, consider a trajectory starting from an arbitrary point $\mathbf{x}_o$ in $\Omega_l$. The trajectory must stay in $\Omega_l$ all the time, because $\dot{V} \leq 0$ implies that $V[\mathbf{x}(t)] \leq V[\mathbf{x}(0)] < l$ for all $t \geq 0$. In addition, because $V(\mathbf{x})$ is continuous in $\mathbf{x}$ (since it is differentiable with respect to $\mathbf{x}$) over the bounded region $\Omega_l$, it is lower bounded in that region; therefore, since we just noticed that the trajectory remains in $\Omega_l$, $V[\mathbf{x}(t)]$ remains lower bounded for all $t \geq 0$. Furthermore, the facts that $\mathbf{f}$ is continuous, $V$ has continuous partial derivatives, and the region $\Omega_l$ is bounded, imply that $\dot{V}$ is uniformly continuous. Therefore, $V[\mathbf{x}(t)]$ satisfies the three conditions ($V$ lower bounded; $\dot{V} \leq 0$; $\dot{V}$ uniformly continuous) of Barbalat's lemma. As a result, $\dot{V}[\mathbf{x}(t)] \to 0$, which implies that all system trajectories starting from within $\Omega_l$ converge to the set $\mathbf{R}$.

The second part of the proof [see, *e.g.*, Hahn, 1968] involves showing that the trajectories cannot converge to just anywhere in the set $\mathbf{R}$: they must converge to the largest invariant set $\mathbf{M}$ within $\mathbf{R}$. This can be proven by showing that any bounded trajectory of an autonomous system converges to an invariant set (the so-called positive limit set of the trajectory), and then simply noticing that this set is a subset of the largest invariant set $\mathbf{M}$.                         □

Note that the asymptotic stability result in the local Lyapunov theorem can be viewed a special case of the above invariant set theorem, where the set $\mathbf{M}$ consists only of the origin.

Let us now illustrate applications of the invariant set theorem using some examples. The first example shows how to conclude asymptotic stability for problems which elude the local Lyapunov theorem. The second example shows how to determine a domain of attraction, an issue which was not specifically addressed before. The third example shows the convergence of system trajectories to a limit cycle.

**Example 3.11: Asymptotic stability of the mass-damper-spring system**

For the system (3.13), one can only draw conclusion of marginal stability using the energy function (3.14) in the local equilibrium point theorem, because $\dot{V}$ is only negative semi-definite according to (3.15). Using the invariant set theorem, however, we can show that the system is actually asymptotically stable. To do this, we only have to show that the set $\mathbf{M}$ contains only one point.

The set $\mathbf{R}$ is defined by $\dot{x} = 0$, *i.e.*, the collection of states with zero velocity, or the whole horizontal axis in the phase plane $(x, \dot{x})$. Let us show that the largest invariant set $\mathbf{M}$ in this set $\mathbf{R}$ contains only the origin. Assume that $\mathbf{M}$ contains a point with a nonzero position $x_1$, then, the acceleration at that point is $\ddot{x} = -(k_o/m)x - (k_1/m)x^3 \neq 0$. This implies that the trajectory will immediately move out of the set **R** and thus also out of the set **M**, a contradiction to the definition. ☐

### Example 3.12: Domain of Attraction

Consider again the system in Example 3.8. For $l = 2$, the region $\Omega_2$, defined by $V(\mathbf{x}) = x_1^2 + x_2^2 < 2$, is bounded. The set **R** is simply the origin **0**, which is an invariant set (since it is an equilibrium point). All the conditions of the local invariant set theorem are satisfied and, therefore, any trajectory starting within the circle converges to the origin. Thus, a domain of attraction is explicitly determined by the invariant set theorem. ☐

### Example 3.13: Attractive Limit Cycle

Consider the system

$$\dot{x}_1 = x_2 - x_1^7 [x_1^4 + 2 x_2^2 - 10]$$

$$\dot{x}_2 = -x_1^3 - 3 x_2^5 [x_1^4 + 2 x_2^2 - 10]$$

Notice first that the set defined by $x_1^4 + 2 x_2^2 = 10$ is invariant, since

$$\frac{d}{dt}(x_1^4 + 2 x_2^2 - 10) = -(4 x_1^{10} + 12 x_2^6)(x_1^4 + 2 x_2^2 - 10)$$

which is zero on the set. The motion *on* this invariant set is described (equivalently) by *either* of the equations

$$\dot{x}_1 = x_2$$

$$\dot{x}_2 = -x_1^3$$

Therefore, we see that the invariant set actually represents a *limit cycle*, along which the state vector moves clockwise.

Is this limit cycle actually attractive? Let us define as a Lyapunov function candidate

$$V = (x_1^4 + 2 x_2^2 - 10)^2$$

which represents a measure of the "distance" to the limit cycle. For any arbitrary positive number $l$, the region $\Omega_l$, which surrounds the limit cycle, is bounded. Using our earlier calculation, we immediately obtain

$$\dot{V} = -8 (x_1^{10} + 3 x_2^6)(x_1^4 + 2 x_2^2 - 10)^2$$

Thus $\dot{V}$ is strictly negative, except if

$$x_1^4 + 2 x_2^2 = 10 \qquad \text{or} \qquad x_1^{10} + 3 x_2^6 = 0$$

in which case $\dot{V} = 0$. The first equation is simply that defining the limit cycle, while the second equation is verified only at the origin. Since both the limit cycle and the origin are invariant sets, the set $\mathbf{M}$ simply consists of their union. Thus, all system trajectories starting in $\Omega_l$ converge either to the limit cycle, or to the origin (Figure 3.15).

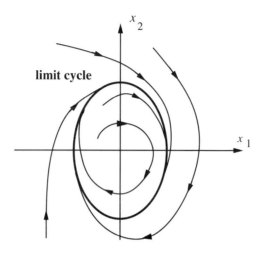

**Figure 3.15 :** Convergence to a limit cycle

Moreover, the equilibrium point at the origin can actually be shown to be *unstable*. However, this result cannot be obtained from linearization, since the linearized system $(\dot{x}_1 = x_2 , \dot{x}_2 = 0)$ is only marginally stable. Instead, and more astutely, consider the region $\Omega_{100}$, and note that while the origin $\mathbf{0}$ does not belong to $\Omega_{100}$, every other point in the region enclosed by the limit cycle is in $\Omega_{100}$ (in other words, the origin corresponds to a local *maximum* of $V$). Thus, while the expression of $\dot{V}$ is the same as before, *now the set $\mathbf{M}$ is just the limit cycle*. Therefore, reapplication of the invariant set theorem shows that any state trajectory starting from the region within the limit cycle, excluding the origin, actually converges to the limit cycle. In particular, this implies that the equilibrium point at the origin is unstable.                            □

Example 3.11 actually represents a very common application of the invariant set theorem: conclude asymptotic stability of an equilibrium point for systems with negative *semi*-definite $\dot{V}$. The following corollary of the invariant set theorem is more specifically tailored to such applications:

**Corollary**    *Consider the autonomous system (3.2), with $\mathbf{f}$ continuous, and let $V(\mathbf{x})$ be a scalar function with continuous partial derivatives. Assume that in a certain neighborhood $\Omega$ of the origin*

• $V(\mathbf{x})$ *is locally positive definite*

• $\dot{V}$ *is negative semi-definite*

• *the set* $\mathbf{R}$ *defined by* $\dot{V}(\mathbf{x}) = 0$ *contains no trajectories of (3.2) other than the trivial trajectory* $\mathbf{x} \equiv \mathbf{0}$

*Then, the equilibrium point* $\mathbf{0}$ *is asymptotically stable. Furthermore, the largest connected region of the form* $\mathbf{\Omega}_l$ *(defined by* $V(\mathbf{x}) < l$ *) within* $\mathbf{\Omega}$ *is a domain of attraction of the equilibrium point.*

Indeed, the largest invariant set $\mathbf{M}$ in $\mathbf{R}$ then contains only the equilibrium point $\mathbf{0}$. Note that

• The above corollary replaces the negative definiteness condition on $\dot{V}$ in Lyapunov's local asymptotic stability theorem by a negative *semi*-definiteness condition on $\dot{V}$, combined with a third condition on the trajectories within $\mathbf{R}$.

• The largest connected region of the form $\mathbf{\Omega}_l$ within $\mathbf{\Omega}$ is *a* domain of attraction of the equilibrium point, but not necessarily *the* whole domain of attraction, because the function $V$ is not unique.

• The set $\mathbf{\Omega}$ itself is not necessarily a domain of attraction. Actually, the above theorem does not guarantee that $\mathbf{\Omega}$ is invariant: some trajectories starting in $\mathbf{\Omega}$ but outside of the largest $\mathbf{\Omega}_l$ may actually end up outside $\mathbf{\Omega}$.

### GLOBAL INVARIANT SET THEOREMS

The above invariant set theorem and its corollary can be simply extended to a global result, by enlarging the involved region to be the whole space and requiring the radial unboundedness of the scalar function $V$.

**Theorem 3.5 (Global Invariant Set Theorem)** *Consider the autonomous system (3.2), with* $\mathbf{f}$ *continuous, and let* $V(\mathbf{x})$ *be a scalar function with continuous first partial derivatives. Assume that*

• $\dot{V}(\mathbf{x}) \leq 0$ *over the whole state space*

• $V(\mathbf{x}) \to \infty$ *as* $\|\mathbf{x}\| \to \infty$

*Let* $\mathbf{R}$ *be the set of all points where* $\dot{V}(\mathbf{x}) = 0$, *and* $\mathbf{M}$ *be the largest invariant set in* $\mathbf{R}$. *Then all solutions globally asymptotically converge to* $\mathbf{M}$ *as* $t \to \infty$.

For instance, the above theorem shows that the limit cycle convergence in

Example 3.13 is actually global: all system trajectories converge to the limit cycle (unless they start exactly at the origin, which is an unstable equilibrium point).

Because of the importance of this theorem, let us present an additional (and very useful) example.

**Example 3.14: A class of second-order nonlinear systems**

Consider a second-order system of the form

$$\ddot{x} + b(\dot{x}) + c(x) = 0$$

where $b$ and $c$ are continuous functions verifying the sign conditions

$$\dot{x}\, b(\dot{x}) > 0 \quad \text{for } \dot{x} \neq 0$$

$$x\, c(x) > 0 \quad \text{for } x \neq 0$$

The dynamics of a mass-damper-spring system with nonlinear damper and spring can be described by equations of this form, with the above sign conditions simply indicating that the otherwise arbitrary functions $b$ and $c$ actually represent "damping" and "spring" effects. A nonlinear R-L-C (resistor-inductor-capacitor) electrical circuit can also be represented by the above dynamic equation (Figure 3.16). Note that if the functions $b$ and $c$ are actually *linear* ( $b(\dot{x}) = \alpha_1\, \dot{x}$ , $c(x) = \alpha_o\, x$ ), the above sign conditions are simply the *necessary and sufficient* conditions for the system's stability (since they are equivalent to the conditions $\alpha_1 > 0$, $\alpha_o > 0$).

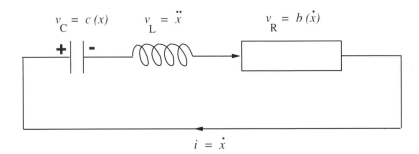

**Figure 3.16 :** A nonlinear R-L-C circuit

Together with the continuity assumptions, the sign conditions on the functions $b$ and $c$ imply that $b(0) = 0$ and $c(0) = 0$ (Figure 3.17). A positive definite function for this system is

$$V = \frac{1}{2}\dot{x}^2 + \int_o^x c(y)\, dy$$

which can be thought of as the sum of the kinetic and potential energy of the system.

Differentiating $V$, we obtain

$$\dot{V} = \dot{x}\,\ddot{x} + c(x)\,\dot{x} = -\dot{x}\,b(\dot{x}) - \dot{x}\,c(x) + c(x)\,\dot{x} = -\dot{x}\,b(\dot{x}) \le 0$$

which can be thought of as representing the power dissipated in the system. Furthermore, by hypothesis, $\dot{x}\,b(\dot{x}) = 0$ only if $\dot{x} = 0$. Now $\dot{x} = 0$ implies that

$$\ddot{x} = -c(x)$$

which is nonzero as long as $x \ne 0$. Thus the system cannot get "stuck" at an equilibrium value other than $x = 0$; in other words, with **R** being the set defined by $\dot{x} = 0$, the largest invariant set **M** in **R** contains only one point, namely $[x = 0, \dot{x} = 0]$. Use of the local invariant set theorem indicates that the origin is a locally asymptotically stable point.

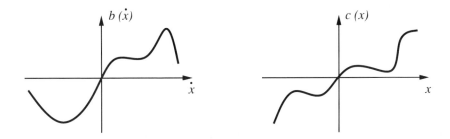

**Figure 3.17** : The functions $b(\dot{x})$ and $c(x)$

Furthermore, if the integral $\int_{o}^{x} c(r)\,dr$ is unbounded as $|x| \to \infty$, then $V$ is a radially unbounded function and the equilibrium point at the origin is globally asymptotically stable, according to the global invariant set theorem.

For instance, the system

$$\ddot{x} + \dot{x}^3 + x^5 = x^4 \sin^2 x$$

is globally asymptotically convergent to $x = 0$ (while, again, its linear approximation would be inconclusive, even about its local stability).      $\square$

The relaxation of the positive definiteness requirement on the function $V$, as compared with Lyapunov's direct method, also allows one to use a single Lyapunov-like function to describe systems with multiple equilibria.

**Example 3.15: Multimodal Lyapunov Function**

Consider the system

$$\ddot{x} + |x^2 - 1| \dot{x}^3 + x = \sin\frac{\pi x}{2}$$

For this system, we study, similarly to Example 3.14, the Lyapunov function

$$V = \frac{1}{2}\dot{x}^2 + \int_0^x (y - \sin\frac{\pi y}{2})\, dy$$

This function has two minima, at $x = \pm 1$; $\dot{x} = 0$, and a local *maximum* in $x$ (a saddle point in the state-space) at $x = 0$; $\dot{x} = 0$. As in Example 3.14, the time-derivative of $V$ is (without calculations)

$$\dot{V} = -\,|x^2 - 1|\,\dot{x}^4$$

*i.e*, the virtual power "dissipated" by the system. Now

$$\dot{V} = 0 \quad \Rightarrow \quad \dot{x} = 0 \quad \text{or} \quad x = \pm 1$$

Let us consider each of these cases:

$$\dot{x} = 0 \quad \Rightarrow \quad \ddot{x} = \sin\frac{\pi x}{2} - x \neq 0 \quad \text{except if} \quad x = 0 \quad \text{or} \quad x = \pm 1$$

$$x = \pm 1 \quad \Rightarrow \quad \ddot{x} = 0$$

Thus, the invariant set theorem indicates that the system converges globally to $(x = 1; \dot{x} = 0)$ or $(x = -1; \dot{x} = 0)$, or to $(x = 0; \dot{x} = 0)$. The first two of these equilibrium points are stable, since they correspond to local mimina of $V$ (note again that linearization is inconclusive about their stability). By contrast, the equilibrium point $(x = 0; \dot{x} = 0)$ is unstable, as can be shown from linearization ($\ddot{x} \approx (\pi/2 - 1)\, x$), or simply by noticing that because that point is a local maximum of $V$ along the $x$ axis, any small deviation in the $x$ direction will drive the trajectory away from it. $\quad\square$

As noticed earlier, several Lyapunov functions may exist for a given system, and therefore several associated invariant sets may be derived. The system then converges to the (necessarily non-empty) *intersection* of the invariant sets $M_i$, which may give a more precise result than that obtained from any of the Lyapunov functions taken separately.   Equivalently, one can notice that the sum of two Lyapunov functions for a given system is also a Lyapunov function, whose set $\mathbf{R}$ is the intersection of the individual sets $\mathbf{R}_i$.

## 3.5   System Analysis Based on Lyapunov's Direct Method

With so many theorems and so many examples presented in the last section, one may feel confident enough to attack practical nonlinear control problems. However, the theorems all make a basic assumption:   an explicit Lyapunov function is somehow known. The question is therefore how to find a Lyapunov function for a specific problem. Yet, there is no general way of finding Lyapunov functions for nonlinear systems. This is a fundamental drawback of the direct method. Therefore, faced with specific systems, one has to use experience, intuition, and physical insights to search for an appropriate Lyapunov function. In this section, we discuss a number of techniques which can facilitate the otherwise blind search of Lyapunov functions.

We first show that, not surprisingly, Lyapunov functions can be systematically found to describe stable *linear* systems. Next, we discuss two of many mathematical methods that may be used to help finding a Lyapunov function for a given nonlinear system. We then consider the use of physical insights, which, when applicable, represents by far the most powerful and elegant way of approaching the problem, and is closest in spirit to the original intuition underlying the direct method. Finally, we discuss the use of Lyapunov functions in transient performance analysis.

## 3.5.1 Lyapunov Analysis of Linear Time-Invariant Systems

Stability analysis for linear time-invariant systems is well known. It is interesting, however, to develop Lyapunov functions for such systems. First, this allows us to describe both linear and nonlinear systems using *a common language*, allowing shared insights between the two classes. Second, as we shall detail later on, Lyapunov functions are "additive", like energy. In other words, Lyapunov functions for combinations of subsystems may be derived by adding the Lyapunov functions of the subsystems. Since nonlinear control systems may include linear components (whether in plant or in controller), we should be able to describe linear systems in the Lyapunov formalism.

We first review some basic results on matrix algebra, since the development of Lyapunov functions for linear systems will make extensive use of quadratic forms.

**SYMMETRIC, SKEW-SYMMETRIC, AND POSITIVE DEFINITE MATRICES**

**Definition 3.10** *A square matrix* $\mathbf{M}$ *is symmetric if* $\mathbf{M} = \mathbf{M}^T$ *(in other words, if* $\forall\, i, j \;\; M_{ij} = M_{ji}$ *). A square matrix* $\mathbf{M}$ *is skew-symmetric if* $\mathbf{M} = -\mathbf{M}^T$ *(i.e, if* $\forall\, i, j \;\; M_{ij} = -M_{ji}$ *).*

An interesting fact is that any square $n \times n$ matrix $\mathbf{M}$ can be represented as the sum of a symmetric matrix and a skew-symmetric matrix. This can be shown by the following decomposition

$$\mathbf{M} = \frac{\mathbf{M} + \mathbf{M}^T}{2} + \frac{\mathbf{M} - \mathbf{M}^T}{2}$$

where the first term on the left side is symmetric and the second term is skew-symmetric.

Another interesting fact is that the quadratic function associated with a skew-symmetric matrix is always zero. Specifically, let $\mathbf{M}$ be a $n \times n$ skew-symmetric matrix and $\mathbf{x}$ an arbitrary $n \times 1$ vector. Then the definition of a skew-symmetric matrix implies that

$$\mathbf{x}^T \mathbf{M} \mathbf{x} = - \mathbf{x}^T \mathbf{M}^T \mathbf{x}$$

Since $\mathbf{x}^T \mathbf{M}^T \mathbf{x}$ is a scalar, the right-hand side of the above equation can be replaced by its transpose. Therefore,

$$\mathbf{x}^T \mathbf{M} \mathbf{x} = - \mathbf{x}^T \mathbf{M} \mathbf{x}$$

This shows that

$$\forall \, \mathbf{x} \, , \quad \mathbf{x}^T \mathbf{M} \mathbf{x} = 0 \tag{3.16}$$

In designing some tracking control systems for robots, for instance, this fact is very useful because it can simplify the control law, as we shall see in chapter 9.

Actually, property (3.16) is a *necessary and sufficient* condition for a matrix $\mathbf{M}$ to be skew-symmetric. This can be easily seen by applying (3.16) to the basis vectors $\mathbf{e}_i$ :

$$[ \, \forall \, i \, , \, \mathbf{e}_i^T \mathbf{M}_s \mathbf{e}_i = 0 \, ] \quad \Rightarrow \quad [ \, \forall \, i, \, M_{ii} = 0 \, ]$$

and

$$[ \, \forall \, (i,j) \, , \, (\mathbf{e}_i + \mathbf{e}_j)^T \mathbf{M}_s (\mathbf{e}_i + \mathbf{e}_j) = 0 \, ] \quad \Rightarrow \quad [ \, \forall \, (i,j), \, M_{ii} + M_{ij} + M_{ji} + M_{jj} = 0 \, ]$$

which, using the first result, implies that

$$\forall \, (i,j) \, , \quad M_{ji} = - M_{ij}$$

In our later analysis of linear systems, we will often use quadratic functions of the form $\mathbf{x}^T \mathbf{M} \mathbf{x}$ as Lyapunov function candidates. In view of the above, each quadratic function of this form, whether $\mathbf{M}$ is symmetric or not, is always equal to a quadratic function with a symmetric matrix. Thus, in considering quadratic functions of the form $\mathbf{x}^T \mathbf{M} \mathbf{x}$ as Lyapunov function candidates, one can always assume, without loss of generality, that $\mathbf{M}$ is symmetric.

We are now in a position to introduce the important concept of positive definite matrices.

**Definition 3.11**     *A square $n \times n$ matrix* **M** *is positive definite (p.d.) if*

$$\mathbf{x} \neq \mathbf{0} \quad \Rightarrow \quad \mathbf{x}^T \mathbf{M} \mathbf{x} > 0$$

In other words, a matrix **M** is positive definite if the quadratic function $\mathbf{x}^T \mathbf{M} \mathbf{x}$ is a positive definite function. This definition implies that to every positive definite matrix is associated a positive definite function. Obviously, the converse is not true.

Geometrically, the definition of positive-definiteness can be interpreted as simply saying that the angle between a vector **x** and its image **Mx** is always less than $90^o$ (Figure 3.18).

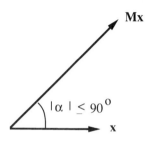

**Figure 3.18 :** Geometric interpretation of the positive-definiteness of a matrix **M**

A *necessary* condition for a square matrix **M** to be *p.d.* is that its diagonal elements be strictly positive, as can be seen by applying the above definition to the basis vectors. A famous matrix algebra result called Sylvester's theorem shows that, *assuming that* **M** *is symmetric*, a *necessary and sufficient* condition for **M** to be *p.d.* is that its principal minors (*i.e.*, $M_{11}$, $M_{11}M_{22} - M_{21} M_{12}$, ... , det **M** ) all be strictly positive; or, equivalently, that all its eigenvalues be strictly positive. In particular, a *p.d.* matrix is always invertible, because the above implies that its determinant is non-zero.

A positive definite matrix **M** can always be decomposed as

$$\mathbf{M} = \mathbf{U}^T \Lambda \mathbf{U} \tag{3.17}$$

where **U** is a matrix of eigenvectors and satisfies $\mathbf{U}^T \mathbf{U} = \mathbf{I}$, and $\Lambda$ is a diagonal matrix containing the eigenvalues of the matrix **M**. Let $\lambda_{min}(\mathbf{M})$ denote the smallest eigenvalue of **M** and $\lambda_{max}(\mathbf{M})$ the largest. Then, it follows from (3.17) that

$$\lambda_{min}(\mathbf{M}) \|\mathbf{x}\|^2 \leq \mathbf{x}^T \mathbf{M} \mathbf{x} \leq \lambda_{max}(\mathbf{M}) \|\mathbf{x}\|^2$$

This is due to the following three facts:

- $\mathbf{x}^T \mathbf{M} \mathbf{x} = \mathbf{x}^T \mathbf{U}^T \Lambda \mathbf{U} \mathbf{x} = \mathbf{z}^T \Lambda \mathbf{z}$, where $\mathbf{U} \mathbf{x} = \mathbf{z}$

- $\lambda_{min}(\mathbf{M}) \, \mathbf{I} \leq \Lambda \leq \lambda_{max}(\mathbf{M}) \, \mathbf{I}$

- $\mathbf{z}^T \mathbf{z} = \|\mathbf{x}\|^2$

The concepts of positive semi-definite, negative definite, and negative semi-definite can be defined similarly. For instance, a square $n \times n$ matrix $\mathbf{M}$ is said to be *positive semi-definite ( p.s.d.)* if

$$\forall \, \mathbf{x} \, , \, \mathbf{x}^T \mathbf{M} \, \mathbf{x} \geq 0$$

By continuity, necessary and sufficient conditions for positive semi-definiteness are obtained by substituting "positive or zero" to "strictly positive" in the above conditions for positive definiteness.  Similarly, a *p.s.d.* matrix is invertible only if it is actually *p.d.* Examples of *p.s.d.* matrices are $n \times n$ matrices of the form $\mathbf{M} = \mathbf{N}^T \mathbf{N}$ where $\mathbf{N}$ is a $m \times n$ matrix. Indeed,

$$\forall \, \mathbf{x} \, , \, \mathbf{x}^T \mathbf{N}^T \mathbf{N} \, \mathbf{x} = (\mathbf{N}\mathbf{x})^T \, (\mathbf{N}\mathbf{x}) \geq 0$$

A matrix inequality of the form

$$\mathbf{M}_1 > \mathbf{M}_2$$

(where $\mathbf{M}_1$ and $\mathbf{M}_2$ are square matrices of the same dimension) means that

$$\mathbf{M}_1 - \mathbf{M}_2 > \mathbf{0}$$

*i.e.*, that the matrix $\mathbf{M}_1 - \mathbf{M}_2$ is positive definite.  Similar notations apply to the concepts of positive semi-definiteness, negative definiteness, and negative semi-definiteness.

A time-varying matrix $\mathbf{M}(t)$ is *uniformly* positive definite if

$$\exists \, \alpha > 0, \ \forall \, t \geq 0, \ \mathbf{M}(t) \geq \alpha \, \mathbf{I}$$

A similar definition applies for uniform negative-definiteness of a time-varying matrix.

## LYAPUNOV FUNCTIONS FOR LINEAR TIME-INVARIANT SYSTEMS

Given a linear system of the form $\dot{\mathbf{x}} = \mathbf{A} \, \mathbf{x}$ , let us consider a quadratic Lyapunov function candidate

$$V = \mathbf{x}^T \mathbf{P} \, \mathbf{x}$$

where **P** is a given symmetric positive definite matrix. Differentiating the positive definite function $V$ along the system trajectory yields another quadratic form

$$\dot{V} = \dot{\mathbf{x}}^T \mathbf{P} \mathbf{x} + \mathbf{x}^T \mathbf{P} \dot{\mathbf{x}} = -\mathbf{x}^T \mathbf{Q} \mathbf{x} \tag{3.18}$$

where

$$\mathbf{A}^T \mathbf{P} + \mathbf{P} \mathbf{A} = -\mathbf{Q} \tag{3.19}$$

The question, thus, is to determine whether the symmetric matrix **Q** defined by the so-called *Lyapunov equation* (3.19) above, is itself *p.d.* If this is the case, then $V$ satisfies the conditions of the basic theorem of section 3.4, and the origin is globally asymptotically stable. However, this "natural" approach may lead to inconclusive result, *i.e.*, **Q** may be not positive definite even for stable systems.

**Example 3.17**: Consider a second-order linear system whose **A** matrix is

$$A = \begin{bmatrix} 0 & 4 \\ -8 & -12 \end{bmatrix}$$

If we take **P** = **I**, then

$$-Q = \mathbf{P}\mathbf{A} + \mathbf{A}^T\mathbf{P} = \begin{bmatrix} 0 & -4 \\ -4 & -24 \end{bmatrix}$$

The matrix **Q** is not positive definite. Therefore, no conclusion can be drawn from the Lyapunov function on whether the system is stable or not.  ☐

A more useful way of studying a given linear system using scalar quadratic functions is, instead, to derive a positive definite matrix **P** from a given positive definite matrix **Q**, *i.e.*,

• choose a positive definite matrix **Q**

• *solve* for **P** from the Lyapunov equation (3.19)

• check whether **P** is *p.d*

If **P** is *p.d.*, then $\mathbf{x}^T\mathbf{P}\mathbf{x}$ is a Lyapunov function for the linear system and global asymptotical stability is guaranteed. Unlike the previous approach of going from a given **P** to a matrix **Q**, this technique of going from a given **Q** to a matrix **P** always leads to conclusive results for stable linear systems, as seen from the following theorem.

**Theorem 3.6** *A necessary and sufficient condition for a LTI system $\dot{\mathbf{x}} = \mathbf{A}\,\mathbf{x}$ to be strictly stable is that, for any symmetric p.d. matrix $\mathbf{Q}$, the unique matrix $\mathbf{P}$ solution of the Lyapunov equation (3.19) be symmetric positive definite.*

**Proof**: The above discussion shows that the condition is sufficient, thus we only need to show that it is also necessary. We first show that given any symmetric *p.d.* matrix $\mathbf{Q}$, there exists a symmetric *p.d.* matrix $\mathbf{P}$ verifying (3.19). We then show that for a given $\mathbf{Q}$, the matrix $\mathbf{P}$ is actually unique.

Let $\mathbf{Q}$ be a given symmetric positive definite matrix, and let

$$\mathbf{P} = \int_0^\infty \exp(\mathbf{A}^T t)\, \mathbf{Q} \exp(\mathbf{A}\, t)\, dt \tag{3.20}$$

One can easily show that this integral exists if and only if $\mathbf{A}$ is strictly stable. Also note that the matrix $\mathbf{P}$ thus defined is symmetric and positive definite, since $\mathbf{Q}$ is. Furthermore, we have

$$-\mathbf{Q} = \int_{t=0}^\infty d[\, \exp(\mathbf{A}^T t)\, \mathbf{Q} \exp(\mathbf{A}\, t)\, ]$$

$$= \int_{t=0}^\infty [\, \mathbf{A}^T \exp(\mathbf{A}^T t)\, \mathbf{Q} \exp(\mathbf{A}\, t) + \exp(\mathbf{A}^T t)\, \mathbf{Q} \exp(\mathbf{A}\, t)\, \mathbf{A}\, ]\, dt$$

$$= \mathbf{A}^T \mathbf{P} + \mathbf{P}\,\mathbf{A}$$

where the first equality comes from the stability of $\mathbf{A}$ (which implies that $\exp(\mathbf{A}\infty) = \mathbf{0}$), the second from differentiating the exponentials explicitly, and the third from the fact that $\mathbf{A}$ is constant and therefore can be taken out of the integrals.

The uniqueness of $\mathbf{P}$ can be verified similarly by noting that another solution $\mathbf{P}_1$ of the Lyapunov equation would necessarily verify

$$\mathbf{P}_1 = -\int_{t=0}^\infty d[\, \exp(\mathbf{A}^T t)\, \mathbf{P}_1 \exp(\mathbf{A}\, t)\, ]$$

$$= -\int_{t=0}^\infty \exp(\mathbf{A}^T t)\, (\, \mathbf{A}^T \mathbf{P}_1 + \mathbf{P}_1\, \mathbf{A}\, ) \exp(\mathbf{A}\, t)\, ]\, dt$$

$$= \int_0^\infty \exp(\mathbf{A}^T t)\, \mathbf{Q} \exp(\mathbf{A}\, t)\, dt = \mathbf{P}$$

An alternate proof of uniqueness is the elegant original proof given by Lyapunov, which makes direct use of fundamental algebra results. The Lyapunov equation (3.19) can be interpreted as defining a linear map from the $n^2$ components of the matrix $\mathbf{P}$ to the $n^2$ components of the matrix $\mathbf{Q}$, where $\mathbf{P}$ and $\mathbf{Q}$ are arbitrary (not necessarily symmetric *p.d.*) square matrices. Since (3.20) actually shows the existence of a solution $\mathbf{P}$ for *any* square matrix $\mathbf{Q}$, the range of this linear map is full, and therefore its null-space is reduced to $\mathbf{0}$. Thus, for any $\mathbf{Q}$, the solution $\mathbf{P}$ is unique.                                                                                                      ◻

The above theorem shows that *any* positive definite matrix $\mathbf{Q}$ can be used to determine the stability of a linear system. A simple choice of $\mathbf{Q}$ is the identity matrix.

**Example 3.18:** Consider again the second-order system of Example 3.17. Let us take $\mathbf{Q} = \mathbf{I}$ and denote $\mathbf{P}$ by

$$\mathbf{P} = \begin{bmatrix} p_{11} & p_{12} \\ p_{21} & p_{22} \end{bmatrix}$$

where, due to the symmetry of $\mathbf{P}$, $p_{21} = p_{12}$. Then the Lyapunov equation is

$$\begin{bmatrix} p_{11} & p_{12} \\ p_{12} & p_{22} \end{bmatrix} \begin{bmatrix} 0 & 4 \\ -8 & -12 \end{bmatrix} + \begin{bmatrix} 0 & -8 \\ 4 & -12 \end{bmatrix} \begin{bmatrix} p_{11} & p_{12} \\ p_{12} & p_{22} \end{bmatrix} = \begin{bmatrix} -1 & 0 \\ 0 & -1 \end{bmatrix}$$

whose solution is

$$p_{11} = 5/16 , \; p_{12} = p_{22} = 1/16$$

The corresponding matrix

$$\mathbf{P} = \frac{1}{16}\begin{bmatrix} 5 & 1 \\ 1 & 1 \end{bmatrix}$$

is positive definite, and therefore the linear system is globally asymptotically stable. Note that we have solved for $\mathbf{P}$ directly, without using the more cumbersome expression (3.20).     ◻

Even though the choice $\mathbf{Q} = \mathbf{I}$ is motivated by computational simplicity, it has a surprising property: the resulting Lyapunov analysis allows us to get the best estimate of the state convergence rate, as we shall see in section 3.5.5.

## 3.5.2 Krasovskii's Method

Let us now come back to the problem of finding Lyapunov functions for general nonlinear systems. Krasovskii's method suggests a simple form of Lyapunov function candidate for autonomous nonlinear systems of the form (3.2), namely, $V = \mathbf{f}^T \mathbf{f}$. The basic idea of the method is simply to check whether this particular choice indeed leads to a Lyapunov function.

**Theorem 3.7 (Krasovskii)**   *Consider the autonomous system defined by (3.2), with the equilibrium point of interest being the origin. Let $\mathbf{A}(\mathbf{x})$ denote the Jacobian matrix of the system, i.e.,*

$$\mathbf{A}(\mathbf{x}) = \frac{\partial \mathbf{f}}{\partial \mathbf{x}}$$

*If the matrix $\mathbf{F} = \mathbf{A} + \mathbf{A}^T$ is negative definite in a neighborhood $\Omega$, then the equilibrium point at the origin is asymptotically stable. A Lyapunov function for this system is*

$$V(\mathbf{x}) = \mathbf{f}^T(\mathbf{x})\, \mathbf{f}(\mathbf{x})$$

*If $\Omega$ is the entire state space and, in addition, $V(\mathbf{x}) \to \infty$ as $\|\mathbf{x}\| \to \infty$, then the equilibrium point is globally asymptotically stable.*

**Proof:** First, let us prove that the negative definiteness of $\mathbf{F}$ implies that $\mathbf{f}(\mathbf{x}) \neq \mathbf{0}$ for $\mathbf{x} \neq \mathbf{0}$.

Since the square matrix $\mathbf{F}(\mathbf{x})$ is negative definite for non-zero $\mathbf{x}$, one can show that the Jacobian matrix $\mathbf{A}(\mathbf{x})$ is invertible, by contradiction. Indeed, assume that $\mathbf{A}$ is singular. Then one can find a non-zero vector $\mathbf{y}_o$ such that $\mathbf{A}(\mathbf{x})\mathbf{y}_o = \mathbf{0}$. Since

$$\mathbf{y}_o^T \mathbf{F} \mathbf{y}_o = 2 \mathbf{y}_o^T \mathbf{A}\, \mathbf{y}_o$$

the singularity of $\mathbf{A}$ implies that $\mathbf{y}_o^T \mathbf{A}\, \mathbf{y}_o = 0$, which contradicts the assumed negative definiteness of $\mathbf{F}$.

The invertibility and continuity of $\mathbf{A}$ guarantee that the function $\mathbf{f}(\mathbf{x})$ can be *uniquely* inverted. This implies that the dynamic system (3.2) has only one equilibrium point in $\Omega$ (otherwise different equilibrium points would correspond to the same value of $\mathbf{f}$), i.e., that $\mathbf{f}(\mathbf{x}) \neq \mathbf{0}$ for $\mathbf{x} \neq \mathbf{0}$.

We can now show the asymptotic stability of the origin. Given the above result, the scalar function $V(\mathbf{x}) = \mathbf{f}^T(\mathbf{x})\, \mathbf{f}(\mathbf{x})$ is positive definite. Using the fact that $\dot{\mathbf{f}} = \mathbf{A}\mathbf{f}$, the derivative of $V$ can be written

$$\dot{V}(\mathbf{x}) = \mathbf{f}^T \dot{\mathbf{f}} + \dot{\mathbf{f}}^T \mathbf{f} = \mathbf{f}^T \mathbf{A}\, \mathbf{f} + \mathbf{f}^T \mathbf{A}^T\, \mathbf{f} = \mathbf{f}^T \mathbf{F} \mathbf{f}$$

The negative definiteness of $\mathbf{F}$ implies the negative definiteness of $\dot{V}$. Therefore, according to Lyapunov's direct method, the equilibrium state at the origin is asymptotically stable. The global asymptotic stability of the system is guaranteed by the global version of Lyapunov's direct method. □

Let us illustrate the use of Krasovskii's theorem on a simple example.

**Example 3.19:** Consider the nonlinear system

$$\dot{x}_1 = -6x_1 + 2x_2$$

$$\dot{x}_2 = 2x_1 - 6x_2 - 2x_2{}^3$$

We have

$$A = \frac{\partial \mathbf{f}}{\partial \mathbf{x}} = \begin{bmatrix} -6 & 2 \\ 2 & -6 - 6x_2{}^2 \end{bmatrix} \qquad \mathbf{F} = \mathbf{A} + \mathbf{A}^T = \begin{bmatrix} -12 & 4 \\ 4 & -12 - 12x_2{}^2 \end{bmatrix}$$

The matrix **F** is easily shown to be negative definite over the whole state space. Therefore, the origin is asymptotically stable, and a Lyapunov function candidate is

$$V(\mathbf{x}) = \mathbf{f}^T(\mathbf{x})\,\mathbf{f}(\mathbf{x}) = (-6x_1 + 2x_2)^2 + (2x_1 - 6x_2 - 2x_2{}^3)^2$$

Since $V(\mathbf{x}) \to \infty$ as $\|\mathbf{x}\| \to \infty$, the equilibrium state at the origin is globally asymptotically stable. □

While the use of the above theorem is very straightforward, its applicability is limited in practice, because the Jacobians of many systems do not satisfy the negative definiteness requirement. In addition, for systems of high order, it is difficult to check the negative definiteness of the matrix **F** for *all* **x**.

An immediate generalization of Krasovskii's theorem is as follows:

**Theorem 3.8 (Generalized Krasovskii Theorem)** *Consider the autonomous system defined by (3.2), with the equilibrium point of interest being the origin, and let* $\mathbf{A}(\mathbf{x})$ *denote the Jacobian matrix of the system. Then, a sufficient condition for the origin to be asymptotically stable is that there exist two symmetric positive definite matrices* **P** *and* **Q**, *such that* $\forall \mathbf{x} \neq \mathbf{0}$, *the matrix*

$$\mathbf{F}(\mathbf{x}) = \mathbf{A}^T \mathbf{P} + \mathbf{P} \mathbf{A} + \mathbf{Q}$$

*is negative semi-definite in some neighborhood* $\Omega$ *of the origin. The function* $V(\mathbf{x}) = \mathbf{f}^T \mathbf{P} \mathbf{f}$ *is then a Lyapunov function for the system. If the region* $\Omega$ *is the whole state space, and if in addition,* $V(\mathbf{x}) \to \infty$ *as* $\|\mathbf{x}\| \to \infty$, *then the system is globally asymptotically stable.*

**Proof**: This theorem can be proven similarly. The positive definiteness of $V(\mathbf{x})$ can be derived as before. Furthermore, the derivative of $V$ can be computed as

$$\dot{V} = \frac{\partial V}{\partial \mathbf{x}} \mathbf{f}(\mathbf{x}) = \mathbf{f}^T \mathbf{P} \mathbf{A}(\mathbf{x}) \mathbf{f} + \mathbf{f}^T \mathbf{P} \mathbf{A}^T(\mathbf{x}) \mathbf{P} \mathbf{f} = \mathbf{f}^T \mathbf{F} \mathbf{f} - \mathbf{f}^T \mathbf{Q} \mathbf{f}$$

Because $\mathbf{F}$ is negative semi-definite and $\mathbf{Q}$ is positive definite, $\dot{V}$ is negative definite and the equilibrium point at the origin is asymptotically stable. If $V(\mathbf{x}) \to \infty$ as $\|\mathbf{x}\| \to \infty$, the global version of Lyapunov's direct method indicates the global asymptotic stability of the system.    □

## 3.5.3 The Variable Gradient Method

The variable gradient method is a formal approach to constructing Lyapunov functions. It involves assuming a certain form for the gradient of an unknown Lyapunov function, and then finding the Lyapunov function itself by integrating the assumed gradient. For low order systems, this approach sometimes leads to the successful discovery of a Lyapunov function.

To start with, let us note that a scalar function $V(\mathbf{x})$ is related to its gradient $\nabla V$ by the integral relation

$$V(\mathbf{x}) = \int_0^{\mathbf{x}} \nabla V \, d\mathbf{x}$$

where $\nabla V = \{\partial V/\partial x_1, \ldots, \partial V/\partial x_n\}^T$. In order to recover a unique scalar function $V$ from the gradient $\nabla V$, the gradient function has to satisfy the so-called curl conditions

$$\frac{\partial \nabla V_i}{\partial x_j} = \frac{\partial \nabla V_j}{\partial x_i} \qquad (i, j = 1, 2, \ldots, n)$$

Note that the $i^{\text{th}}$ component $\nabla V_i$ is simply the directional derivative $\partial V/\partial x_i$. For instance, in the case $n = 2$, the above simply means that

$$\frac{\partial \nabla V_1}{\partial x_2} = \frac{\partial \nabla V_2}{\partial x_1}$$

The principle of the variable gradient method is to assume a specific form *for the gradient* $\nabla V$, instead of assuming a specific form for the Lyapunov function $V$ itself. A simple way is to assume that the gradient function is of the form

$$\nabla V_i = \sum_{j=1}^{n} a_{ij} x_j \qquad (3.21)$$

where the $a_{ij}$'s are coefficients to be determined.    This leads to the following procedure for seeking a Lyapunov function $V$ :

- assume that $\nabla V$ is given by (3.21) (or another form)

- solve for the coefficients $a_{ij}$ so as to satisfy the curl equations

- restrict the coefficients in (3.21) so that $\dot{V}$ is negative semi-definite (at least locally)

- compute $V$ from $\nabla V$ by integration;

- check whether $V$ is positive definite

Since satisfaction of the curl conditions implies that the above integration result is independent of the integration path, it is usually convenient to obtain $V$ by integrating along a path which is parallel to each axis in turn, *i.e.*,

$$V(\mathbf{x}) = \int_o^{x_1} \nabla V_1(x_1,0,...,0) \, dx_1 \; + \; \int_o^{x_2} \nabla V_2(x_1,x_2,0,...,0) dx_2 + \; ...$$

$$+ \; \int_o^{x_n} \nabla V_n(x_1,x_2,...,x_n) dx_n$$

**Example 3.20**: Let us use the variable gradient method to find a Lyapunov function for the nonlinear system

$$\dot{x}_1 = - 2x_1$$

$$\dot{x}_2 = - 2x_2 + 2x_1 x_2^2$$

We assume that the gradient of the undetermined Lyapunov function has the following form

$$\nabla V_1 = a_{11} x_1 + a_{12} x_2$$

$$\nabla V_2 = a_{21} x_1 + a_{22} x_2$$

The curl equation is

$$\frac{\partial \nabla V_1}{\partial x_2} = \frac{\partial \nabla V_2}{\partial x_1}$$

$$a_{12} + x_2 \frac{\partial a_{12}}{\partial x_2} = a_{21} + x_1 \frac{\partial a_{21}}{\partial x_1}$$

If the coefficients are chosen to be

$$a_{11} = a_{22} = 1, a_{12} = a_{21} = 0$$

which leads to

$$\nabla V_1 = x_1 \qquad \nabla V_2 = x_2$$

then $\dot{V}$ can be computed as

$$\dot{V} = \nabla V \dot{\mathbf{x}} = -2x_1^2 - 2x_2^2(1 - x_1 x_2)$$

Thus, $\dot{V}$ is locally negative definite in the region $(1 - x_1 x_2) > 0$. The function $V$ can be computed as

$$V(\mathbf{x}) = \int_o^{x_1} x_1 \, dx_1 + \int_o^{x_2} x_2 \, dx_2 = \frac{x_1^2 + x_2^2}{2} \tag{3.22}$$

This is indeed positive definite, and therefore the asymptotic stability is guaranteed.

Note that (3.22) is not the only Lyapunov function obtainable by the variable gradient method. For example, by taking

$$a_{11} = 1, \ a_{12} = x_2^2$$

$$a_{21} = 3x_2^2, \ a_{22} = 3$$

we obtain the positive definite function

$$V = \frac{x_1^2}{2} + \frac{3}{2}x_2^2 + x_1 x_2^3 \tag{3.23}$$

whose derivative is

$$\dot{V} = -2x_1^2 - 6x_2^2 - 2x_2^2 (x_1 x_2 - 3x_1^2 x_2^2)$$

One easily verifies that $\dot{V}$ is a locally negative definite function (noting that the quadratic terms are dominant near the origin), and therefore, (3.23) represents another Lyapunov function for the system. □

## 3.5.4 Physically Motivated Lyapunov Functions

The Lyapunov functions in the above sections 3.5.1-3.5.3, and in a number of examples earlier in section 3.4, have been obtained from a *mathematical* point of view, *i.e.*, we examined the mathematical features of the given differential equations and searched for Lyapunov function candidates $V$ that can make $\dot{V}$ negative. We did not pay much attention to where the dynamic equations came from and what properties the physical systems had. However, this purely mathematical approach, though effective for simple systems, is often of little use for complicated dynamic equations. On the other hand, if engineering insight and physical properties are properly exploited, an elegant and powerful Lyapunov analysis may be possible for very complex systems.

**Example 3.21: Global asymptotic stability of a robot position controller**

A fundamental task in robotic applications is for robot manipulators to transfer objects from one point to another, the so-called robot position control problem. In the last decade, engineers had been routinely using P.D. (proportional plus derivative) controllers to control robot arms. However, there was no theoretical justification for the stability of such control systems, because the dynamics of a robot is highly nonlinear.

A robot arm consists a number of links connected by rotational or translational joints, with the last link equipped with some end-effector (Figure 3.19). The dynamics of an $n$-link robot arm can be expressed by a set of $n$ equations,

$$\mathbf{H}(\mathbf{q})\ddot{\mathbf{q}} + \mathbf{b}(\mathbf{q}, \dot{\mathbf{q}}) + \mathbf{g}(\mathbf{q}) = \mathbf{\tau} \tag{3.24}$$

where $\mathbf{q}$ is an $n$-dimensional vector describing the joint positions of the robot, $\mathbf{\tau}$ is the vector of input torques, $\mathbf{g}$ is the vector of gravitational torques, $\mathbf{b}$ represents the Coriolis and centripetal forces caused by the motion of the links, and $\mathbf{H}$ the $n \times n$ inertia matrix of the robot arm. Consider a controller simply composed of a P.D. term and a gravity compensation term

$$\mathbf{\tau} = -\mathbf{K}_D \dot{\mathbf{q}} - \mathbf{K}_P \mathbf{q} + \mathbf{g}(\mathbf{q}) \tag{3.25}$$

where $\mathbf{K}_D$ and $\mathbf{K}_P$ are constant positive definite $n \times n$ matrices. It is almost impossible to use trial-and error to search for a Lyapunov function for the closed loop dynamics defined by (3.24) and (3.25), because (3.24) contains hundreds of terms for the 5-link or 6-link robot arms commonly found in industry. Therefore, it seems very difficult to show that $\dot{\mathbf{q}} \rightarrow \mathbf{0}$ and $\mathbf{q} \rightarrow \mathbf{0}$.

With the aid of physical insights, however, a Lyapunov function can be successfully found for such complex robotic systems. First, note that the inertia matrix $\mathbf{H}(\mathbf{q})$ is positive definite for any $\mathbf{q}$. Second, the P.D. control term can be interpreted as mimicking a combination of dampers and springs. This suggests the following Lyapunov function candidate

$$V = \frac{1}{2} [\dot{\mathbf{q}}^T \mathbf{H} \dot{\mathbf{q}} + \mathbf{q}^T \mathbf{K}_p \mathbf{q}]$$

where the first term represents the kinetic energy of the manipulator, and the second term denotes the "artificial potential energy" associated with the virtual spring in the control law (3.25).

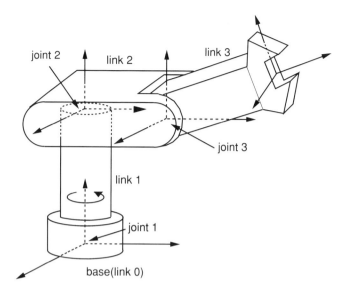

**Figure 3.19 :** A robot manipulator

In computing the derivative of this function, one can use the energy theorem in mechanics, which states that the rate of change of kinetic energy in a mechanical system is equal to the power provided by the external forces. Therefore,

$$\dot{V} = \dot{\mathbf{q}}^T (\boldsymbol{\tau} - \mathbf{g}) + \dot{\mathbf{q}}^T \mathbf{K}_P \mathbf{q}$$

Substitution of the control law (3.25) in the above equation then leads to

$$\dot{V} = -\dot{\mathbf{q}}^T \mathbf{K}_D \dot{\mathbf{q}}$$

Since the arm cannot get "stuck" at any position such that $\mathbf{q} \neq \mathbf{0}$ (which can be easily shown by noting that acceleration is non-zero in such situations), the robot arm must settle down at $\dot{\mathbf{q}} = \mathbf{0}$ and $\mathbf{q} = \mathbf{0}$, according to the invariant set theorem. Thus, the system is actually *globally* asymptotically stable.                                                                    □

Two lessons can be learned from this practical example. The first is that one should use as many physical properties as possible in analyzing the behavior of a system. The second lesson is that physical concepts like energy may lead us to some uniquely powerful choices of Lyapunov functions. Physics will play a major role in the development of the multi-input nonlinear controllers of chapter 9.

## 3.5.5 Performance Analysis

In the preceding sections, we have been primarily concerned with using Lyapunov functions for stability analysis. But sometimes Lyapunov functions can further provide estimates of the transient performance of stable systems. In particular, they can allow us to estimate the convergence rate of asymptotically stable linear or nonlinear systems. In this section, we first derive a simple lemma on differential inequalities. We then show how Lyapunov analysis may be used to determine the convergence rates of linear and nonlinear systems.

### A SIMPLE CONVERGENCE LEMMA

**Lemma**: *If a real function W(t) satisfies the inequality*

$$\dot{W}(t) + \alpha W(t) \leq 0 \tag{3.26}$$

*where $\alpha$ is a real number. Then*

$$W(t) \leq W(0) e^{-\alpha t}$$

**Proof**: Let us define a function $Z(t)$ by

$$Z(t) = \dot{W} + \alpha W \tag{3.27}$$

Equation (3.26) implies that $Z(t)$ is non-positive. The solution of the first-order equation (3.27) is

$$W(t) = W(0) e^{-\alpha t} + \int_{o}^{t} e^{-\alpha(t-r)} Z(t) \, dr$$

Because the second term in the right-hand-side of the above equation is non-positive, one has

$$W(t) \leq W(0) e^{-\alpha t} \qquad \qquad \square$$

The above lemma implies that, if $W$ is a non-negative function, the satisfaction of (3.26) guarantees the exponential convergence of $W$ to zero. In using Lyapunov's direct method for stability analysis, it is sometimes possible to manipulate $\dot{V}$ into the form of (3.26). In such a case, the exponential convergence of $V$ and the convergence rate can be inferred and, in turn, the exponential convergence rate of the state may then be determined. In later chapters, we will provide examples of using this lemma for estimating convergence rates of nonlinear or adaptive control systems.

### ESTIMATING CONVERGENCE RATES FOR LINEAR SYSTEMS

Now let us evaluate the convergence rate of a stable linear system based on the Lyapunov analysis described in section 3.5.1. Let us denote the largest eigenvalue of the matrix $\mathbf{P}$ by $\lambda_{max}(\mathbf{P})$, the smallest eigenvalue of $\mathbf{Q}$ by $\lambda_{min}(\mathbf{Q})$, and their ratio $\lambda_{min}(\mathbf{Q})/\lambda_{max}(\mathbf{P})$ by $\gamma$. The positive definiteness of $\mathbf{P}$ and $\mathbf{Q}$ implies that these scalars are all strictly positive. Since matrix theory shows that

$$\mathbf{P} \le \lambda_{max}(\mathbf{P})\,\mathbf{I} \qquad\qquad \lambda_{min}(\mathbf{Q})\,\mathbf{I} \le \mathbf{Q}$$

we have

$$\mathbf{x}^T \mathbf{Q}\mathbf{x} \ge \frac{\lambda_{min}(\mathbf{Q})}{\lambda_{max}(\mathbf{P})} \mathbf{x}^T[\lambda_{max}(\mathbf{P})\mathbf{I}]\mathbf{x} \ge \gamma V$$

This and (3.18) imply that

$$\dot{V} \le -\gamma V$$

This, according to lemma, means that

$$\mathbf{x}^T \mathbf{P}\mathbf{x} \le V(0)\,e^{-\gamma t}$$

This, together with the fact $\mathbf{x}^T \mathbf{P}\mathbf{x} \ge \lambda_{min}(\mathbf{P})\|\mathbf{x}(t)\|^2$, implies that the state $\mathbf{x}$ converges to the origin with a rate of at least $\gamma/2$.

One might naturally wonder how this convergence rate estimate varies with the choice of $\mathbf{Q}$, and how it relates to the familiar notion of dominant pole in linear theory. An interesting result is that *the convergence rate estimate is largest for* $\mathbf{Q} = \mathbf{I}$. Indeed, let $\mathbf{P}_o$ be the solution of the Lyapunov equation corresponding to $\mathbf{Q} = \mathbf{I}$ :

$$\mathbf{A}^T \mathbf{P}_o + \mathbf{P}_o\,\mathbf{A} = -\mathbf{I}$$

and let $\mathbf{P}$ be the solution corresponding to some other choice of $\mathbf{Q}$

$$\mathbf{A}^T \mathbf{P} + \mathbf{P}\,\mathbf{A} = -\mathbf{Q}_1$$

Without loss of generality, we can assume that $\lambda_{min}(\mathbf{Q}_1) = 1$, since rescaling $\mathbf{Q}_1$ will rescale $\mathbf{P}$ by the same factor, and therefore will not affect the value of the corresponding $\gamma$. Subtracting the above two equations yields

$$\mathbf{A}^T (\mathbf{P} - \mathbf{P}_o) + (\mathbf{P} - \mathbf{P}_o)\,\mathbf{A} = -(\mathbf{Q}_1 - \mathbf{I})$$

Now since $\lambda_{min}(\mathbf{Q}_1) = 1 = \lambda_{max}(\mathbf{I})$, the matrix $(\mathbf{Q}_1 - \mathbf{I})$ is positive semi-definite, and hence the above equation implies that $(\mathbf{P} - \mathbf{P}_o)$ is positive semi-definite. Therefore

$$\lambda_{max}(\mathbf{P}) \ge \lambda_{max}(\mathbf{P}_o)$$

Since $\lambda_{min}(\mathbf{Q}_1) = 1 = \lambda_{min}(\mathbf{I})$, the convergence rate estimate

$$\gamma = \lambda_{min}(\mathbf{Q})/\lambda_{max}(\mathbf{P})$$

corresponding to $\mathbf{Q} = \mathbf{I}$ is larger than (or equal to) that corresponding to $\mathbf{Q} = \mathbf{Q}_1$.

If the stable matrix $\mathbf{A}$ is symmetric, then the meaning of this "optimal" value of $\gamma$, corresponding to the choice $\mathbf{Q} = \mathbf{I}$, can be interpreted easily. Indeed, all eigenvalues of $\mathbf{A}$ are then real, and furthermore $\mathbf{A}$ is diagonalizable, *i.e.*, there exists a change of state coordinates such that in these coordinates $\mathbf{A}$ is diagonal. One immediately verifies that, in these coordinates, the matrix $\mathbf{P} = -1/2\,\mathbf{A}^{-1}$ verifies the Lyapunov equation for $\mathbf{Q} = \mathbf{I}$, and that therefore the corresponding $\gamma/2$ is simply the absolute value of the *dominant pole* of the linear system. Furthermore, $\gamma$ is obviously independent of the choice of state coordinates.

## ESTIMATING CONVERGENCE RATES FOR NONLINEAR SYSTEMS

The estimation of convergence rate for nonlinear systems also involves manipulating the expression of $\dot{V}$ so as to obtain an explicit estimate of $V$. The difference lies in that, for nonlinear systems, $V$ and $\dot{V}$ are not necessarily quadratic functions of the states.

**Example 3.22**: Consider again the system in Example 3.8. Given the chosen Lyapunov function candidate $V = \|\mathbf{x}\|^2$, the derivative $\dot{V}$, can be written

$$\dot{V} = 2V(V-1)$$

that is,

$$\frac{dV}{V(1-V)} = -2\,dt$$

The solution of this equation is easily found to be

$$V(\mathbf{x}) = \frac{\alpha e^{-2t}}{1 + \alpha e^{-2t}}$$

where

$$\alpha = \frac{V(0)}{1 - V(0)}$$

If $\|\mathbf{x}(0)\|^2 = V(0) < 1$, *i.e.*, if the trajectory starts inside the unit circle, then $\alpha > 0$, and

$$V(t) < \alpha e^{-2t}$$

This implies that the norm $\|\mathbf{x}(t)\|$ of the state vector converges to zero exponentially, with a rate of 1.

However, if the trajectory starts outside the unit circle, *i.e.*, if $V(0) > 1$, then $\alpha < 0$, so that

$V(t)$ and therefore $\|\mathbf{x}\|$ tend to infinity in a finite time (the system is said to exhibit finite escape time, or "explosion").                                                                                 ◻

# 3.6   Control Design Based on Lyapunov's Direct Method

The previous sections have been concerned with using Lyapunov's direct method for system analysis. In doing the analysis, we have implicitly presumed that certain control laws have been chosen for the systems. However, in many control problems, the task is to find an appropriate control law for a given plant. In the following, we briefly discuss how to apply Lyapunov's direct method for designing stable control systems. Many of the controller design methods we will describe in chapters 6-9 are actually based on Lyapunov concepts.

There are basically two ways of using Lyapunov's direct method for control design, and both have a trial and error flavor. The first technique involves hypothesizing one form of control law and then finding a Lyapunov function to justify the choice. The second technique, conversely, requires hypothesizing a Lyapunov function candidate and then finding a control law to make this candidate a real Lyapunov function.

We saw an application of the first technique in the robotic P.D. control example in section 3.5.4, where a P.D. controller is chosen based on physical intuition, and a Lyapunov function is found to show the global asymptotic convergence of the resulting closed-loop system. The second technique can be illustrated on the following simple example.

**Example 3.23: Regulator Design**

Consider the problem of stabilizing the system

$$\ddot{x} - \dot{x}^3 + x^2 = u$$

*i.e.*, to bring it to equilibrium at $x \equiv 0$. Based on Example 3.14, it is sufficient to choose a continuous control law $u$ of the form

$$u = u_1(\dot{x}) + u_2(x)$$

where

$$\dot{x}(\dot{x}^3 + u_1(\dot{x})) < 0 \quad \text{for } \dot{x} \neq 0$$

$$x(x^2 - u_2(x)) > 0 \quad \text{for } x \neq 0$$

The above inequalities also imply that globally stabilizing controllers can be designed even in the presence of some uncertainty on the dynamics. For instance, the system

$$\ddot{x} + \alpha_1 \dot{x}^3 + \alpha_2 x^2 = u$$

where the constants $\alpha_1$ and $\alpha_2$ are unknown, but such that $\alpha_1 > -2$ and $|\alpha_2| < 5$, can be globally stabilized using the control law

$$u = -2 \dot{x}^3 - 5(x + x^3)$$                                        □

For some classes of nonlinear systems, systematic design procedures have been developed based on these above two techniques, as can be seen in chapter 7 on the sliding control design methodology, in chapter 8 on adaptive control, and in chapter 9 on physically-motivated designs.

Finally, it is important to notice that, just as a nonlinear system may be globally asymptotically stable while its linear approximation is only marginally stable, a nonlinear system may be controllable while its linear approximation is not. Consider for instance the system

$$\ddot{x} + \dot{x}^5 = x^2 u$$

This system can be made to converge asymptotically to zero simply by letting $u = -x$. However, its linear approximation at $x = 0$ and $u = 0$ is $\ddot{x} \approx 0$, and therefore is uncontrollable!

# 3.7  Summary

Stability is a fundamental issue in control system analysis and design. Various concepts of stability, such as Lyapunov stability, asymptotic stability, exponential stability, and global asymptotic or exponential stability, must be defined in order to accurately characterize the complex and rich stability behaviors exhibited by nonlinear systems.

Since analytical solutions of nonlinear differential equations usually cannot be obtained, the two methods of Lyapunov are of major importance in the determination of nonlinear system stability.

The linearization method is concerned with the small motion of nonlinear systems around equilibrium points. It is mainly of theoretical value, justifying the use of linear control for the design and analysis of weakly nonlinear physical systems.

The direct method, based on so-called Lyapunov functions, is not restricted to small motions. In principle, it is applicable to essentially all dynamic systems, whether linear or nonlinear, continuous-time or discrete-time , of finite or infinite order, and in small or large motion. However, the method suffers from the common difficulty of finding a Lyapunov function for a given system. Since there is no generally effective approach for finding Lyapunov functions, one has to use trial-and-error, experience, and intuition to search for appropriate Lyapunov functions. Physical properties (such as energy conservation) and physical insights may be exploited in formulating Lyapunov functions, and may lead to uniquely powerful choices. Simple mathematical techniques, such as, *e.g.*, Krasovskii's method or the variable gradient method, can also be of help.

Generally, the application of Lyapunov theory to *controller design* is more easily rewarding. This is because, in design, one often has the freedom to deliberately modify the dynamics (by designing the controller) in such a way that a chosen scalar function becomes a Lyapunov function for the closed-loop system. In the second part of the book, we will see many applications of Lyapunov theory to the construction of effective nonlinear control systems.

## 3.8  Notes and References

In Lyapunov's original work [Lyapunov, 1892], the linearization method (which, today, is sometimes incorrectly referred to as the first method) is simply given as an example of application of the direct (or second) method. The first method was the so-called method of exponents, which is used today in the analysis of chaos.

Many references can be found on Lyapunov theory, *e.g.*, [La Salle and Lefschetz, 1961] (on which the invariant set theorems are based), [Kalman and Bertram, 1960; Hahn, 1967; Yoshizawa, 1966]. An inspiring discussion of the role of scalar summarizing functions in science, along with a very readable introduction to elementary Lyapunov theory, can be found in [Luenberger, 1979].

Examples 3.1 and 3.13 are adapted from [Luenberger, 1979]. Examples 3.3 and 3.8 are adapted from [Vidyasagar, 1978]. The counterexample in Figure 3.12 is from [Hahn, 1967]. Example 3.14 is adapted from [La Salle and Lefschetz, 1961; Vidyasagar, 1978]. The variable gradient method in subsection 3.5.3 is adapted from [Ogata, 1970]. The robotic example of section 3.5.4 is based on [Takegaki and Arimoto, 1981]. The remark on the "optimal" choice of **Q** in section 3.5.5 is from [Vidyasagar, 1982]. A detailed study of Krasovskii's theorems can be found in [Krasovskii, 1963].

## 3.9 Exercises

**3.1**    The norm used in the definitions of stability need not be the usual Euclidian norm. If the state-space is of finite dimension $n$ (*i.e.*, the state vector has $n$ components), stability and its type are independent of the choice of norm (all norms are "equivalent"), although a particular choice of norm may make analysis easier. For $n = 2$, draw the unit balls corresponding to the following norms:

(i)    $\| \mathbf{x} \|^2 = (x_1)^2 + (x_2)^2$    (Euclidian norm)

(ii)    $\| \mathbf{x} \|^2 = (x_1)^2 + 5\,(x_2)^2$

(iii)    $\| \mathbf{x} \| = |x_1| + |x_2|$

(iv)    $\| \mathbf{x} \| = \text{Sup}(\,|x_1|\,,|x_2|\,)$

Recall that a ball $\mathbf{B}(\mathbf{x}_o, R)$, of center $\mathbf{x}_o$ and radius $R$, is the set of $\mathbf{x}$ such that $\| \mathbf{x} - \mathbf{x}_o\| \le R$, and that the unit ball is $\mathbf{B}(\mathbf{0}, 1)$.

**3.2**    For the following systems, find the equilibrium points and determine their stability. Indicate whether the stability is asymptotic, and whether it is global.

(*a*)    $\dot{x} = -x^3 + \sin^4 x$

(*b*)    $\dot{x} = (5 - x)^5$

(*c*)    $\ddot{x} + \dot{x}^5 + x^7 = x^2 \sin^8 x \cos^2 3x$

(*d*)    $\ddot{x} + (x - 1)^4 \dot{x}^7 + x^5 = x^3 \sin^3 x$

(*e*)    $\ddot{x} + (x - 1)^2 \dot{x}^7 + x = \sin(\pi x/2)$

**3.3**    For the Van der Pol oscillator of Example 3.3, demonstrate the existence of a limit cycle using the linearization method.

**3.4**    This exercise, adapted from [Hahn, 1967], provides an example illustrating the motivation of the radial unboundedness condition in Theorem 3.3. Consider the second-order system

$$\dot{x}_1 = -\frac{6x_1}{z^2} + 2x_2$$

$$\dot{x}_2 = -\frac{2(x_1 + x_2)}{z^2}$$

with $z = 1 + x_1^2$. On the hyperbola $x_2^h = 2/(x_1 - \sqrt{2}\,)$, the system trajectory slope is

$$\frac{\dot{x}_2}{\dot{x}_1} = \frac{-1}{1 + 2^{3/2}x_1 + 2x_1^2}$$

while the slope of the hyperbola is

$$\frac{dx_2^h}{dx_1} = \frac{-1}{1 - 2^{3/2}x_1 + x_1^2/2}$$

Note that for $x_1 > \sqrt{2}$, the first expression is larger than the second, implying that the trajectories cannot cut the branch of the hyperbola which lies in the first quadrant, in the direction toward the axes (since on the hyperbola we have $\dot{x}_1 > 0$ if $x_1 > \sqrt{2}$). Thus, there are trajectories which do not tend toward the origin, indicating the lack of global asymptotic stability. Use the scalar function

$$V(\mathbf{x}) = \frac{x_1^2}{z} + x_2^2$$

to analyze the stability of the above system.

**3.5**    Determine regions of attraction of the pendulum, using as Lyapunov functions the pendulum's total energy, and the modified Lyapunov function of page 67. Comment on the two results.

**3.6**    Show that given a constant matrix $\mathbf{M}$ and any time-varying vector $\mathbf{x}$, the time-derivative of the scalar $\mathbf{x}^T \mathbf{M} \mathbf{x}$ can be written

$$\frac{d}{dt} \mathbf{x}^T \mathbf{M} \mathbf{x} = \mathbf{x}^T (\mathbf{M} + \mathbf{M}^T) \dot{\mathbf{x}} = \dot{\mathbf{x}}^T (\mathbf{M} + \mathbf{M}^T) \mathbf{x}$$

and that, if $\mathbf{M}$ is symmetric, it can also be written

$$\frac{d}{dt} \mathbf{x}^T \mathbf{M} \mathbf{x} = 2 \mathbf{x}^T \mathbf{M} \dot{\mathbf{x}} = 2 \dot{\mathbf{x}}^T \mathbf{M} \mathbf{x}$$

**3.7**    Consider an $n \times n$ matrix $\mathbf{M}$ of the form $\mathbf{M} = \mathbf{N}^T\mathbf{N}$, where $\mathbf{N}$ is a $m \times n$ matrix. Show that $\mathbf{M}$ is *p.d.* if, and only if, $m \geq n$ and $\mathbf{N}$ has full rank.

**3.8**    Show that if $\mathbf{M}$ is a symmetric matrix such that

$$\forall \mathbf{x}, \ \mathbf{x}^T \mathbf{M} \mathbf{x} = 0$$

then $\mathbf{M} = \mathbf{0}$.

**3.9**    Show that if symmetric *p.d.* matrices $\mathbf{P}$ and $\mathbf{Q}$ exist such that

$$\mathbf{A}^T \mathbf{P} + \mathbf{P} \mathbf{A} + 2\lambda \mathbf{P} = -\mathbf{Q}$$

then all the eigenvalues of $\mathbf{A}$ have a real part strictly less than $-\lambda$. (Adapted from [Luenberger,

1979].)

**3.10**    Consider the system

$$A_1 \ddot{y} + A_2 \dot{y} + A_3 y = 0$$

where the $2n \times 1$ vector $x = [y^T \; \dot{y}^T]^T$ is the state, and the $n \times n$ matrices $A_j$ are all symmetric positive definite. Show that the system is globally asymptotically stable, with $0$ as a unique equilibrium point.

**3.11**    Consider the system

$$\dot{x} = A x \qquad y = c^T x$$

Use the invariant set theorem to show that if the system is observable, and if there exists a symmetric *p.d.* matrix $P$ such that

$$A^T P + P A = -c c^T$$

then the system is asymptotically stable.

Can the result be derived using the direct method? (Adapted from [Luenberger, 1979].)

**3.12**    Use Krasovskii's theorem to justify Lyapunov's linearization method.

**3.13**    Consider the system

$$\dot{x} = 4x^2 y - f_1(x)\,(x^2 + 2y^2 - 4)$$
$$\dot{y} = -2x^3 - f_2(y)\,(x^2 + 2y^2 - 4)$$

where the continuous functions $f_1$ and $f_2$ have the same sign as their argument. Show that the system tends towards a limit cycle independent of the explicit expressions of $f_1$ and $f_2$.

**3.14**    The second law of thermodynamics states that the entropy of an isolated system can only increase with time. How does this relate to the notion of a Lyapunov function?

# Chapter 4
# Advanced Stability Theory

In the previous chapter, we studied Lyapunov analysis of autonomous systems. In many practical problems, however, we encounter non-autonomous systems. For instance, a rocket taking off is a non-autonomous system, because the parameters involved in its dynamic equations, such as air temperature and pressure, vary with time. Furthermore, as discussed earlier, determining the stability of a nominal motion for an autonomous system requires the stability analysis of an equivalent non-autonomous system around an equilibrium point. Therefore, stability analysis techniques for non-autonomous systems must be developed. This constitutes the major topic of this chapter.

After extending the concepts of stability and the Lyapunov stability theorems to non-autonomous systems, in sections 4.1 and 4.2, we discuss a number of interesting related topics in advanced stability theory. Some Lyapunov theorems for concluding *instability* of nonlinear systems are provided in section 4.3. Section 4.4 discusses a number of so-called converse theorems asserting the existence of Lyapunov functions. Besides their theoretical interest, these existence theorems can be valuable in some nonlinear system analysis and design problems. In section 4.5, we describe a very simple mathematical result, known as Barbalat's lemma, which can be conveniently used to solve some asymptotic stability problems beyond the treatment of Lyapunov stability theorems, and which we shall use extensively in chapters 8 and 9. In section 4.6, we discuss the so-called *positive real* linear systems and their unique properties, which shall be exploited later in the book, particularly in chapter 8. Section 4.7 describes passivity theory, a convenient way of interpreting, representing, and combining Lyapunov or Lyapunov-like functions. Section 4.8 discusses a special class of nonlinear systems which can be systematically treated by Lyapunov analysis. Section 4.9 studies some non-Lyapunov techniques which can be used to establish boundedness of signals in nonlinear systems. Finally, section 4.10 discusses mathematical conditions for the existence and unicity of solutions of nonlinear differential equations.

# 4.1  Concepts of Stability for Non-Autonomous Systems

The concepts of stability for non-autonomous systems are quite similar to those of autonomous systems. However, due to the dependence of non-autonomous system behavior on initial time $t_o$, the definitions of these stability concepts include $t_o$ explicitly. Furthermore, a new concept, *uniformity*, is necessary to characterize non-autonomous systems whose behavior has a certain consistency for different values of initial time $t_o$. In this section, we concisely extend the stability concepts for autonomous systems to non-autonomous systems, and introduce the new concept of uniformity.

### EQUILIBRIUM POINTS AND INVARIANT SETS

For non-autonomous systems, of the form

$$\dot{\mathbf{x}} = \mathbf{f}(\mathbf{x}, t) \tag{4.1}$$

*equilibrium points* $\mathbf{x}^*$ are defined by

$$\mathbf{f}(\mathbf{x}^*, t) \equiv \mathbf{0} \qquad \forall \, t \geq t_o \tag{4.2}$$

Note that this equation must be satisfied $\forall \, t \geq t_o$, implying that the system should be able to stay at the point $\mathbf{x}^*$ all the time. For instance, one easily sees that the linear time-varying system

$$\dot{\mathbf{x}} = \mathbf{A}(t)\mathbf{x} \tag{4.3}$$

has a unique equilibrium point at the origin $\mathbf{0}$ unless $\mathbf{A}(t)$ is always singular.

**Example 4.1:** The system

$$\dot{x} = -\frac{a(t)x}{1 + x^2} \tag{4.4}$$

has an equilibrium point at $x = 0$. However, the system

$$\dot{x} = -\frac{a(t)x}{1+x^2} + b(t) \tag{4.5}$$

with $b(t) \neq 0$, does not have an equilibrium point. It can be regarded as a system under external input or disturbance $b(t)$. Since Lyapunov theory is mainly developed for the stability of nonlinear systems with respect to initial conditions, such problems of forced motion analysis are more appropriately treated by other methods, such as those in section 4.9.    $\blacksquare$

The definition of invariant set is the same for non-autonomous systems as for autonomous systems. Note that, unlike in autonomous systems, a system trajectory is generally not an invariant set for a non-autonomous system (Exercise 4.1).

## EXTENSIONS OF THE PREVIOUS STABILITY CONCEPTS

Let us now extend the previously defined concepts of stability, instability, asymptotic stability, and exponential stability to non-autonomous systems. The key in doing so is to properly include the initial time $t_o$ in the definitions.

**Definition 4.1**    *The equilibrium point* **0** *is* <u>*stable*</u> *at* $t_o$ *if for any* $R > 0$, *there exists a positive scalar* $r(R, t_o)$    *such that*

$$\|\mathbf{x}(t_o)\| < r \quad \Rightarrow \quad \|\mathbf{x}(t)\| < R \qquad \forall\, t \geq t_o \tag{4.6}$$

*Otherwise, the equilibrium point* **0** *is* <u>*unstable*</u>.

Again, the definition means that we can keep the state in a ball of arbitrarily small radius $R$ by starting the state trajectory in a ball of sufficiently small radius $r$. Definition 4.1 differs from definition 3.3 in that the radius $r$ of the initial ball may depend on the initial time $t_o$.

The concept of asymptotic stability can also be defined for non-autonomous systems.

**Definition 4.2**    *The equilibrium point* **0** *is* <u>*asymptotically stable*</u> *at time* $t_o$ *if*

- *it is stable*

- $\exists\, r(t_o) > 0$   *such that*    $\|\mathbf{x}(t_o)\| < r(t_o)$    $\Rightarrow$    $\|\mathbf{x}(t)\| \to \mathbf{0}$ *as* $t \to \infty$

Here, the asymptotic stability requires that there exists an attractive region for *every* initial time $t_o$ . The size of attractive region and the speed of trajectory convergence may depend on the initial time $t_o$.

The definitions of exponential stability and global asymptotic stability are also straightforward.

**Definition 4.3**    *The equilibrium point* **0** *is <u>exponentially stable</u> if there exist two positive numbers,* $\alpha$ *and* $\lambda$, *such that for sufficiently small* $\mathbf{x}(t_o)$,

$$\|\mathbf{x}(t)\| \leq \alpha \|\mathbf{x}_o\| \, e^{-\lambda(t-t_o)} \qquad \forall \, t \geq t_o$$

**Definition 4.4**    *The equilibrium point* **0** *is <u>globally asymptotically stable</u> if* $\forall \, \mathbf{x}(t_o)$

$$\mathbf{x}(t) \rightarrow \mathbf{0} \qquad \text{as } t \rightarrow \infty$$

### Example 4.2: A first-order linear time-varying system

Consider the first-order system

$$\dot{x}(t) = -a(t)x(t)$$

Its solution is

$$x(t) = x(t_o) \exp\left[-\int_{t_o}^{t} a(r)\,dr\right]$$

Thus, the system is stable if $a(t) \geq 0$, $\forall \, t \geq t_o$. It is asymptotically stable if $\int_{o}^{\infty} a(r)\,dr = +\infty$. It is exponentially stable if there exists a strictly positive number $T$ such that $\forall \, t \geq 0$, $\int_{t}^{t+T} a(r)\,dr \geq \gamma$, with $\gamma$ being a positive constant.

For instance,

- The system $\dot{x} = -x/(1+t)^2$ is stable (but not asymptotically stable).

- The system $\dot{x} = -x/(1+t)$ is asymptotically stable.

- The system $\dot{x} = -tx$ is exponentially stable.

Another interesting example is the system

$$\dot{x}(t) = -\frac{x}{1+\sin x^2}$$

whose solution can be expressed as

$$x(t) = x(t_o) \exp\left[\int_{t_o}^{t} \frac{-1}{1+\sin x^2(r)}\,dr\right]$$

Since

$$\int_{t_o}^{t} \frac{1}{1+\sin x^2(r)}\,dr \geq \frac{t-t_o}{2}$$

the system is exponentially convergent with rate 1/2 .                              □

## UNIFORMITY IN STABILITY CONCEPTS

The previous concepts of Lyapunov stability and asymptotic stability for non-autonomous systems both indicate the important effect of initial time. In practice, it is usually desirable for the system to have a certain uniformity in its behavior regardless of when the operation starts. This motivates us to consider the definitions of uniform stability and uniform asymptotic stability. *As we shall see in later chapters, non-autonomous systems with uniform properties have some desirable ability to withstand disturbances.* It is also useful to point out that, because the behavior of autonomous systems is independent of the initial time, all the stability properties of an autonomous system are uniform.

**Definition 4.5**    *The equilibrium point* **0** *is locally <u>uniformly stable</u> if the scalar r in Definition 4.1 can be chosen independently of $t_o$ , i.e., if $r = r(R)$.*

The intuitive reason for introducing the concept of uniform stability is to rule out systems which are "less and less stable" for larger values of $t_o$ . Similarly, the definition of uniform asymptotic stability also intends to restrict the effect of the initial time $t_o$ on the state convergence pattern.

**Definition 4.6**    *The equilibrium point at the origin is locally <u>uniformly asymptotically stable</u> if*

- *it is uniformly stable*

- *There exists a ball of attraction* $\mathbf{B}_{R_o}$ *, whose radius is independent of $t_o$ , such that any system trajectory with initial states in* $\mathbf{B}_{R_o}$ *converges to* **0** *uniformly in $t_o$*

By uniform convergence in terms of $t_o$ , we mean that for all $R_1$ and $R_2$ satisfying $0 < R_2 < R_1 \leq R_o$ , $\exists T(R_1, R_2) > 0$, such that, $\forall t_o \geq 0$,

$$\|\mathbf{x}(t_o)\| < R_1 \quad \Rightarrow \quad \|\mathbf{x}(t)\| < R_2 \qquad \forall t \geq t_o + T(R_1, R_2)$$

*i.e.*, the state trajectory, starting from within a ball $\mathbf{B}_{R_1}$, will converge into a smaller ball $\mathbf{B}_{R_2}$ after a time period $T$ which is independent of $t_o$ .

By definition, uniform asymptotic stability always implies asymptotic stability. The converse is generally not true, as illustrated by the following example.

**Example 4.3:** Consider the first-order system

$$\dot{x} = - \frac{x}{1 + t}$$

This system has the general solution

$$x(t) = \frac{1 + t_o}{1 + t} \, x(t_o)$$

This solution asymptotically converges to zero. But the convergence is not uniform. Intuitively, this is because a larger $t_o$ requires a longer time to get close to the origin.      □

Using Definition 4.3, one can easily prove that exponential stability always implies uniform asymptotic stability.

The concept of globally uniformly asymptotic stability can be defined by replacing the ball of attraction $\mathbf{B}_{R_o}$ by the whole state space.

# 4.2 Lyapunov Analysis of Non-Autonomous Systems

We now extend the Lyapunov analysis results of chapter 3 to the stability analysis of non-autonomous systems. Although many of the ideas in chapter 3 can be similarly applied to the non-autonomous case, the conditions required in the treatment of non-autonomous systems are more complicated and more restrictive. We start with the description of the direct method. We then apply the direct method to the stability analysis of linear time-varying systems. Finally, we discuss the linearization method for non-autonomous nonlinear systems.

## 4.2.1 Lyapunov's Direct Method for Non-Autonomous Systems

The basic idea of the direct method, *i.e.*, concluding the stability of nonlinear systems using scalar Lyapunov functions, can be similarly applied to non-autonomous systems. Besides more mathematical complexity, a major difference in non-autonomous systems is that the powerful La Salle's theorems do not apply. This drawback will partially be compensated by a simple result in section 4.5 called Barbalat's lemma.

### TIME-VARYING POSITIVE DEFINITE FUNCTIONS AND DECRESCENT FUNCTIONS

When studying non-autonomous systems using Lyapunov's direct method, scalar functions with explicit time-dependence $V(t, \mathbf{x})$ may have to be used, while in autonomous system analysis time-invariant functions $V(\mathbf{x})$ suffice. We now introduce a simple definition of positive definiteness for such scalar functions.

**Definition 4.7**    *A scalar time-varying function $V(\mathbf{x}, t)$ is* locally positive definite *if $V(\mathbf{0}, t) = 0$ and there exists a time-invariant positive definite function $V_o(\mathbf{x})$ such that*

$$\forall \, t \geq t_o \,, \quad V(\mathbf{x}, t) \geq V_o(\mathbf{x}) \tag{4.7}$$

Thus, a time-variant function is locally positive definite if it *dominates a time-invariant* locally positive definite function. Globally positive definite functions can be defined similarly.

Other related concepts can be defined in the same way, in a local or a global sense. A function $V(\mathbf{x}, t)$ is negative definite if $- V(\mathbf{x}, t)$ is positive definite; $V(\mathbf{x}, t)$ is positive semi-definite if $V(\mathbf{x}, t)$ dominates a time-invariant positive semi-definite function; $V(\mathbf{x}, t)$ is negative semi-definite if $- V(\mathbf{x}, t)$ is positive semi-definite.

In Lyapunov analysis of non-autonomous systems, the concept of decrescent functions is also necessary.

**Definition 4.8**    *A scalar function $V(\mathbf{x}, t)$ is said to be* decrescent *if $V(\mathbf{0}, t) = 0$, and if there exists a time-invariant positive definite function $V_1(\mathbf{x})$ such that*

$$\forall \, t \geq 0 \,, \quad V(\mathbf{x}, t) \leq V_1(\mathbf{x})$$

In other words, a scalar function $V(\mathbf{x}, t)$ is decrescent if it is *dominated by a time-invariant positive definite function*.

**Example 4.4:** A simple example of a time-varying positive definite function is

$$V(\mathbf{x}, t) \,=\, (1 + \sin^2 t)\,(x_1{}^2 + x_2{}^2)$$

because it dominates the function $V_o(\mathbf{x}) = x_1{}^2 + x_2{}^2$. This function is also decrescent because it is dominated by the function $V_1(\mathbf{x}) = 2\,(x_1{}^2 + x_2{}^2)$.    □

Given a time-varying scalar function $V(\mathbf{x}, t)$, its derivative along a system trajectory is

$$\frac{dV}{dt} \,=\, \frac{\partial V}{\partial t} + \frac{\partial V}{\partial \mathbf{x}} \dot{\mathbf{x}} \,=\, \frac{\partial V}{\partial t} + \frac{\partial V}{\partial \mathbf{x}} f(\mathbf{x}, t) \tag{4.8}$$

## LYAPUNOV THEOREM FOR NON-AUTONOMOUS SYSTEM STABILITY

The main Lyapunov stability results for non-autonomous systems can be summarized by the following theorem.

**Theorem 4.1 (Lyapunov theorem for non-autonomous systems)**

**Stability**: *If, in a ball* $\mathbf{B}_{R_o}$ *around the equilibrium point* **0**, *there exists a scalar function* $V(\mathbf{x}, t)$ *with continuous partial derivatives such that*

    *1. V is positive definite*

    *2. $\dot{V}$ is negative semi-definite*

*then the equilibrium point* **0** *is stable in the sense of Lyapunov.*

**Uniform stability and uniform asymptotic stability**: *If, furthermore,*

    *3. V is decrescent*

*then the origin is uniformly stable. If condition 2 is strengthened by requiring that $\dot{V}$ be negative definite, then the equilibrium point is uniformly asymptotically stable.*

**Global uniform asymptotic stability**: *If the ball $B_{R_o}$ is replaced by the whole state space, and condition 1, the strengthened condition 2, condition 3, and the condition*

    *4. $V(\mathbf{x}, t)$ is radially unbounded*

*are all satisfied, then the equilibrium point at* **0** *is globally uniformly asymptotically stable.*

Similarly to the case of autonomous systems, if, in a certain neighborhood of the equilibrium point, $V$ is positive definite and $\dot{V}$, its derivative along the system trajectories, is negative semi-definite, then $V$ is called a Lyapunov function for the non-autonomous system.

The proof of this important theorem, which we now detail, is rather technical. Hurried readers may skip it in a first reading, and go directly to Example 4.5.

In order to prove the above theorem, we first translate the definitions of positive definite functions and decrescent functions in terms of the so-called class-K functions.

**Definition 4.9**     *A continuous function* $\alpha: \mathbf{R}^+ \to \mathbf{R}^+$ *is said to be of* <u>*class K*</u> *(or to belong to class K), if*

    • $\alpha(0) = 0$

    • $\alpha(p) > 0 \quad \forall p > 0$

    • $\alpha$ *is non-decreasing*

The following lemma indicates the relation of positive definite and decrescent functions to class K functions.

**Lemma 4.1**: *A function $V(\mathbf{x}, t)$ is locally (or globally) positive definite if, and only if, there exists a function $\alpha$ of class K such that $V(\mathbf{0}, t) = 0$ and*

$$V(\mathbf{x}, t) \geq \alpha(\|\mathbf{x}\|) \tag{4.9}$$

*$\forall\, t \geq 0$ and $\forall\, \mathbf{x} \in \mathbf{B}_{R_o}$ (or the whole state space).*

*A function $V(\mathbf{x}, t)$ is locally (or globally) decrescent if and only if there exists a class K function $\beta$ such that $V(\mathbf{0}, t) = 0$ and*

$$V(\mathbf{x}, t) \leq \beta(\|\mathbf{x}\|) \tag{4.10}$$

*$\forall\, t \geq 0$ and $\forall\, \mathbf{x} \in \mathbf{B}_{R_0}$ (or in the whole state space).*

**Proof**: Let us prove the positive definite function part first. Sufficiency is obvious from the definition, because $\alpha(\|\mathbf{x}\|)$ itself is a scalar time-invariant positive definite function. We now consider necessity, *i.e.*, assume that there exists a time-invariant positive function $V_o(\mathbf{x})$ such that $V(\mathbf{x}, t) \geq V_o(\mathbf{x})$, and show that a function $\alpha$ of class K exists such that (4.9) holds. Let us define

$$\alpha(p) = \inf_{p \,\leq\, \|\mathbf{x}\| \,\leq\, R} V_o(\mathbf{x}) \tag{4.11}$$

Then, $\alpha(0) = 0$, $\alpha$ is continuous and non-decreasing. Because $V_o(\mathbf{x})$ is a continuous function and non-zero except at $\mathbf{0}$, $\alpha(p) > 0$ for $p > 0$. Therefore, $\alpha$ is a class K function. Because of (4.11), (4.9) is satisfied.

The second part of the lemma can be proven similarly, with the function $\beta$ defined by

$$\beta(p) = \sup_{0 \,\leq\, \|\mathbf{x}\| \,\leq\, p} V_1(\mathbf{x}) \tag{4.12}$$

where $V_1(\mathbf{x})$ is the time-invariant positive function in Definition 4.8.    $\square$

Given the above lemma, we can now restate Theorem 4.1 as follows:

**Theorem 4.1**    *Assume that, in a neighborhood of the equilibrium point $\mathbf{0}$, there exists a scalar function $V(\mathbf{x}, t)$ with continuous first order derivatives and a class-K function $\alpha$ such that, $\forall \mathbf{x} \neq \mathbf{0}$*

*1. $V(\mathbf{x}, t) \geq \alpha(\|\mathbf{x}\|) > 0$*

*2a. $\dot{V}(\mathbf{x}, t) \leq 0$*

*then the origin $\mathbf{0}$ is Lyapunov stable. If, furthermore, there is a scalar class-K function $\beta$ such that*

*3. $V(\mathbf{x}, t) \leq \beta(\|\mathbf{x}\|)$*

*then* **0** *is uniformly stable. If conditions 1 and 3 are satisfied and condition 2a is replaced by condition 2b*

$$2b. \ \dot{V} \leq - \gamma(\|\mathbf{x}\|) < 0$$

*with* $\gamma$ *being another class-K function, then* **0** *is uniformly asymptotically stable. If conditions 1, 2b and 3 are satisfied in the whole state space, and*

$$\lim_{\mathbf{x} \to \infty} \ \alpha(\|\mathbf{x}\|) \to \infty$$

*then* **0** *is globally uniformly asymptotically stable.*

**Proof**· We derive the three parts of the theorem in sequence.

**Lyapunov stability**: To establish Lyapunov stability, we must show that given $R > 0$, there exists $r > 0$ such that (4.6) is satisfied. Because of conditions 1 and 2a,

$$\alpha(\|\mathbf{x}(t)\|) \leq V[\mathbf{x}(t), t] \leq V[\mathbf{x}(t_o), t_o] \qquad \forall t \geq t_o \tag{4.13}$$

Because $V$ is continuous in terms of $\mathbf{x}$ and $V(\mathbf{0}, t_o) = 0$, we can find $r$ such that

$$\|\mathbf{x}(t_o)\| < r \quad => \quad V(\mathbf{x}(t_o), t_o) < \alpha(R)$$

This means that if $\|\mathbf{x}(t_o)\| < r$, then $\alpha(\|\mathbf{x}(t)\|) < \alpha(R)$, and, accordingly, $\|\mathbf{x}(t)\| < R$ , $\forall t \geq t_o$.

**Uniform stability and uniform asymptotic stability**: From conditions 1 and 3,

$$\alpha(\|\mathbf{x}(t)\|) \leq V(\mathbf{x}(t), t) \leq \beta(\|\mathbf{x}(t)\|)$$

For any $R > 0$, there exists $r(R) > 0$ such that $\beta(r) < \alpha(R)$ (Figure 4.1). Let the initial condition $\mathbf{x}(t_o)$ be chosen such that $\|\mathbf{x}(t_o)\| < r$. Then

$$\alpha(R) > \beta(r) \geq V[\mathbf{x}(t_o), t_o] \geq V[\mathbf{x}(t), t] \geq \alpha(\|\mathbf{x}(t)\|)$$

This implies that

$$\forall \ t \geq t_o , \quad \|\mathbf{x}(t)\| < R$$

Uniform stability is asserted because $r$ is independent of $t_o$ .

In establishing uniform asymptotic stability, the basic idea is that if $\mathbf{x}$ does not converge to the origin, then it can be shown that there is a positive number $a$ such that $- \dot{V} [\mathbf{x}(t), t] \geq a > 0$ . This implies that

$$V[\mathbf{x}(t), t] - V[\mathbf{x}(t_o), t_o] = \int_{t_o}^{t} \dot{V} \, dt \leq - (t - t_0) a$$

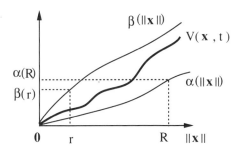

**Figure 4.1** : A positive definite and decrescent function

and thus, that

$$0 \le V[\mathbf{x}(t), t] \le V[\mathbf{x}(t_o), t_o] - (t - t_0)a$$

which leads to a contradiction for large $t$. Let us now detail the proof.

Let $\|\mathbf{x}(t_o)\| \le r$, with $r$ obtained as before. Let $\mu$ be be any positive constant such that $0 < \mu < \|\mathbf{x}(t_o)\|$. We can find another positive constant $\delta(\mu)$ such that $\beta(\delta) < \alpha(\mu)$. Define $\varepsilon = \gamma(\delta)$ and set

$$T = T(\mu, r) = \frac{\beta}{\varepsilon}$$

Then, if $\|\mathbf{x}(t)\| > \mu$ for all $t$ in the period $t_o \le t \le t_1 \equiv t_o + T$, we have

$$0 < \alpha(\mu) \le V[\mathbf{x}(t_1), t_1] \le V[\mathbf{x}(t_o, t_o] - \int_{t_o}^{t_1} \gamma(\|\mathbf{x}(s)\|)\, ds \le V[\mathbf{x}(t_o), t_o)] - \int_{t_o}^{t_1} \gamma(\delta)\, ds$$

$$\le V[\mathbf{x}(t_o), t_o] - (t_1 - t_o)\varepsilon \le \beta(r) - T\varepsilon = 0$$

This is a contradiction, and so there must exist $t_2 \in [t_o, t_1]$ such that $\|\mathbf{x}(t_2)\| \le \delta$. Thus, for all $t \ge t_2$.

$$\alpha(\|\mathbf{x}(t)\|) \le V[\mathbf{x}(t), t] \le V[\mathbf{x}(t_2), t_2] \le \beta(\delta) < \alpha(\mu)$$

As a result,

$$\|\mathbf{x}(t)\| < \mu \qquad \forall t \ge t_o + T \ge t_2$$

which shows uniform asymptotic stability.

**Global uniform asymptotic stability**: Since $\alpha(\cdot)$ is radially unbounded, $R$ can be found such that $\beta(r) < \alpha(R)$ for any $r$. In addition, $r$ can be made arbitrarily large. Hence, the origin $\mathbf{x} = \mathbf{0}$ is globally uniformly asymptotically stable.   $\square$

**Example 4.5: Global Asymptotic Stability**

Consider the system defined by

$$\dot{x}_1(t) = - x_1(t) - e^{-2t} x_2(t)$$

$$\dot{x}_2(t) = x_1(t) - x_2(t)$$

To determine the stability of the equilibrium point at **0**, let us choose the following scalar function

$$V(\mathbf{x}, t) = x_1^2 + (1 + e^{-2t}) x_2^2$$

This function is positive definite, because it dominates the time-invariant positive function $x_1^2 + x_2^2$. It is also decrescent, because it is dominated by the time-invariant positive definite function $x_1^2 + 2x_2^2$. Furthermore,

$$\dot{V}(\mathbf{x}, t) = - 2[ x_1^2 - x_1 x_2 + x_2^2(1 + 2e^{-2t})]$$

This shows that

$$\dot{V} \leq - 2(x_1^2 - x_1 x_2 + x_2^2) = - (x_1 - x_2)^2 - x_1^2 - x_2^2$$

Thus, $\dot{V}$ is negative definite, and therefore, the point **0** is globally asymptotically stable.   ☐

Stability results for non-autonomous systems can be less intuitive than those for autonomous systems, and therefore, particular care is required in applying the above theorem, as we now illustrate with two simple examples.

For autonomous systems, the origin is guaranteed to be asymptotically stable if $V$ is positive definite and $\dot{V}$ is negative definite. Therefore, one might be tempted to conjecture that the same conditions are sufficient to guarantee the asymptotic stability of the system. However, this intuitively appealing statement is *incorrect*, as demonstrated by the following counter-example.

**Example 4.6: Importance of the decrescence condition**

Let $g(t)$ be a continuously-differentiable function which coincides with the function $e^{-t/2}$ except around some peaks where it reaches the value 1. Specifically, $g^2(t)$ is shown in Figure 4.2. There is a peak for each integer value of $t$. The width of the peak corresponding to abcissa $n$ is assumed to be smaller than $(1/2)^n$. The infinite integral of $g^2$ thus satisfies

$$\int_0^\infty g^2(r)\,dr < \int_0^\infty e^{-r}\,dr + \sum_{n=1}^\infty \frac{1}{2^n} = 2$$

and therefore, the scalar function

$$V(\mathbf{x}, t) = \frac{x^2}{g^2(t)} [\, 3 - \int_o^t g^2(r)\, dr\, ] \tag{4.14}$$

is positive definite ( $V(\mathbf{x}, t) > x^2$ ).

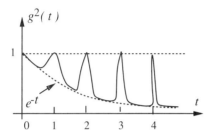

**Figure 4.2 :** The function $g^2(t)$

Now consider the first-order differential equation

$$\dot{x} = \frac{\dot{g}(t)}{g(t)} x \tag{4.15}$$

If we choose $V(\mathbf{x}, t)$ in (4.14) as the Lyapunov function candidate, we easily find that

$$\dot{V} = -x^2$$

*i.e.*, $\dot{V}$ *is negative definite.* Yet, the general solution of (4.15) is

$$x(t) = \frac{g(t)}{g(t_o)} x(t_o)$$

and hence *the origin is not asymptotically stable.*                               □

Since the positive definiteness of $V$ and the negative semi-definiteness of $\dot{V}$ are already sufficient to guarantee the Lyapunov stability of the origin, one may wonder what additional property the negative definiteness of $\dot{V}$ does provide. It can be shown that, if $\dot{V}$ is negative definite, then an infinite sequence $t_i$'s ($i = 1, 2, ...$) can be found such that the corresponding state values $\mathbf{x}(t_i)$ converge to zero as $i \to \infty$ (a result of mostly theoretical interest).

Another illustration that care is required before jumping to conclusions involves the following interesting second-order dynamics

$$\ddot{x} + c(t)\dot{x} + k_o x = 0 \tag{4.16}$$

which can represent a mass-spring-damper system (of mass 1), where $c(t) \geq 0$ is a time-varying damping coefficient, and $k_o$ is a spring constant. Physical intuition may suggest that the equilibrium point (0,0) is asymptotically stable as long as the damping

$c(t)$ remains larger than a strictly positive constant (implying constant dissipation of energy), as is the case for autonomous nonlinear mass-spring-damper systems. However, *this is not necessarily true*. Indeed, consider the system

$$\ddot{x} + (2 + e^t)\dot{x} + x = 0$$

One easily verifies that, for instance, with the initial condition $x(0) = 2$, $\dot{x}(0) = -1$, the solution is $x(t) = 1 + e^{-t}$, which tends to $x = 1$ instead! Here the damping increases so fast that the system gets "stuck" at $x = 1$.

Let us study the asymptotic stability of systems of the form of (4.16), using a Lyapunov analysis.

### Example 4.7: Asymptotic stability with time-varying damping

Lyapunov stability of the system (although not its asymptotic stability) can be easily established using the mechanical energy of the system as a Lyapunov function. Let us now use a different Lyapunov function to determine sufficient conditions for the asymptotic stability of the origin for the system (4.16). Consider the following positive definite function

$$V(\mathbf{x}, t) = \frac{(\dot{x} + \alpha x)^2}{2} + \frac{b(t)}{2} x^2$$

where $\alpha$ is any positive constant smaller than $\sqrt{k_o}$ , and

$$b(t) = k_o - \alpha^2 + \alpha c(t)$$

$\dot{V}$ can be easily computed as

$$\dot{V} = [\alpha - c(t)]\dot{x}^2 + \frac{\alpha}{2} [\dot{c}(t) - 2k_o]x^2$$

Thus, if there exist positive numbers $\alpha$ and $\beta$ such that

$$c(t) > \alpha > 0 \qquad \dot{c}(t) \le \beta < 2k_o$$

then $\dot{V}$ is negative definite. Assuming in addition that $c(t)$ is upper bounded (guaranteeing the decrescence of $V$), the above conditions imply the asymptotic convergence of the system.

It can be shown [Rouche, *et al.*, 1977] that, actually, the technical assumption that $c(t)$ is upper bounded is not necessary. Thus, for instance, the system

$$\ddot{x} + (2 + 8t)\dot{x} + 5x = 0$$

is asymptotically stable.                                          □

## 4.2.2 Lyapunov Analysis of Linear Time-Varying Systems

None of the standard approaches for analyzing linear time-invariant systems (*e.g.*, eigenvalue determination) applies to linear time-varying systems. Thus, it is of interest to consider the application of Lyapunov's direct method for studying the stability of linear time-varying systems. Indeed, a number of results are available in this regard. In addition, in view of the relation between the stability of a non-autonomous system and that of its (generally time-varying) linearization, to be discussed in section 4.2.3, such results on linear time-varying systems can also be of practical importance for local stability analysis of nonlinear non-autonomous systems.

Consider linear time-varying systems of the form

$$\dot{\mathbf{x}} = \mathbf{A}(t)\,\mathbf{x} \tag{4.17}$$

Since LTI systems are asymptotically stable if their eigenvalues all have negative real parts, one might be tempted to conjecture that system (4.17) will be stable if at *any* time $t \geq 0$, the eigenvalues of $A(t)$ all have negative real parts. If this were indeed the case, it would make the analysis of linear time-varying systems very easy. However, this conjecture is *not true*.

Consider for instance the system

$$\begin{bmatrix} \dot{x}_1 \\ \dot{x}_2 \end{bmatrix} = \begin{bmatrix} -1 & e^{2t} \\ 0 & -1 \end{bmatrix} \begin{bmatrix} x_1 \\ x_2 \end{bmatrix} \tag{4.18}$$

The eigenvalues of the matrix $\mathbf{A}(t)$ are both equal to $-1$ at all times. Yet, solving first for $x_2$ and then substituting in the $\dot{x}_1$ equation, one sees that the system verifies

$$x_2 = x_2(0)\,e^{-t} \qquad\qquad \dot{x}_1 + x_1 = x_2(0)\,e^{t}$$

and therefore is unstable, since $x_1$ can be viewed as the output of a first-order filter whose input $x_2(0)\,e^t$ tends to infinity.

A simple result, however, is that the time-varying system (4.17) *is* asymptotically stable if the eigenvalues of the *symmetric* matrix $\mathbf{A}(t) + \mathbf{A}^T(t)$ (all of which are real) remain strictly in the left-half complex plane

$$\exists\,\lambda > 0\,,\,\forall\,i\,,\,\forall\,t \geq 0\,,\,\lambda_i(\mathbf{A}(t) + \mathbf{A}^T(t)) \leq -\lambda \tag{4.19}$$

This can be readily shown using the Lyapunov function $V = \mathbf{x}^T\mathbf{x}$, since

$$\dot{V} = \mathbf{x}^T \dot{\mathbf{x}} + \dot{\mathbf{x}}^T \mathbf{x} = \mathbf{x}^T \left( \mathbf{A}(t) + \mathbf{A}^T(t) \right) \mathbf{x} \le -\lambda \, \mathbf{x}^T \mathbf{x} = -\lambda \, V$$

so that

$$\forall \, t \ge 0 \, , \quad 0 \le \mathbf{x}^T \mathbf{x} = V(t) \le V(0) \, e^{-\lambda t}$$

and therefore $\mathbf{x}$ tends to zero exponentially.

Of course, the above result also applies in case the matrix $\mathbf{A}$ depends explicitly on the state. It is also important to notice that the result provides a *sufficient* condition for asymptotic stability (some asymptotically stable systems do not verify (4.19), see Exercise 4.8).

A large number of more specialized theorems are available to determine the stability of classes of linear time-varying systems. We now give two examples of such results. The first result concerns "perturbed" linear systems, *i.e.*, systems of the form (4.17), where the matrix $\mathbf{A}(t)$ is a sum of a constant stable matrix and some "small" time-varying matrix. The second result is a more technical property of systems such that the matrix $\mathbf{A}(t)$ maintains all its eigenvalues in the left-half plane *and* satisfies certain smoothness conditions .

**Perturbed linear systems**

Consider a linear time-varying system of the form

$$\dot{\mathbf{x}} = \left( \mathbf{A}_1 + \mathbf{A}_2(t) \right) \mathbf{x} \tag{4.20}$$

where the matrix $\mathbf{A}_1$ is constant and Hurwitz (*i.e.*, has all its eigenvalues strictly in the left-half plane), and the time-varying matrix $\mathbf{A}_2(t)$ is such that

$$\mathbf{A}_2(t) \to \mathbf{0} \quad \text{as} \quad t \to \infty$$

and

$$\int_o^\infty \| \mathbf{A}_2(t) \| \, dt < \infty \quad \text{(\textit{i.e}, the integral exists and is finite)}$$

Then the system (4.20) is globally exponentially stable.

**Example 4.8:** Consider the system

$$\dot{x}_1 = -\left( 5 + x_2{}^5 + x_3{}^8 \right) x_1$$

$$\dot{x}_2 = -x_2 + 4 \, x_3{}^2$$

$$\dot{x}_3 = -(2 + \sin t) \, x_3$$

Since $x_3$ tends to zero exponentially, so does $x_3{}^2$, and therefore, so does $x_2$. Applying the above result to the first equation, we conclude that the system is globally exponentially stable.    $\square$

**Sufficient smoothness conditions on the $A(t)$ matrix**

Consider the linear system (4.17), and assume that at *any* time $t \geq 0$, the eigenvalues of $A(t)$ all have negative real parts

$$\exists\, \alpha > 0\,,\, \forall\, i\,,\, \forall\, t \geq 0\,,\, \lambda_i[\mathbf{A}(t)] \leq -\alpha \tag{4.21}$$

If, *in addition*, the matrix $\mathbf{A}(t)$ remains *bounded*, and

$$\int_o^\infty \mathbf{A}^T(t)\,\mathbf{A}(t)\,dt < \infty \quad (i.e,\text{ the integral exists and is finite})$$

then the system is globally exponentially stable.

# 4.2.3  * The Linearization Method for Non-Autonomous Systems

Lyapunov's linearization method can also be developed for non-autonomous systems. Let a non-autonomous system be described by (4.1) and $\mathbf{0}$ be an equilibrium point. Assume that $\mathbf{f}$ is continuously differentiable with respect to $\mathbf{x}$. Let us denote

$$\mathbf{A}(t) = \left(\frac{\partial \mathbf{f}}{\partial \mathbf{x}}\right)_{\mathbf{x}=\mathbf{0}} \tag{4.22}$$

Then for any *fixed* time $t$ (*i.e.*, regarding $t$ as a parameter), a Taylor expansion of $\mathbf{f}$ leads to

$$\dot{\mathbf{x}} = \mathbf{A}(t)\mathbf{x} + \mathbf{f}_{h.o.t.}(\mathbf{x},t)$$

If $\mathbf{f}$ can be well approximated by $\mathbf{A}(t)\mathbf{x}$ for any time $t$, *i.e.*,

$$\lim_{\|\mathbf{x}\| \to \mathbf{0}} \sup \frac{\|\,\mathbf{f}_{h.o.t.}(\mathbf{x},t)\,\|}{\|\mathbf{x}\|} = 0 \qquad \forall\, t \geq 0 \tag{4.23}$$

then the system

$$\dot{\mathbf{x}} = \mathbf{A}(t)\mathbf{x} \tag{4.24}$$

is said to be the linearization (or linear approximation) of the nonlinear non-autonomous system (4.1) around equilibrium point $\mathbf{0}$.

Note that

- The Jacobian matrix **A** thus obtained from a non-autonomous nonlinear system is generally time-varying, contrary to what happens for autonomous nonlinear systems. But in some cases **A** is constant. For example, the nonlinear system $\dot{x} = -x + x^2/t$ leads to the linearized system $\dot{x} = -x$.

- Our later results require that the uniform convergence condition (4.23) be satisfied. Some non-autonomous systems may not satisfy this condition, and Lyapunov's linearization method cannot be used for such systems. For example, (4.23) is *not* satisfied for the system $\dot{x} = -x + tx^2$.

Given a non-autonomous system satisfying condition (4.23), we can assert its (local) stability if its linear approximation is uniformly asymptotically stable, as stated in the following theorem:

**Theorem 4.2**   *If the linearized system (with condition (4.23) satisfied) is uniformly asymptotically stable, then the equilibrium point* **0** *of the original non-autonomous system is also uniformly asymptotically stable.*

Note that the linearized time-varying system must be *uniformly* asymptotically stable in order to use this theorem. If the linearized system is only asymptotically stable, no conclusion can be drawn about the stability of the original nonlinear system. Counter-examples can easily verify this point.

Unlike Lyapunov's linearization method for autonomous systems, the above theorem does not relate the instability of the linearized time-varying system to that of the nonlinear system. There does exist a simple result which infers the instability of a non-autonomous system from that of its linear approximation, but it is applicable only to non-autonomous systems whose linear approximations are time-invariant.

**Theorem 4.3**   *If the Jacobian matrix* $\mathbf{A}(t)$ *is constant,* $\mathbf{A}(t) = \mathbf{A}_o$ *, and if (4.23) is satisfied, then the instability of the linearized system implies that of the original non-autonomous nonlinear system, i.e., (4.1) is unstable if one or more of the eigenvalues of* $\mathbf{A}_o$ *has a positive real part.*

# 4.3   *\* Instability Theorems

The preceding study of Lyapunov's direct method is concerned with providing sufficient conditions for stability of nonlinear systems. This section provides some instability theorems based on Lyapunov's direct method.

Note that, for *autonomous* systems, one might think that the conclusive results provided by Lyapunov's linearization method are sufficient for the study of instability.

However, in some cases, instability theorems based on the direct method may be advantageous. Indeed, if the linearization method fails (*i.e.*, if the linearized system is marginally stable), these theorems may be used to determine the instability of the nonlinear system. Another advantage may be convenience, since the theorems do not require linearization of the system equations.

We state the theorems directly in a non-autonomous setting. For autonomous systems, the conditions simplify in a straightforward fashion.

**Theorem 4.4 (First instability theorem)**    *If, in a certain neighborhood $\Omega$ of the origin, there exists a continuously differentiable, decrescent scalar function $V(\mathbf{x}, t)$ such that*

- $V(\mathbf{0}, t) = 0 \quad \forall\, t \geq t_o$

- $V(\mathbf{x}, t_o)$ *can assume strictly positive values arbitrarily close to the origin*

- $\dot{V}(\mathbf{x}, t)$ *is positive definite (locally in $\Omega$)*

*then the equilibrium point $\mathbf{0}$ at time $t_o$ is unstable.*

Note that the second condition is weaker than requiring the positive definiteness of $V$. For example, the function $V(\mathbf{x}) = x_1^2 - x_2^2$ is obviously not positive definite, but it can assume positive values arbitrarily near the origin ($V(\mathbf{x}) = x_1^2$ along the line $x_2 = 0$).

**Example 4.9:** Consider the system

$$\dot{x}_1 = 2x_2 + x_1(x_1^2 + 2x_2^4) \tag{4.26}$$

$$\dot{x}_2 = -2x_1 + x_2(x_1^2 + x_2^4) \tag{4.27}$$

Linearization of this system yields $\dot{x}_1 = 2x_2$ and $\dot{x}_2 = -2x_1$. The eigenvalues of this system are $+2j$ and $-2j$, indicating the inability of Lyapunov's linearization method for this system. However, if we take

$$V = \frac{1}{2}(x_1^2 + x_2^2)$$

its derivative is

$$\dot{V} = (x_1^2 + x_2^2)(x_1^2 + x_2^4)$$

Because of the positive definiteness of $V$ and $\dot{V}$, the above theorem indicates the instability of the system.    $\square$

**Theorem 4.5 (Second instability theorem)** *If, in a certain neighborhood* $\Omega$ *of the origin, there exists a continuously differentiable, decrescent scalar function* $V(\mathbf{x}, t)$ *satisfying*

- $V(\mathbf{0}, t_o) = 0$ *and* $V(\mathbf{x}, t_o)$ *can assume strictly positive values arbitrarily close to the origin*

- $\dot{V}(\mathbf{x}, t) - \lambda V(\mathbf{x}, t) \geq 0 \quad \forall\, t \geq t_o \quad \forall\, \mathbf{x} \in \Omega$

*with* $\lambda$ *being a strictly positive constant, then the equilibrium point* $\mathbf{0}$ *at time* $t_o$ *is unstable.*

**Example 4.10:** Consider the system described by

$$x_1 = x_1 + 3x_2 \sin^2 x_2 + 5x_1 x_2^2 \sin^2 x_1 \tag{4.28}$$

$$\dot{x}_2 = 3x_1 \sin^2 x_2 + x_2 - 5x_1^2 x_2 \cos^2 x_1 \tag{4.29}$$

Let us consider the function $V(\mathbf{x}) = (1/2)(x_1^2 - x_2^2)$ , which was shown earlier to assume positive values arbitrarily near the origin. Its derivative is

$$\dot{V} = x_1^2 - x_2^2 + 5x_1^2 x_2^2 = 2V + 5x_1^2 x_2^2$$

Thus, the second instability theorem shows that the equilibrium point at the origin is unstable. Of course, in this particular case, the instability could be predicted more easily by the linearization method. $\qquad \blacksquare$

In order to apply the above two theorems, $\dot{V}$ is required to satisfy certain conditions at all points in the neighborhood $\Omega$. The following theorem (Cetaev's theorem) replaces theses conditions by a boundary condition on a subregion in $\Omega$.

**Theorem 4.6 (Third instability theorem)** *Let* $\Omega$ *be a neighborhood of the origin. If there exists a scalar function* $V(\mathbf{x}, t)$ *with continuous first partial derivatives, decrescent in* $\Omega$, *and a region* $\Omega_I$ *in* $\Omega$, *such that*

- $V(\mathbf{x}, t)$ *and* $\dot{V}(\mathbf{x}, t)$ *are positive definite in* $\Omega_I$

- *The origin is a boundary point of* $\Omega_I$

- *At the boundary points of* $\Omega_I$ *within* $\Omega$, $V(\mathbf{x}, t) = 0$ *for all* $t \geq t_o$

*then the equilibrium point* $\mathbf{0}$ *at time* $t_o$ *is unstable.*

The geometrical meaning of this theorem can be seen from Figure 4.3. Let us illustrate the use of this theorem on a simple example.

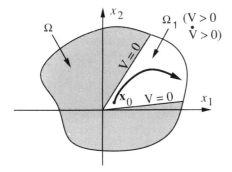

**Figure 4.3** : Geometrical interpretation of the third instability theorem

**Example 4.11:** Consider the system

$$\dot{x}_1 = x_1{}^2 + x_2{}^3$$

$$\dot{x}_2 = -x_2 + x_1{}^3$$

The linearization of this system leads to a pole at the origin and a pole at $-1$. Therefore, Lyapunov's linearization method cannot be used to determine the stability of this nonlinear system. Now let us take the function $V = x_1 - x_2{}^2/2$. Its derivative is

$$\dot{V} = \dot{x}_1 - x_2\dot{x}_2 = x_1{}^2 + x_2{}^2 + x_2{}^3 - x_1{}^3 x_2$$

Examining $V$ and $\dot{V}$ and using Cetaev's theorem, one can show the instability of the origin.     □

# 4.4  * Existence of Lyapunov Functions

In the previous Lyapunov theorems, the existence of Lyapunov functions is always assumed, and the objective is to deduce the stability properties of the systems from the properties of the Lyapunov functions. In view of the common difficulty in finding Lyapunov functions, one may naturally wonder whether Lyapunov functions always exist for stable systems. A number of interesting results concerning the existence of Lyapunov functions, called *converse Lyapunov theorems*, have been obtained in this regard. For many years, these theorems were thought to be of no practical value because, like the previously described theorems, they do not tell us how to generate Lyapunov functions for a system to be analyzed, but only represent comforting reassurances in the search for Lyapunov functions. In the past few years, however, there has been a resurgence of interest in these results. The reason is that a subsystem of a nonlinear system may be known to possess some stability properties, and the converse theorems allow us to construct a Lyapunov function for the subsystem, which may subsequently lead to the generation of a Lyapunov function for the whole system. In particular, the converse theorems can be used in connection with stability analysis of feedback linearizable systems and robustness analysis of adaptive control systems.

## THE CONVERSE THEOREMS

There exists a converse theorem for essentially every Lyapunov stability theorem (stability, uniform stability, asymptotic stability, uniform asymptotic stability, global uniform asymptotic stability and instability). We now present three of the converse theorems.

**Theorem 4.7 (stability)** *If the origin of (4.1) is stable, there exists a positive definite function V(**x**, t) with a non-positive derivative.*

This theorem indicates the existence of a Lyapunov function for every stable system.

**Theorem 4.8 (uniform asymptotic stability)** *If the equilibrium point at the origin is uniformly asymptotically stable, there exists a positive definite and decrescent function V(**x**, t) with a negative definite derivative.*

This theorem is theoretically important because it will later be useful in establishing robustness of uniform asymptotic stability to persistent disturbance.

The next theorem on exponential stability has more practical value than the second theorem, because its use may allow us to explicitly estimate convergence rates in some nonlinear systems.

**Theorem 4.9 (exponential stability)** *If the vector function **f**(**x**, t) in (4.1) has continuous and bounded first partial derivatives with respect to **x** and t, for all **x** in a ball $\mathbf{B}_r$ and for all $t \geq 0$, then the equilibrium point at the origin is exponentially stable if, and only if, there exists a function V(**x**, t) and strictly positive constants $\alpha_1, \alpha_2, \alpha_3, \alpha_4$ such that $\forall\, \mathbf{x} \in \mathbf{B}_r$, $\forall\, t \geq 0$*

$$\alpha_1 \|\mathbf{x}\|^2 \leq V(\mathbf{x}, t) \leq \alpha_2 \|\mathbf{x}\|^2 \tag{4.30}$$

$$\dot{V} \leq -\alpha_3 \|\mathbf{x}\|^2 \tag{4.31}$$

$$\|\frac{\partial V}{\partial \mathbf{x}}\| \leq \alpha_4 \|\mathbf{x}\| \tag{4.32}$$

The proofs of the converse theorems typically assume that the solution of the system is available, and then construct a Lyapunov function based on the assumed solution [Hahn, 1967]. Proof of Theorem 4.9 can be found in [Bodson, 1986; Sastry and Bodson, 1989].

# 4.5 Lyapunov-Like Analysis Using Barbalat's Lemma

For autonomous systems, the invariant set theorems are powerful tools to study stability, because they allow asymptotic stability conclusions to be drawn even when $\dot{V}$ is only negative *semi*-definite. However, the invariant set theorems are not applicable to non-autonomous systems. Therefore, *asymptotic* stability analysis of non-autonomous systems is generally much harder than that of autonomous systems, since it is usually very difficult to find Lyapunov functions with a *negative definite* derivative. An important and simple result which partially remedies this situation is Barbalat's lemma. Barbalat's lemma is a purely mathematical result concerning the asymptotic properties of functions and their derivatives. When properly used for dynamic systems, particularly non-autonomous systems, it may lead to the satisfactory solution of many asymptotic stability problems.

## 4.5.1 Asymptotic Properties of Functions and Their Derivatives

Before discussing Barbalat's lemma itself, let us clarify a few points concerning the asymptotic properties of functions and their derivatives. Given a differentiable function $f$ of time $t$, the following three facts are important to keep in mind.

- $\dot{f} \to 0 \;\not\Rightarrow\; f$ converges

  The fact that $\dot{f}(t) \to 0$ does <u>not</u> imply that $f(t)$ has a limit as $t \to \infty$ .

  Geometrically, a diminishing derivative means flatter and flatter slopes. However, it does not necessarily imply that the function approaches a limit. Consider, for instance, the rather benign function $f(t) = \sin(\log t)$ . While

  $$\dot{f}(t) = \frac{\cos(\log t)}{t} \to 0 \text{ as } t \to \infty$$

  the function $f(t)$ keeps oscillating (slower and slower). The function $f(t)$ may even be unbounded, as with $f(t) = \sqrt{t}\ \sin(\log t)$ . Note that functions of the form $\log t$ , $\sin t$ , $e^{\alpha t}$ , and combinations thereof, are quite easy to find in dynamic system responses.

- $f$ converges $\;\not\Rightarrow\; \dot{f} \to 0$

  The fact that $f(t)$ has a finite limit as $t \to \infty$ does <u>not</u> imply that $\dot{f}(t) \to 0$.

  For instance, while the function $f(t) = e^{-t} \sin(e^{2t})$ tends to zero, its derivative $\dot{f}$ is unbounded. Note that this is not linked to the frequent sign changes of the function.  Indeed, with $f(t) = e^{-t} \sin^2(e^{2t}) \geq 0$, $\dot{f}$ is still unbounded.

- If $f$ is lower bounded and decreasing ($\dot{f} \leq 0$), then it converges to a limit.

    This is a standard result in calculus.  However, it does not say whether the slope of the curve will diminish or not.

## 4.5.2  Barbalat's Lemma

Now, given that a function tends towards a finite limit, what additional requirement can guarantee that its derivative actually converges to zero?  Barbalat's lemma indicates that the derivative itself should have some smoothness.  More precisely, we have

**Lemma 4.2 (Barbalat)**    *If the differentiable function $f(t)$ has a finite limit as $t \to \infty$, and if $\dot{f}$ is uniformly continuous, then $\dot{f}(t) \to 0$ as $t \to \infty$.*

Before proving this result, let us define what we mean by *uniform* continuity. Recall that a function $g(t)$ is continuous on $[0, \infty)$ if

$$\forall\, t_1 \geq 0, \forall\, R > 0\,, \exists\, \eta(R, t_1) > 0\,, \forall\, t \geq 0, \quad |t - t_1| < \eta \quad \Rightarrow \quad |g(t) - g(t_1)| < R$$

A function $g$ is said to be *uniformly continuous* on $[0, \infty)$ if

$$\forall\, R > 0\,, \exists\, \eta(R) > 0\,, \forall\, t_1 \geq 0, \forall\, t \geq 0, \quad |t - t_1| < \eta \quad \Rightarrow \quad |g(t) - g(t_1)| < R$$

In other words, $g$ is *uniformly* continuous if one can always find an $\eta$ which does not depend on the specific point $t_1$  – and in particular, such that $\eta$ does not shrink as $t_1 \to \infty$, as shall be important when proving the lemma.  Note that $t$ and $t_1$ play a symmetric role in the definition of uniform continuity.

Uniform continuity of a function is often awkward to assert directly from the above definition.  A more convenient approach is to examine the function's derivative. Indeed, a very simple *sufficient* condition for a differentiable function to be uniformly continuous is that *its derivative be bounded*. This can be easily seen from the finite difference theorem

$$\forall\, t, \forall\, t_1\,, \quad \exists\, t_2 \text{ (between } t \text{ and } t_1) \quad \text{such that} \quad g(t) - g(t_1) = \dot{g}(t_2)\,(t - t_1)$$

and therefore, if $R_1 > 0$ is an upper bound on the function $|\dot{g}|$, one can always use $\eta = R/R_1$ independently of $t_1$ to verify the definition of uniform continuity.

Let us now prove Barbalat's lemma, by contradiction.

**Proof of Barbalat's lemma**: Assume that $\dot{f}(t)$ does not approach zero as $t \to \infty$. Then $\exists\, \varepsilon_o > 0, \forall\, T > 0, \exists\, t > T, \; |\dot{f}(t)| \geq \varepsilon_o$. Therefore, we can get an *infinite* sequence of $t_i$'s (such that $t_i \to \infty$ as $i \to \infty$) such that $|\dot{f}(t_i)| \geq R_o$. Since $\dot{f}(t)$ is assumed to be *uniformly* continuous, $\exists\, \eta > 0$, such that for any $t'$ and $t''$ satisfying $|t' - t''| < \eta$

$$|\dot{f}(t') - \dot{f}(t'')| < \frac{\varepsilon_o}{2}$$

This implies that for any $t$ within the $\eta$-neighborhood of $t_i$ (*i.e.*, such that $|t - t_i| < \eta$ )

$$|\dot{f}(t)| > \frac{\varepsilon_o}{2}$$

Hence, for all $t_i$,

$$\left| \int_{t_i - \eta}^{t_i + \eta} \dot{f}(t)\, dt \right| = \int_{t_i - \eta}^{t_i + \eta} |\dot{f}|(t)\, dt \geq \frac{\varepsilon_o}{2} 2\eta = \varepsilon_o \eta$$

where the left equality comes from the fact that $\dot{f}$ keeps a constant sign over the integration interval, due to the continuity of $\dot{f}$ and the bound $|\dot{f}(t)| > \varepsilon_o/2 > 0$.

This result would contradict the known fact that the integral $\int_o^t \dot{f}(r)\, dr$ has a limit (equal to $f(\infty) - f(0)$ ) as $t \to \infty$.    □

Given the simple sufficient condition for uniform continuity mentioned earlier, an immediate and practical corollary of Barbalat's lemma can be stated as follows: *if the differentiable function $f(t)$ has a finite limit as $t \to \infty$, and is such that $\ddot{f}$ exists and is bounded, then $\dot{f}(t) \to 0$ as $t \to \infty$.*

The following example illustrates how to assert the uniform continuity of signals in control systems.

**Example 4.12:** Consider a strictly stable linear system whose input is bounded. Then the system output is uniformly continuous.

Indeed, write the system in the standard form

$$\dot{\mathbf{x}} = \mathbf{A}\,\mathbf{x} + \mathbf{B}\,\mathbf{u}$$

$$\mathbf{y} = \mathbf{C}\,\mathbf{x}$$

Since **u** is bounded and the linear system is strictly stable, thus the state **x** is bounded. This in turn implies from the first equation that $\dot{\mathbf{x}}$ is bounded, and therefore from the second equation that

$\dot{\mathbf{y}} = \mathbf{C}\,\dot{\mathbf{x}}$ is bounded. Thus the system output $\mathbf{y}$ is uniformly continuous.      ◻

## USING BARBALAT'S LEMMA FOR STABILITY ANALYSIS

To apply Barbalat's lemma to the analysis of dynamic systems, one typically uses the following immediate corollary, which looks very much like an invariant set theorem in Lyapunov analysis:

**Lemma 4.3 ("Lyapunov-Like Lemma")**      *If a scalar function $V(\mathbf{x}, t)$ satisfies the following conditions*

- $V(\mathbf{x}, t)$ *is lower bounded*

- $\dot{V}(\mathbf{x}, t)$ *is negative semi-definite*

- $\dot{V}(\mathbf{x}, t)$ *is uniformly continuous in time*

*then $\dot{V}(\mathbf{x}, t) \to 0$ as $t \to \infty$.*

Indeed, $V$ then approaches a finite limiting value $V_\infty$, such that $V_\infty \leq V(\mathbf{x}(0), 0)$ (this does not require uniform continuity). The above lemma then follows from Barbalat's lemma.

To illustrate this procedure, let us consider the asymptotic stability analysis of a simple adaptive control system.

**Example 4.13:** As we shall detail in chapter 8, the closed-loop error dynamics of an adaptive control system for a first-order plant with one unknown parameter is

$$\dot{e} = -e + \theta\, w(t)$$

$$\dot{\theta} = -e\, w(t)$$

where $e$ and $\theta$ are the two states of the closed-loop dynamics, representing tracking error and parameter error, and $w(t)$ is a bounded continuous function (in the general case, the dynamics has a similar form but with $e$, $\theta$, and $w(t)$ replaced by vector quantities). Let us analyze the asymptotic properties of this system.

Consider the lower bounded function

$$V = e^2 + \theta^2$$

Its derivative is

$$\dot{V} = 2e(-e + \theta w) + 2\theta(-e w(t)) = -2e^2 \leq 0$$

This implies that $V(t) \leq V(0)$, and therefore, that $e$ and $\theta$ are bounded. But the invariant set theorems cannot be used to conclude the convergence of $e$, because the dynamics is non-autonomous.

To use Barbalat's lemma, let us check the uniform continuity of $\dot{V}$. The derivative of $\dot{V}$ is

$$\ddot{V} = -4e(-e+\theta w)$$

This shows that $\ddot{V}$ is bounded, since $w$ is bounded by hypothesis, and $e$ and $\theta$ were shown above to be bounded. Hence, $\dot{V}$ is uniformly continuous. Application of Barbalat's lemma then indicates that $e \to 0$ as $t \to \infty$.

Note that, although $e$ converges to zero, the system is not *asymptotically stable*, because $\theta$ is only guaranteed to be bounded. $\qquad\square$

The analysis in the above example is quite similar to a Lyapunov analysis based on invariant set theorems. Such an analysis based on Barbalat's lemma shall be called a *Lyapunov-like* analysis. It presents two subtle but important differences with Lyapunov analysis, however. The first is that the function $V$ can simply be a lower bounded function of $\mathbf{x}$ and $t$ instead of a positive definite function. The second difference is that the derivative $\dot{V}$ must be shown to be uniformly continuous, in addition to being negative or zero. This is typically done by proving that $\ddot{V}$ is bounded. Of course, in using the Lyapunov-like lemma for stability analysis, the primary difficulty is still the proper choice of the scalar function $V$.

# 4.6 Positive Linear Systems

In the analysis and design of nonlinear systems, it is often possible and useful to decompose the system into a linear subsystem and a nonlinear subsystem. If the transfer function (or transfer matrix) of the linear subsystem is so-called *positive real*, then it has important properties which may lead to the generation of a Lyapunov function for the whole system. In this section, we study linear systems with positive real transfer functions or transfer matrices, and their properties. Such systems, called *positive linear systems*, play a central role in the analysis and design of many nonlinear control problems, as will be seen later in the book.

## 4.6.1 PR and SPR Transfer Functions

We consider rational transfer functions of $n^{\text{th}}$-order single-input single-output linear systems, represented in the form

$$h(p) = \frac{b_m p^m + b_{m-1} p^{m-1} + ... + b_o}{p^n + a_{n-1} p^{n-1} + ... + a_o}$$

The coefficients of the numerator and denominator polynomials are assumed to be real numbers and $n \geq m$. The difference $n-m$ between the order of the denominator and that of the numerator is called the *relative degree* of the system.

**Definition 4.10**    *A transfer function $h(p)$ is <u>positive real</u> if*

$$Re[h(p)] \geq 0 \quad for\ all \quad Re[p] \geq 0 \tag{4.33}$$

*It is <u>strictly positive real</u> if $h(p-\varepsilon)$ is positive real for some $\varepsilon > 0$.*

Condition (4.33), called the positive real condition, means that $h(p)$ always has a positive (or zero) real part when $p$ has positive (or zero) real part. Geometrically, it means that the rational function $h(p)$ maps every point in the closed right half (*i.e.*, including the imaginary axis) of the complex plane into the closed right half of the $h(p)$ plane. The concept of positive real functions originally arose in the context of circuit theory, where the transfer function of a passive network (passive in the sense that no energy is generated in the network, *e.g.*, a network consisting of only inductors, resistors, and capacitors) is rational and positive real. In section 4.7, we shall reconcile the PR concept with passivity.

**Example 4.14: A strictly positive real function**

Consider the rational function

$$h(p) = \frac{1}{p + \lambda}$$

which is the transfer function of a first-order system, with $\lambda > 0$. Corresponding to the complex variable $p = \sigma + j\omega$,

$$h(p) = \frac{1}{(\sigma + \lambda) + j\omega} = \frac{\sigma + \lambda - j\omega}{(\sigma + \lambda)^2 + \omega^2}$$

Obviously, $Re[h(p)] \geq 0$ if $\sigma \geq 0$. Thus, $h(p)$ is a positive real function. In fact, one can easily see that $h(p)$ is strictly positive real, for example by choosing $\varepsilon = \lambda/2$ in Definition 4.9.    □

For higher-order transfer functions, it is often difficult to use the definition directly in order to test the positive realness condition, because this involves checking the positivity condition over the entire right-half of the complex plane. The following theorem can simplify the algebraic complexity.

**Theorem 4.10**    *A transfer function $h(p)$ is strictly positive real (SPR) if and only if*

  *i) $h(p)$ is a strictly stable transfer function*

  *ii) the real part of $h(p)$ is strictly positive along the $j\omega$ axis, i.e.,*

$$\forall \, \omega \geq 0 \quad \mathrm{Re}[h(j\omega)] > 0 \qquad\qquad (4.34)$$

The proof of this theorem is presented in the next section, in connection with the so-called passive systems.

The above theorem implies simple *necessary* conditions for asserting whether a given transfer function $h(p)$ is SPR:

- $h(p)$ is strictly stable

- The Nyquist plot of $h(j\omega)$ lies entirely in the right half complex plane. Equivalently, the phase shift of the system in response to sinusoidal inputs is always less than $90^o$

- $h(p)$ has relative degree 0 or 1

- $h(p)$ is strictly minimum-phase (*i.e.*, all its zeros are strictly in the left-half plane)

The first and second conditions are immediate from the theorem. The last two conditions can be derived from the second condition simply by recalling the procedure for constructing Bode or Nyquist frequency response plots (systems with relative degree larger than 1 and non-minimum phase systems have phase shifts larger than $90^o$ at high frequencies, or, equivalently have parts of the Nyquist plot lying in the left-half plane).

  **Example 4.15: SPR and non-SPR transfer functions**

    Consider the following systems

$$h_1(p) = \frac{p-1}{p^2 + a\,p + b}$$

$$h_2(p) = \frac{p+1}{p^2 - p + 1}$$

$$h_3(p) = \frac{1}{p^2 + a\,p + b}$$

$$h_4(p) = \frac{p+1}{p^2 + p + 1}$$

The transfer functions $h_1$, $h_2$, and $h_3$ are not SPR, because $h_1$ is non-minimum phase, $h_2$ is unstable, and $h_3$ has relative degree larger than 1.

Is the (strictly stable, minimum-phase, and of relative degree 1) function $h_4$ actually SPR? We have

$$h_4(j\omega) = \frac{j\omega + 1}{-\omega^2 + j\omega + 1} = \frac{[j\omega + 1][-\omega^2 - j\omega + 1]}{[1 - \omega^2]^2 + \omega^2}$$

(where the second equality is obtained by multiplying numerator and denominator by the complex conjugate of the denominator) and thus

$$\mathrm{Re}[\, h_4(j\omega)\,] = \frac{-\omega^2 + 1 + \omega^2}{[1 - \omega^2]^2 + \omega^2} = \frac{1}{[1 - \omega^2]^2 + \omega^2}$$

which shows that $h_4$ is SPR (since it is also strictly stable). Of course, condition (4.34) can also be checked directly on a computer. □

The basic difference between PR and SPR transfer functions is that PR transfer functions may tolerate poles on the $j\omega$ axis, while SPR functions cannot.

**Example 4.16:** Consider the transfer function of an integrator,

$$h(p) = \frac{1}{p}$$

Its value corresponding to $p = \sigma + j\omega$ is

$$h(p) = \frac{\sigma - j\omega}{\sigma^2 + \omega^2}$$

One easily sees from Definition 4.9 that $h(p)$ is PR but not SPR. □

More precisely, we have the following result, which complements Theorem 4.10.

**Theorem 4.11** *A transfer function $h(p)$ is positive real if, and only if,*

*i) $h(p)$ is a stable transfer function*

*(ii) The poles of $h(p)$ on the $j\omega$ axis are simple (i.e., distinct) and the associated residues are real and non-negative*

*iii) $\mathrm{Re}[h(j\omega)] \geq 0$ for any $\omega \geq 0$ such that $j\omega$ is not a pole of $h(p)$*

## 4.6.2 The Kalman-Yakubovich Lemma

If a transfer function of a system is SPR, there is an important mathematical property associated with its state-space representation, which is summarized in the celebrated Kalman-Yakubovich (KY) lemma.

**Lemma 4.4 (Kalman-Yakubovich)** *Consider a controllable linear time-invariant system*

$$\dot{\mathbf{x}} = \mathbf{A}\mathbf{x} + \mathbf{b}u$$

$$y = \mathbf{c}^T\mathbf{x}$$

*The transfer function*

$$h(p) = \mathbf{c}^T[p\mathbf{I} - \mathbf{A}]^{-1}\mathbf{b} \tag{4.35}$$

*is strictly positive real if, and only if, there exist positive definite matrices* $\mathbf{P}$ *and* $\mathbf{Q}$ *such that*

$$\mathbf{A}^T\mathbf{P} + \mathbf{P}\mathbf{A} = -\mathbf{Q} \tag{4.36a}$$

$$\mathbf{P}\mathbf{b} = \mathbf{c} \tag{4.36b}$$

The proof of this lemma is presented in section 4.7 in connection with passivity in linear systems. Beyond its mathematical statement, which shall be extensively used in chapter 8 (Adaptive Control), the KY lemma has important physical interpretations and uses in generating Lyapunov functions, as discussed in section 4.7.

The KY lemma can be easily extended to PR systems. For such systems, it can be shown that there exist a positive definite matrix $\mathbf{P}$ and a positive *semi*-definite matrix $\mathbf{Q}$ such that (4.36a) and (4.36b) are verified. The usefulness of this result is that it is applicable to transfer functions containing a pure integrator ( $1/p$ in the frequency-domain), of which we shall see many in chapter 8 when we study adaptive controller design. The Kalman-Yakubovich lemma is also referred to as the *positive real lemma*.

In the KY lemma, the involved system is required to be asymptotically stable and completely controllable. A modified version of the KY lemma, relaxing the controllability condition, can be stated as follows:

**Lemma 4.5 (Meyer-Kalman-Yakubovich)** *Given a scalar $\gamma \geq 0$, vectors **b** and **c**, an asymptotically stable matrix **A**, and a symmetric positive definite matrix **L**, if the transfer function*

$$H(\mathbf{p}) = \frac{\gamma}{2} + \mathbf{c}^T [p\mathbf{I} - \mathbf{A}]^{-1} \mathbf{b}$$

*is SPR, then there exist a scalar $\varepsilon > 0$, a vector **q**, and a symmetric positive definite matrix **P** such that*

$$\mathbf{A}^T \mathbf{P} + \mathbf{P} \mathbf{A} = -\mathbf{q}\mathbf{q}^T - \varepsilon \mathbf{L}$$

$$\mathbf{P}\mathbf{b} = \mathbf{c} + \sqrt{\gamma}\ \mathbf{q}$$

This lemma is different from Lemma 4.4 in two aspects. First, the involved system now has the output equation

$$y = \mathbf{c}^T \mathbf{x} + \frac{\gamma}{2} u$$

Second, the system is only required to be stabilizable (but not necessarily controllable).

## 4.6.3 Positive Real Transfer Matrices

The concept of positive real transfer function can be generalized to rational positive real matrices. Such generalization is useful for the analysis and design of multi-input-multi-output nonlinear control systems.

**Definition 4.11** *An m×m transfer matrix **H**(p) is called PR if*

    *$\mathbf{H}(p)$ has elements which are analytic for $\text{Re}(p) > 0$;*

    *$\mathbf{H}(p) + \mathbf{H}^T(p^*)$ is positive semi-definite for $\text{Re}(p) > 0$.*

*where the asterisk $^*$ denotes the complex conjugate transpose. $\mathbf{H}(p)$ is SPR if $\mathbf{H}(p - \varepsilon)$ is PR for some $\varepsilon > 0$.*

The Kalman-Yakubovich lemma and the Meyer-Kalman-Yakubovich lemma can be easily extended to positive real transfer matrices.

# 4.7   The Passivity Formalism

As we saw earlier, Lyapunov functions are generalizations of the notion of energy in a dynamic system. Thus, intuitively, we expect Lyapunov functions to be "additive", *i.e.*, Lyapunov functions for combinations of systems to be derived by simply adding the Lyapunov functions describing the subsystems. Passivity theory formalizes this intuition, and derives simple rules to describe combinations of subsystems or "blocks" expressed in a Lyapunov-like formalism.   It also represents an approach to constructing Lyapunov functions or Lyapunov-like functions for feedback control purposes.

As a motivation, recall first that the dynamics of state-determined physical systems, whether linear or nonlinear, satisfy energy-conservation equations of the form

$$\frac{d}{dt} [\text{Stored Energy}] = [\text{External Power Input}] + [\text{Internal Power Generation}]$$

These equations actually form the basis of modeling techniques such as bond-graphs. The external power input term can be represented as the scalar product $\mathbf{y}^T\mathbf{u}$ of an input ("effort" or "flow") $\mathbf{u}$, and a output ("flow" or "effort") $\mathbf{y}$.

In the following, we shall more generally consider systems which verify equations of the form

$$\dot{V}_1(t) = \mathbf{y}_1^T \mathbf{u}_1 - g_1(t) \tag{4.37}$$

where $V_1(t)$ and $g_1(t)$ are scalar functions of time, $\mathbf{u}_1$ is the system input, and $\mathbf{y}_1$ is its output. Note that, from a mathematical point of view, the above form is quite general (given an arbitrary system, of input $\mathbf{u}_1(t)$ and output $\mathbf{y}_1(t)$, we can let, for instance, $g_1(t) \equiv 0$ and $V_1(t) = \int_o^t \mathbf{y}_1^T(r) \mathbf{u}_1(r) \, dr$). It is the physical or "Lyapunov-like" properties that $V_1(t)$ and $g_1(t)$ may have, and how they are transmitted through combinations with similar systems, that we shall be particularly interested in.

## 4.7.1   Block Combinations

Assume that we couple a system in the form (4.37), or *power form*, to one verifying the similar equation

$$\dot{V}_2(t) = \mathbf{y}_2^T \mathbf{u}_2 - g_2(t)$$

In a feedback configuration, namely $\mathbf{u}_2 = \mathbf{y}_1$ and $\mathbf{u}_1 = -\mathbf{y}_2$ (Figure 4.4), assuming of course that the vectors $\mathbf{u}_i$ and $\mathbf{y}_j$ are all of the same dimension. We then have

$$\frac{d}{dt}[\,V_1(t) + V_2(t)\,] = -[\,g_1(t) + g_2(t)\,] \qquad (4.38)$$

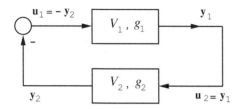

**Figure 4.4** : Two blocks of the form (4.37), in a feedback configuration

Let us assume that the function $V_1 + V_2$ is lower bounded (*e.g.*, is positive). Then, using the same reasoning as in section 4.5, we have

- If $\forall\, t \geq 0$, $g_1(t) + g_2(t) \geq 0$, then the function $V_1 + V_2$ is upper bounded, and

$$\int_o^\infty [\,g_1(t) + g_2(t)\,]\, dt < \infty$$

- If in addition, the function $g_1 + g_2$ is uniformly continuous, then $[\,g_1(t) + g_2(t)\,] \to 0$ as $t \to \infty$.

- In particular, if $g_1(t)$ and $g_2(t)$ are both non-negative and uniformly continuous, then they both tend to zero as $t \to \infty$

Note that an explicit expression of $V_1 + V_2$ is not needed in the above results. More generally, without any assumption on the sign of $V_1 + V_2$ or $g_1 + g_2$, we can state that

- If $V_1 + V_2$ has a finite limit as $t \to \infty$, and if $g_1 + g_2$ is uniformly continuous, then $[\,g_1(t) + g_2(t)\,] \to 0$ as $t \to \infty$.

A system verifying an equation of the form (4.37) with $V_1$ lower bounded and $g_1 \geq 0$ is said to be *passive* (or to be a *passive mapping* between $\mathbf{u}_1$ and $\mathbf{y}_1$). Furthermore, a passive system is said to be *dissipative* if

$$\int_o^\infty \mathbf{y}_1{}^T(t)\,\mathbf{u}_1(t)\, dt \neq \mathbf{0} \quad => \quad \int_o^\infty g_1(t)\, dt > 0$$

**Example 4.17:** The nonlinear mass-spring-damper system

$$m\ddot{x} + x^2\dot{x}^3 + x^7 = F$$

represents a dissipative mapping from external force $F$ to velocity $\dot{x}$, since

$$\frac{d}{dt}\left(\frac{1}{2}m\dot{x}^2 + \frac{1}{8}x^8\right) = \dot{x}F - x^2\dot{x}^4$$

Of course, here $V_1$ is simply the total (kinetic plus potential) energy stored in the system, and $g_1$ is the dissipated power. □

**Example 4.18:** Consider the system (Figure 4.5)

$$\dot{x} + \lambda(t)\,x = u \tag{4.39}$$

$$y = h(x)$$

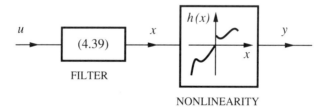

FILTER

NONLINEARITY

**Figure 4.5 :** A passive single-input single-output system

where the function $h$ is of the same sign as its argument, although not necessarily continuous, and $\lambda(t) \geq 0$. The mapping $u \to y$ is passive, since

$$\frac{d}{dt}\int_0^x h(\xi)\,d\xi = h(x)\dot{x} = yu - \lambda(t)\,h(x)\,x$$

with $\int_0^x h(\xi)\,d\xi \geq 0$ and $\lambda(t)\,h(x)\,x \geq 0$ for all $x$. The mapping is actually dissipative if $\lambda(t)$ is not identically zero.

Of course, the function $\lambda(t)$ may actually be of the form $\lambda[x(t)]$. For instance, the system

$$\dot{x} + x^3 = u$$

$$y = x - \sin^2 x$$

is a dissipative mapping from $u$ to $y$. □

A particularly convenient aspect of formalizing the construction of Lyapunov-like functions as above, is that *parallel and feedback combinations of systems in the power form are still in the power form*. Indeed, it is straightforward to verify that, for both the parallel combination and the feedback combination (Figure 4.6), one has

$$\mathbf{y}^T \mathbf{u} = \mathbf{y}_1{}^T \mathbf{u}_1 + \mathbf{y}_2{}^T \mathbf{u}_2$$

Namely, for the parallel combination

$$\mathbf{y}^T \mathbf{u} = (\mathbf{y}_1 + \mathbf{y}_2)^T \mathbf{u} = \mathbf{y}_1{}^T \mathbf{u} + \mathbf{y}_2{}^T \mathbf{u} = \mathbf{y}_1{}^T \mathbf{u}_1 + \mathbf{y}_2{}^T \mathbf{u}_2$$

and for the feedback combination

$$\mathbf{y}^T \mathbf{u} = \mathbf{y}_1{}^T (\mathbf{u}_1 + \mathbf{y}_2) = \mathbf{y}_1{}^T \mathbf{u}_1 + \mathbf{y}_1{}^T \mathbf{y}_2 = \mathbf{y}_1{}^T \mathbf{u}_1 + \mathbf{u}_2{}^T \mathbf{y}_2$$

Incidentally, this result is a particular case of what is known in circuit theory as *Tellegen's power conservation theorem*. Thus, we have, for the overall system

$$V = V_1 + V_2 \qquad\qquad g = g_1 + g_2$$

By induction, *any combination* of feedback and/or parallel combinations of systems in the power form can also be described in the power form, with the corresponding $V$ and $g$ simply being equal to the *sum* of the individual $V_i$ and $g_i$.

$$V = \sum_i V_i \qquad\qquad g = \sum_i g_i$$

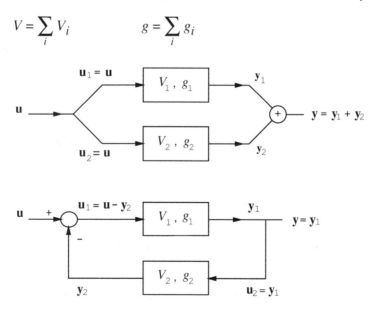

**Figure 4.6 :** Parallel and feedback combinations

The power of this simple result is, of course, that it does not require that the subsystems be linear.

Note that, assuming that $V$ is lower bounded, the overall system can be passive while some of its components may be "active" ( $g_i < 0$ ): for the system to be passive, the *sum* of the $g_i$ has to be positive, *i.e.*, there should be more overall "power dissipation" in the system than there is "power generation". Also, note that the passivity of a block is preserved if its input or its output is multiplied by a strictly positive constant (an input gain or an output gain), since this simply multiplies the associated $V_i$ and $g_i$ by the same constant. Thus we have, more generally

$$V = \sum_i \alpha_i V_i \qquad\qquad g = \sum_i \alpha_i \ g_i$$

where $\alpha_i$ is the product of the input gain and the output gain for block $i$.

> **Example 4.19:** Consider again the adaptive control system of Example 4.13. The fact that
>
> $$\frac{1}{2}\frac{d}{dt} e^2 \ = \ e\,\dot{e} \ = \ e\,\theta\,w(t) - e^2$$
>
> can be interpreted as stating that the mapping $\theta\,w(t) \rightarrow e$ is dissipative. Furthermore, using the parameter adaptation law
>
> $$\dot{\theta} = -\,e\,w(t) \qquad\qquad\qquad\qquad\qquad\qquad (4.40)$$
>
> then corresponds to inserting a passive feedback block between $e$ and $-\theta\,w(t)$, since
>
> $$\frac{1}{2}\frac{d}{dt} \theta^2 \ = \ -\theta\,w(t)\,e \qquad\qquad\qquad\qquad\qquad \square$$

Note that, for a passive system, we may always assume that $g \equiv 0$, simply by adding $\int_o^t g(r)\,dr$ to the original $V$. Hence, the definition of passivity is often written as

$$\exists\,\alpha > -\infty\,, \ \forall\,t \geq 0\,, \qquad \int_o^t \mathbf{y}^T(r)\,\mathbf{u}(r)\,dr \ \geq \ \alpha \qquad\qquad (4.41)$$

which simply says that there exists a lower bounded $V$ such that $g \equiv 0$.

Also, note that the power form is expressed in terms of the *dot-product* $\mathbf{y}^T\mathbf{u}$. Therefore, if $\mathbf{u}_a$ and $\mathbf{y}_a$ are other choices of inputs and outputs for the system such that $\mathbf{y}_a^T\,\mathbf{u}_a \ = \ \mathbf{y}^T\mathbf{u}$ at all times, then they satisfy the same passivity properties as $\mathbf{u}$ and $\mathbf{y}$. For instance, if the mapping $\mathbf{u} \rightarrow \mathbf{y}$ is passive, so is the mapping $\mathbf{Au} \rightarrow \mathbf{A}^{-T}\mathbf{y}$, where the matrix $\mathbf{A}$ is any (perhaps time-varying) invertible matrix. In particular, passivity is conserved through orthogonal transformations ( $\mathbf{A}\,\mathbf{A}^T = \mathbf{I}$ ). Furthermore, note that the dimension of the vectors $\mathbf{u}_a$ and $\mathbf{y}_a$ is not necessarily the same as that of the vectors $\mathbf{u}$ and $\mathbf{y}$ .

## 4.7.2 Passivity in Linear Systems

An important practical feature of the passivity formulation is that it is easy to characterize passive *linear* systems. This allows linear blocks to be straightforwardly incorporated or added in a nonlinear control problem formulated in terms of passive mappings.

As we shall now show, a strictly stable linear SISO system is passive if, and only if,

$$\forall \, \omega \geq 0 \, , \, \text{Re}[h(j\omega)] \geq 0 \tag{4.42}$$

where $h$ is the transfer function of the system (which we shall assume to be rational) and Re refers to the real part of a complex number. Geometrically, condition (4.42) can also be written as (Figure 4.7)

$$\forall \, \omega \geq 0 \, , \, | \text{Arg}\, h(j\omega) | \, \leq \, \frac{\pi}{2} \tag{4.43}$$

Thus, we see that a strictly stable linear SISO system is passive if, and only if its *phase shift* in response to a sinusoidal input is always *less than (or equal to) 90°.*

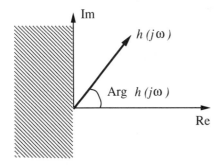

**Figure 4.7** : Geometric interpretation of the passivity condition for a linear SISO system

**Proof of condition (4.42):** The proof is based on Parseval's theorem, which relates the time-domain and frequency-domain expressions of a signal's squared norm or "energy", as well as those of the correlation of two signals.

Consider a strictly stable linear SISO system, of transfer function $y(p)/u(p) = h(p)$, initially at rest ( $y \equiv 0$ ) at $t = 0$. Let us apply to this system an arbitrary control input between $t = 0$ and some positive time $t$, and no input afterwards. Recalling expression (4.41) of passivity, we compute

$$\int_0^t y(r)\, u(r)\, dr \;=\; \int_{-\infty}^{\infty} y(r)\, u(r)\, dr \;=\; \frac{1}{2\pi} \int_{-\infty}^{\infty} y(j\omega)\, u^*(j\omega)\, d\omega$$

where the first equality comes from the fact that $u$ is zero after time $t$ (and, of course, both $u$ and $y$ are zero for negative time), and the second equality comes from Parseval's theorem, with the superscript $^*$ referring to complex conjugation. Since $y(j\omega) = h(j\omega)\, u(j\omega)$, we thus have

$$\int_0^t y(r)\, u(r)\, dr \;=\; \frac{1}{2\pi} \int_{-\infty}^{\infty} h(j\omega)\, |u(j\omega)|^2\, d\omega$$

Now since $h$ is the transfer function of a real system, its coefficients are real, and thus $h(-j\omega) = [h(j\omega)]^*$. Hence,

$$\int_0^t y(r)\, u(r)\, dr \;=\; \frac{1}{\pi} \int_0^{\infty} \mathrm{Re}[h(j\omega)]\, |u(j\omega)|^2\, d\omega \qquad (4.44)$$

Given expression (4.41) of passivity, equation (4.44) shows that (4.42) is a *sufficient* condition for the system to be passive. Indeed, taking an arbitrary input $u$, the integral $\int_0^t y(r)\, u(r)\, dr$ does not depend on the values of $u$ at times later than $t$ (so that our earlier assumption that $u$ is zero after time $t$ is not restrictive).

Equation (4.44) also shows that (4.42) is a *necessary* condition for the system to be passive. Indeed, if (4.44) was not verified, then there would be a finite interval in $\omega$ over which $\mathrm{Re}[h(j\omega)] < 0$, because $h$ is continuous in $\omega$. The integral could then be made arbitrarily negative by choosing $|u(j\omega)|$ large enough over this finite interval. $\qquad\square$

Note that we have assumed that $h(p)$ is *strictly* stable, so as to guarantee the existence of the frequency-domain integrals in the above proof. Actually, using standard results in complex variable analysis, the proof can be extended easily to the case where $h(p)$ has perhaps some poles on the $j\omega$ axis, provided that these poles be simple (*i.e.*, distinct) and that the associated residues be non-negative. As discussed earlier in section 4.6, systems belonging to this more general class and verifying condition (4.42) are called *positive real (PR) systems*. Thus, a linear single-input system is passive if (and only if) it is positive real.

Condition (4.42) can also be formally stated as saying that the Nyquist plot of $h$ is in the right half-plane. Similarly, if the Nyquist plot of a strictly stable (or PR) linear system of transfer function $h$ is strictly in the right half-plane (except perhaps for $\omega = \infty$), *i.e*, if

$$\forall \, \omega \geq 0 \,, \, \text{Re}[h(j\omega)] > 0 \tag{4.45}$$

then the system is actually dissipative. As discussed in section 4.6, strictly stable linear systems verifying (4.45) are called *strictly positive real (SPR) systems*.

It can also be shown that, more generally, a strictly stable linear MIMO system is passive if, and only if,

$$\forall \, \omega \geq 0 \,, \, \mathbf{H}(j\omega) + \mathbf{H}^T(-j\omega) \, \geq \, \mathbf{0}$$

where $\mathbf{H}$ is the transfer matrix of the system. It is dissipative if

$$\forall \, \omega \geq 0 \,, \, \mathbf{H}(j\omega) + \mathbf{H}^T(-j\omega) \, > \, \mathbf{0}$$

## THE KALMAN-YAKUBOVICH LEMMA

For linear systems, the closeness of the concepts of stability and passivity can be understood easily by considering the Lyapunov equations associated with the systems, as we now show. The discussion also provides a more intuitive perspective on the KY lemma of section 4.6.2, in the light of the passivity formalism.

Recall from our discussion of Lyapunov functions for linear systems (section 3.5.1) that, for any strictly stable linear system of the form $\dot{\mathbf{x}} = \mathbf{A}\,\mathbf{x}$, one has

$$\forall \, \mathbf{Q} \; symmetric \; p.d. \,, \; \exists \, \mathbf{P} \; symmetric \; p.d. \,, \; such \; that \quad \mathbf{A}^T \, \mathbf{P} + \mathbf{P} \, \mathbf{A} = - \, \mathbf{Q} \tag{4.46}$$

an algebraic matrix equation which we referred to as the Lyapunov equation for the linear system. Letting

$$V = \frac{1}{2} \mathbf{x}^T \mathbf{P} \mathbf{x}$$

yields

$$\dot{V} = -\frac{1}{2} \mathbf{x}^T \mathbf{Q} \mathbf{x}$$

Consider now a linear system, *strictly stable in open-loop*, in the standard form

$$\dot{\mathbf{x}} = \mathbf{A}\,\mathbf{x} + \mathbf{B}\,\mathbf{u} \qquad \mathbf{y} = \mathbf{C}\mathbf{x}$$

The Lyapunov equation (4.46) is verified for this system, since it is only related to the system's stability, as characterized by the matrix $\mathbf{A}$, and is independent of the input and output matrices $\mathbf{B}$ and $\mathbf{C}$. Thus, with the same definition of $V$ as above, $\dot{V}$ now simply contains an extra term associated with the input $\mathbf{u}$

$$\dot{V} = \mathbf{x}^T \, \mathbf{P} \, (\mathbf{Ax} + \mathbf{Bu}) \; = \; \mathbf{x}^T \mathbf{PBu} - \frac{1}{2} \mathbf{x}^T \mathbf{Qx} \tag{4.47}$$

Since $\mathbf{y} = \mathbf{Cx}$, we see that (4.47) defines a dissipative mapping between $\mathbf{u}$ and $\mathbf{y}$, provided that the matrices $\mathbf{B}$ and $\mathbf{C}$ be related by

$$\mathbf{C} = \mathbf{B}^T \, \mathbf{P}$$

This result, known as the *Kalman-Yakubovich (KY) lemma*, shows the closeness of the passivity and stability concepts, given compatible choices of inputs and outputs. Since the Lyapunov equation (4.46) can be satisfied for any arbitrary symmetric *p.d.* matrix $\mathbf{Q}$, the KY lemma states that given any open-loop strictly stable linear system, one can construct an infinity of dissipative input-output maps simply by using compatible choices of inputs and outputs. In particular, given the system's physical inputs and the associated matrix $\mathbf{B}$, one can choose an infinity of outputs from which the linear system will look dissipative.

In the single-input case, and given our earlier discussion of frequency-domain characterizations of the passivity of linear systems, the KY lemma can be equivalently stated as Lemma 4.4. Note that the controllability condition in that frequency-domain formulation simply ensures that the transfer function $h(p)$ completely characterizes the linear system defined by $(\mathbf{A}, \mathbf{b}, \mathbf{c})$ (since $\mathbf{P}$ is symmetric positive definite and $(\mathbf{A}, \mathbf{b})$ is controllable, thus $(\mathbf{A}, \mathbf{c}^T) = (\mathbf{A}, \mathbf{b}^T \mathbf{P})$ is observable). Also, as noted earlier, the KY lemma can be extended to PR systems, for which it can be shown that there exist a positive definite matrix $\mathbf{P}$ and a positive *semi*-definite matrix $\mathbf{Q}$ such that (4.36a) and (4.36b) are verified. The main usefulness of this result is that it is applicable to transfer functions containing a pure integrator, which are common in adaptive controller design.

**Example 4.20:** The passivity of the adaptation law (4.40) of Example 4.19 can also be shown directly by noticing that the integrator structure

$$\dot{\theta} = - \, e \, w(t)$$

implies that the mapping $- e \, w(t) \; \rightarrow \; \theta$ is passive, and therefore that the mapping $e \rightarrow - w(t) \, \theta$ is also passive (since $\theta \, [- e \, w(t)] = [- w(t) \, \theta] \, e$ ).

Furthermore, the passivity interpretation shows that the integrator in the above update law can be replaced by *any PR transfer function*, while still guaranteeing that the tracking error $e$ tends to zero. Indeed, since the dissipation term $g_2$ is simply zero using the original update law, the KY lemma shows that, with the modified update law, there exists a symmetric positive definite matrix $\mathbf{P}$ and a symmetric positive *semi*-definite matrix $\mathbf{Q}$ (which, in this simple first-order case, are simply scalars $P > 0$ and $Q \geq 0$ ) such that

$$V = e^2 + P\,\theta^2 + \int_o^t Q\,[\theta(r)]^2\,dr$$

$$\dot{V} = -2\,e^2$$

The tracking convergence proof can then be completed as before using Barbalat's lemma.

Thus, the passivity interpretation can quickly suggest additional design flexibility.     ☐

## PASSIVITY INTERPRETATION OF SPR SYSTEMS

The passivity interpretation of SPR systems may allow one to quickly determine whether a transfer function is SPR, using physical insights.

Consider for instance the transfer function

$$h_5(p) = \frac{10p}{4p^2 + 5p + 1}$$

We can determine whether $h_5$ is SPR using a procedure similar to that used for the function $h_4$ in Example 4.15. We can also simply notice that $h_5$ can be *interpreted* as the transfer function of a mass-spring-damper system

$$4\ddot{x} + 5\dot{x} + x = 10u$$

$$y = \dot{x}$$

with force as input and velocity as output. Thus $h_5$ is dissipative, and thus SPR (since it is also strictly stable).

Finally, one can easily verify that

- If $h(p)$ is SPR, so is $1/h(p)$

- If $h_1(p)$ and $h_2(p)$ are SPR, so is

$$h(p) = \alpha_1 h_1(p) + \alpha_2 h_2(p)$$

provided that $\alpha_1 \geq 0$ and $\alpha_2 \geq 0$

- If $h_1(p)$ and $h_2(p)$ are SPR, so is

$$h(p) = \frac{h_1(p)}{1 + h_1(p)\,h_2(p)}$$

which is the overall transfer function of the negative feedback system having $h_1(p)$ as the forward path transfer function and $h_2(p)$ as the feedback path transfer function

While these results can be derived directly, they are simply versions specific to the linear single-input case of more general properties of passive systems. The first result simply reflects the fact that the input $u$ and the output $y$ play a symmetric role in the definition of passivity. The last two results represent the linear single-input frequency-domain interpretation of our earlier general discussion on the combination of passive blocks. Actually, if either $h_1(p)$ *or* $h_2(p)$ is SPR, while the other is merely passive, then $h(p)$ is SPR. This allows us to easily construct new SPR transfer functions simply by taking any stable transfer function having a phase shift smaller than $90^o$ at all frequencies, and putting it in feedback or in parallel configuration with any SPR transfer function.

# 4.8  * Absolute Stability

The systems considered in this section have the interesting structure shown in Figure 4.8. The forward path is a linear time-invariant system, and the feedback part is a memoryless nonlinearity, *i.e.*, a nonlinear static mapping.   The equations of such systems can be written as

$$\dot{\mathbf{x}} = \mathbf{A}\mathbf{x} - \mathbf{b}\,\phi(y) \tag{4.48a}$$

$$y = \mathbf{c}^T \mathbf{x} \tag{4.48b}$$

where $\phi$ is some nonlinear function and $G(p) = \mathbf{c}^T [p\mathbf{I} - \mathbf{A}]^{-1}\mathbf{b}$ .   Many systems of practical interest can be represented in this structure.

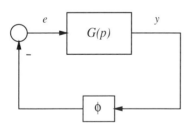

**Figure 4.8 :** System structure in absolute stability problems

### THE ISSUE OF ABSOLUTE STABILITY

The nonlinear system in Figure 4.8 has a special structure.  If the feedback path simply contains a constant gain, *i.e.*, if $\phi(y) = \alpha\, y$, then the stability of the whole system, a linear feedback system, can be simply determined by examining the eigenvalues of the closed-loop system matrix $\mathbf{A} - \alpha\mathbf{b}\mathbf{c}^T$. However, the stability analysis of the whole system with an arbitrary nonlinear feedback function $\phi$ is much more difficult.

In analyzing this kind of system using Lyapunov's direct method, we usually require the nonlinearity to satisfy a so-called sector condition, whose definition is given below.

**Definition 4.12**   *A continuous function* $\phi$ *is said to belong to the <u>sector</u>* $[k_1, k_2]$, *if there exists two non-negative numbers* $k_1$ *and* $k_2$ *such that*

$$y \neq 0 \quad \Rightarrow \quad k_1 \leq \frac{\phi(y)}{y} \leq k_2 \tag{4.49}$$

Geometrically, condition (4.49) implies that the nonlinear function always lies between the two straight lines $k_1 y$ and $k_2 y$, as shown in Figure 4.9. Two properties are implied by equation (4.49). First, it implies that $\phi(0) = 0$. Secondly, it implies that $y\phi(y) \geq 0$, *i.e*, that the graph of $\phi(y)$ lies in the first and third quadrants. Note that in many of later discussions, we will consider the special case of $\phi(y)$ belonging to the sector $[0, k]$, *i.e*, $\exists\, k > 0$, such that

$$0 \leq \phi(y) \leq k\,y \tag{4.50}$$

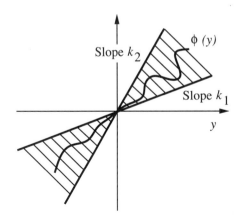

**Figure 4.9 :** The sector condition (4.49)

Assume that the nonlinearity $\phi(y)$ is a function belonging to the sector $[k_1, k_2]$, and that the **A** matrix of the linear subsystem in the forward path is stable (*i.e.*, Hurwitz). What additional constraints are needed to guarantee the stability of the whole system? In view of the fact that the nonlinearity in Figure 4.9 is bounded by the two straight lines, which correspond to constant gain feedback, it may be plausible that the stability of the nonlinear system should have some relation to the stability of constant gain feedback systems. In 1949, the Soviet scientist M.A. Aizerman made the following <u>conjecture</u>: *if the matrix* $[\mathbf{A} - \mathbf{b}\mathbf{c}^T k]$ *is stable for all values of k in* $[k_1, k_2]$, *then the nonlinear system is globally asymptotically stable.*

## POPOV'S CRITERION

Aizerman's is a very interesting conjecture. If it were true, it would allow us to deduce the stability of a nonlinear system by simply studying the stability of linear systems. However, several counter-examples show that this conjecture is false.    After Aizerman, many researchers continued to seek conditions that guarantee the stability of the nonlinear system in Figure 4.8. Popov's criterion imposes additional conditions on the linear subsystem, leading to a sufficient condition for asymptotic stability reminiscent of Nyquist's criterion (a necessary *and* sufficient condition) in linear system analysis.

A number of versions have been developed for Popov's criterion. The following basic version is fairly simple and useful.

**Theorem 4.12 (Popov's Criterion)**    *If the system described by (4.48) satisfies the conditions*

- *the matrix* **A** *is Hurwitz (i.e., has all its eigenvalues strictly in the left half-plane) and the pair* [**A**, **b**] *is controllable*

- *the nonlinearity* $\phi$ *belongs to the sector* $[0, k]$

- *there exists a strictly positive number* $\alpha$ *such that*

$$\forall \omega \geq 0 \qquad \mathrm{Re}\,[(1 + j\alpha\omega)\, G(j\omega)] + \frac{1}{k} \geq \varepsilon \tag{4.51}$$

*for an arbitrarily small* $\varepsilon > 0$, *then the point* **0** *is globally asymptotically stable.*

Inequality (4.51) is called *Popov's inequality*.    The criterion can be proven constructing a Lyapunov function candidate based on the KY lemma.

Let us note the main features of Popov's criterion:

- It only applies to autonomous systems.

- It is restricted to a single memoryless nonlinearity.

- The stability of the nonlinear system may be determined by examining the frequency-response functions of a linear subsystem, without the need of searching for explicit Lyapunov functions.

- It only gives a *sufficient* condition.

The criterion is most easy to apply by using its graphical interpretation. Let

$$G(j\omega) = G_1(\omega) + jG_2(\omega)$$

Then expression (4.51) can be written

$$G_1(\omega) - \alpha\omega G_2(\omega) + \frac{1}{k} \geq \varepsilon \tag{4.52}$$

Now let us construct an *associated transfer function* $\mathbf{W}(j\omega)$, with the same real part as $G(j\omega)$, but an imaginary part equal to $\omega \, \text{Im}(G(j\omega))$, *i.e.*,

$$\mathbf{W}(j\omega) = x + j\,y = G_1(\omega) + j\omega G_2(\omega)$$

Then (4.52) implies that the nonlinear system is guaranteed to be globally asymptotically stable if, in the complex plane having $x$ and $y$ as coordinates, the polar plot of $W(j\omega)$ is (uniformly) below the line $x - \alpha y + (1/k) = 0$ (Figure 4.10). The polar plot of $W$ is called a *Popov plot*. One easily sees the similarity of this criterion to the Nyquist criterion for linear systems. In the Nyquist criterion, the stability of a linear feedback system is determined by examining the position of the polar plot of $G(j\omega)$ relative to the point $(0, -1)$, while in the Popov criterion, the stability of a nonlinear feedback system is determined by checking the position of the associated transfer function $W(j\omega)$ with respect to a line.

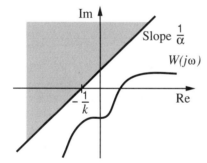

**Figure 4.10 :** The Popov plot

**Example 4.21:** Let us determine the stability of a nonlinear system of the form (4.48) where the linear subsystem is defined by

$$G(j\omega) = \frac{p+3}{p^2 + 7p + 10}$$

and the nonlinearity satisfies condition (4.50).

First, the linear subsystem is strictly stable, because its poles are $-2$ and $-5$. It is also controllable, because there is no pole-zero cancellation. Let us now check the Popov inequality. The frequency response function $G(j\omega)$ is

$$G(j\omega) = \frac{j\omega + 3}{-\omega^2 + 7\omega j + 10}$$

Therefore,

$$G_1 = \frac{4\omega^2 + 30}{\omega^4 + 29\omega^2 + 100}$$

$$G_2 = \frac{-\omega(\omega^2 + 11)}{\omega^4 + 29\omega^2 + 100}$$

Substituting the above into (4.52) leads to

$$4\omega^2 + 30 + \alpha\omega^2(\omega^2 + 11) + (\frac{1}{k} - \varepsilon)(\omega^4 + 29\omega^2 + 100) > 0$$

It is clear that this inequality can be satisfied by any strictly positive number $\alpha$, and any strictly positive number $k$, *i.e.*, $0 < k < \infty$. Thus, the nonlinear system is globally asymptotically stable as long as the nonlinearity belongs to the first and third quadrants. $\square$

## THE CIRCLE CRITERION

A more direct generalization of Nyquist's criterion to nonlinear systems is the circle criterion, whose basic version can be stated as follows.

**Theorem 4.13 (Circle Criterion)**    *If the system (4.48) satisfies the conditions*

> • *the matrix* **A** *has no eigenvalue on the* $j\omega$ *axis, and has* $\rho$ *eigenvalues strictly in the right half-plane;*
>
> • *the nonlinearity* $\phi$ *belongs to the sector* $[k_1, k_2]$ *;*
>
> • *one of the following is true*
>
>> • $0 < k_1 \le k_2$ *, the Nyquist plot of* $G(j\omega)$ *does not enter the disk* **D**$(k_1, k_2)$ *and encircles it* $\rho$ *times counter-clockwise;*
>>
>> • $0 = k_1 < k_2$ *, and the Nyquist plot of* $G(j\omega)$ *stays in the half-plane* Re $p > -1/k_2$ *;*
>>
>> • $k_1 < 0 < k_2$ *, and the Nyquist plot of* $G(j\omega)$ *stays in the interior of the disk* **D**$(k_1, k_2)$ *;*
>>
>> • $k_1 < k_2 < 0$ *, the Nyquist plot of* $-G(j\omega)$ *does not enter the disk* **D**$(-k_1, -k_2)$ *and encircles it* $\rho$ *times counter-clockwise;*

*then the equilibrium point* **0** *of the system is globally asymptotically stable.*

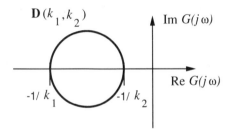

**Figure 4.11** : The circle criterion

Thus we see that, essentially, the critical point $-1/k$ in Nyquist's criterion is replaced in the circle criterion by the circle of Figure 4.11 (which tends towards the point $-1/k_1$ as $k_2$ tends to $k_1$, *i.e.*, as the conic sector gets thinner). Of course, the circle criterion states *sufficient* but not necessary conditions.

The circle criterion can be extended to non-autonomous systems.

# 4.9 * Establishing Boundedness of Signals

In the stability analysis or convergence analysis of nonlinear systems, a frequently encountered problem is that of establishing the boundedness of certain signals. For instance, in order to use Barbalat's lemma, one has to show the uniform continuity of $\dot{f}$, which can be most conveniently shown by proving the boundedness of $\ddot{f}$. Similarly, in studying the effects of disturbances, it is also desirable to prove the boundedness of system signals in the presence of disturbances. In this section, we provide two useful results for such purposes.

### THE BELLMAN-GRONWALL LEMMA

In system analysis, one can often manipulate the signal relations into an integral inequality of the form

$$y(t) \leq \int_o^t a(\tau)\, y(\tau)\, d\tau + b(t) \tag{4.53}$$

where $y(t)$, the variable of concern, appears on both sides of the inequality. The problem is to gain an explicit bound on the magnitude of $y$ from the above inequality. The Bellman-Gronwall lemma can be used for this purpose.

**Lemma 4.6 (Bellman-Gronwall)**    *Let a variable y(t) satisfy (4.53), with a(t), b(t) being known real functions. Then*

$$y(t) \le \int_o^t a(\tau) b(\tau) \exp[\int_\tau^t a(r) dr] d\tau + b(t) \tag{4.54}$$

*If b(t) is differentiable, then*

$$y(t) \le b(0) \exp[\int_o^t a(\tau) d\tau] + \int_o^t \dot{b}(\tau) \exp[\int_\tau^t a(r) dr] d\tau \tag{4.55}$$

*In particular, if b(t) is a constant, we simply have*

$$y(t) \le b(0) \exp[\int_o^t a(\tau) d\tau] \tag{4.56}$$

**Proof**: The proof is based on defining a new variable and transforming the integral inequality into a differential equation, which can be easily solved. Let

$$v(t) = \int_o^t a(\tau) y(\tau) d\tau \tag{4.57}$$

Then differentiation of $v$ and use of (4.53) leads to

$$\dot{v} = a(t) y(t) \le a(t) v(t) + a(t) b(t)$$

Let

$$s(t) = a(t) y(t) - a(t) v(t) - a(t) b(t)$$

which is obviously a non-positive function. Then $v(t)$ satisfies

$$\dot{v}(t) - a(t) v(t) = a(t) b(t) + s(t)$$

Solving this equation with initial condition $v(0) = 0$, yields

$$v(t) = \int_0^t \exp[\int_\tau^t a(r) dr] \, [a(\tau) b(\tau) + s(\tau)] d\tau \tag{4.58}$$

Since $s(\cdot)$ is a non-positive function,

$$v(t) \le \int_o^t \exp[\int_\tau^t a(r) dr] \, a(\tau) b(\tau) d\tau$$

This, together with the definition of $v$ and the inequality (4.53), leads to

$$y(t) \le \int_0^t \exp[\int_\tau^t a(r)\,dr]\, a(\tau)\, b(\tau)\, d\tau + b(t)$$

If $b(t)$ is differentiable, we obtain, by partial integration

$$\int_0^t \exp[\int_\tau^t a(r)\,dr]\, a(\tau)\, b(\tau)\, d\tau = -\, b(\tau)\, \exp[\int_\tau^t a(r)\,dr] \,\Big|_{\tau=o}^{\tau=t} + \int_0^t \dot{b}(\tau)\exp[\int_\tau^t a(r)\,dr]\, d\tau \qquad \square$$

## TOTAL STABILITY

Consider the nonlinear system

$$\ddot{x} + 2\dot{x}^3 + 3x = d(t) \tag{4.59}$$

which represents a non-linear mass-spring-damper system with a disturbance $d(t)$ (which may be due to unmodeled Coulomb friction, motor ripple, parameter variations, *etc*). Is $x$ bounded when the disturbance is bounded? This is the main question addressed by the so-called total stability theory (or stability under persistent disturbances).

In total stability, the systems concerned are described in the form

$$\dot{\mathbf{x}} = \mathbf{f}(\mathbf{x}, t) + \mathbf{g}(\mathbf{x}, t) \tag{4.60}$$

where $\mathbf{g}(\mathbf{x},t)$ is a perturbation term. Our objective is to derive a boundedness condition for the perturbed equation (4.60) from the stability properties of the associated unperturbed system

$$\dot{\mathbf{x}} = \mathbf{f}(\mathbf{x}, t) \tag{4.61}$$

We assume that $\mathbf{x} = 0$ is an equilibrium point for the unperturbed dynamics (4.61), *i.e.*, $\mathbf{f}(\mathbf{0}, t) = 0$. But the origin is not necessarily an equilibrium point for the perturbed dynamics (4.60). The concept of total stability characterizes the ability of a system to withstand small persistent disturbances, and is defined as follows:

**Definition 4.13** *The equilibrium point* $\mathbf{x} = 0$ *for the unperturbed system (4.61) is said to be* <u>totally stable</u> *if for every* $\varepsilon \ge 0$, *two numbers* $\delta_1$ *and* $\delta_2$ *exist such that* $\|\mathbf{x}(t_o)\| < \delta_1$ *and* $\|\mathbf{g}(\mathbf{x},t)\| < \delta_2$ *imply that every solution* $\mathbf{x}(t)$ *of the perturbed system (4.60) satisfies the condition* $\|\mathbf{x}(t)\| < \varepsilon$.

The above definition means that an equilibrium point is totally stable if the state of the perturbed system can be kept arbitrarily close to zero by restricting the initial state and the perturbation to be sufficiently small. Note that total stability is simply a local version (with small input) of BIBO (bounded input bounded output) stability. It is also useful to remark that if the unperturbed system is linear, then total stability is guaranteed by the asymptotic stability of the unperturbed system.

The following theorem is very useful to assert the total stability of a nonlinear system.

**Theorem 4.14**    *If the equilibrium point of (4.61) is uniformly asymptotically stable, then it is totally stable.*

This theorem can be proven by using the converse Lyapunov theorem 4.8. It means that uniformly asymptotically stable systems can withstand small disturbances. Because uniformly asymptotic stability can be asserted by the Lyapunov theorem 4.1, total stability of a system may be similarly established by theorem 4.1. Note that asymptotic stability is not sufficient to guarantee the total stability of a nonlinear system as can be verified by counter-examples. We also point out that exponentially stable systems are always totally stable because exponential stability implies uniform asymptotic stability.

**Example 4.22:** Consider again the system (4.59). Let us analyze the stability of the unperturbed system

$$\ddot{x} + 2\,\dot{x}^3 + 3x = 0$$

first. Using the scalar function

$$V(\mathbf{x}) = \frac{1}{2}\,[\dot{x}^2 + 3x^2]$$

and the invariant set theorem, one easily shows that the equilibrium point is globally asymptotically stable. Because the system is autonomous, the stability is also uniform. Thus, the above theorem shows that the system can withstand small disturbances $d(t)$.    ☐

Total stability guarantees boundedness to only small-disturbance, and requires only local uniform asymptotic stability of the equilibrium point. One might wonder whether the global uniform asymptotic stability can guarantee the boundedness of the state in the presence of large (though still bounded) perturbations. The following counter-example demonstrates that this is not true.

**Example 4.23:** The nonlinear equation

$$\ddot{x} + f(\dot{x}) + x = w(t) \tag{4.62}$$

can be regarded as representing mass-spring-damper system containing nonlinear damping $f(\dot{x})$ and excitation force $w(t)$, where $f$ is a first and third quadrant continuous nonlinear function such that

$$|f(y)| \leq 1 \qquad -\infty < y < \infty$$

as illustrated in Figure 4.12.

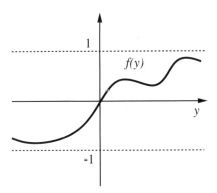

**Figure 4.12 :** A nonlinear damping function

The system is totally stable, because the equilibrium point can be shown to be globally uniformly asymptotically stable using the Lyapunov function $V = (1/2)(\dot{x}^2 + x^2)$. Is the output bounded for bounded input?

Let us consider the response of the system to the excitation force $w(t) = A \sin t$, $A > 8/\pi$. By writing (4.62) as

$$\ddot{x} + x = A \sin t - f(\dot{x})$$

and solving this linear equation with $(A \sin t - f(\dot{x}))$ as input, we obtain

$$x(t) = \frac{A}{2}(\sin t - t \cos t) - \int_0^t \sin(t-\tau)f(\dot{x})\, d\tau \geq \frac{A}{2}(\sin t - t \cos t) - \int_0^t |\sin(t-\tau)|\, d\tau$$

The integral on the right-hand side can be shown to be smaller than $(2/\pi)(1+\varepsilon)t$ for any $t \geq t_o$, and $\varepsilon > 0$ and some $t_o$. At $t_n = (2n+1)\pi$,

$$x(t_n) \geq (2n+1)\pi \left[\frac{A}{2} - \frac{2}{\pi}(1+\varepsilon)\right]$$

Therefore, if we take $A > 8/\pi$ and $\varepsilon = 1/2$, $x(t_n) \to \infty$. $\qquad\qquad \square$

# 4.10 * Existence and Unicity of Solutions

This section discusses the mathematically important question of existence and unicity of solutions of nonlinear differential equations. We first describe a simple and quite general sufficient condition for a nonlinear differential equation to admit a solution, and then simple but conservative conditions for this solution to be unique.

**Theorem 4.15 (Cauchy existence theorem)** *Consider the differential equation* $\dot{\mathbf{x}} = \mathbf{f}(\mathbf{x}, t)$, *with initial condition* $\mathbf{x}(t_o) = \mathbf{x}_o$. *If the function* $\mathbf{f}$ *is continuous in the closed region*

$$|t - t_o| \leq T , \; \|\mathbf{x} - \mathbf{x}_o\| \leq R$$

*where T and R are strictly positive constants, then the equation has at least one solution* $\mathbf{x}(t)$ *which satisfies the initial condition and is continuous over a finite time period* $[t_o, t_1]$ *(where* $t_1 > t_o$ *).*

The above theorem indicates that the continuity of $\mathbf{f}$ is sufficient for the local *existence* of solutions. However, it does not guarantee the *uniqueness* of the solution.

**Example 4.24: An equation with multiple solutions**

Consider the equation

$$\dot{y} = 3 y^{2/3}$$

with initial condition $y(0) = 0$. Two of its solutions are $y(t) \equiv 0$ and $y = t^3$.    ☐

The following theorem gives a *sufficient* condition for the unique existence of a solution.

**Theorem 4.16** *If the function* $\mathbf{f}(\mathbf{x}, t)$ *is continuous in t, and if there exists a strictly positive constant L such that*

$$\|\mathbf{f}(\mathbf{x}_2, t) - \mathbf{f}(\mathbf{x}_1, t)\| \leq L \, \|\mathbf{x}_2 - \mathbf{x}_1\| \tag{4.63}$$

*for all* $\mathbf{x}_1$ *and* $\mathbf{x}_2$ *in a finite neighborhood of the origin and all t in the interval* $[t_o, t_o + T]$ *(with T being a strictly positive constant), then* $\dot{\mathbf{x}} = \mathbf{f}(\mathbf{x}, t)$ *has a unique solution* $\mathbf{x}(t)$ *for sufficiently small initial states and in a sufficiently short time interval.*

Condition (4.63) is called a *Lipschitz condition* and $L$ is known as a *Lipschitz constant*. If (4.63) is verified, then $\mathbf{f}$ is said to be locally Lipschitz in $\mathbf{x}$. If (4.63) is verified for any time $t$, then $\mathbf{f}$ is said to be locally Lipschitz in $\mathbf{x}$ uniformly with respect to $t$. Note that the satisfaction of a Lipschitz condition implies (locally) the continuity of $\mathbf{f}$ in terms of $\mathbf{x}$, as can be easily proven from the definition of continuity. Conversely, if locally $\mathbf{f}$ has a continuous and bounded Jacobian with respect to $\mathbf{x}$, then $\mathbf{f}$ is locally Lipschitz. When (4.63) is satisfied for any $\mathbf{x}_1$ and $\mathbf{x}_2$ in the state space, $\mathbf{f}$ is said to be *globally Lipschitz*. The above theorem can then be extended to guarantee unique existence of a solution in a global sense (*i.e.*, for any initial condition and any time period).

While the condition for existence of solutions, as stated by Cauchy's theorem, is rather benign, the sufficient condition for unicity is quite strong, and, actually, overly conservative. Most results on nonlinear dynamics simply assume that **f** is smooth enough to guarantee existence and unicity of the solutions. Note that this is always the case of good physical system models (at least in classical physics).

Actually, precise mathematical results exist about the relation between the existence of a Lyapunov function for a given system and the existence and unicity of solutions (see, *e.g.*, [Yoshizawa, 1966, 1975]). From a practical point of view, these results essentially mean that the existence of a Lyapunov function to describe a system will guarantee the system's "good behavior" under some mild smoothness assumptions on the dynamics

# 4.11  Summary

Some advanced topics in nonlinear control theory are presented in this chapter. Lyapunov theory for non-autonomous systems is discussed first. Its results are quite similar to those for autonomous systems, although more involved conditions are required. A major difference is that the powerful invariant-set theorem does not apply to non-autonomous system, although Barbalat's lemma can often be a simple and effective substitute. A number of instability theorems are also presented. Such theorems are useful for non-autonomous systems, or for autonomous systems whose linearizations are only marginally stable. Theorems on the existence of Lyapunov functions may be of use in constructing Lyapunov functions for systems part of which is known to have certain stability properties. The passivity formalism is also introduced, as a notationally convenient and physically motivated interpretation of Lyapunov or Lyapunov-like analysis. The chapter also includes some results for establishing the boundedness of signals in nonlinear systems.

# 4.12  Notes and References

A comprehensive yet readable book on Lyapunov analysis of non-autonomous systems is [Hahn, 1967], on which most of the stability definitions in this chapter are based. The definitions and results concerning positive definite and decrescent functions are based on [Hahn, 1967; Vidyasagar, 1978]. The statement and proof of Theorem 4.1 are adapted from [Kalman and Bertram, 1960]. Example 4.5 is adapted from [Vidyasagar, 1978], Example 4.6 from [Massera, 1949], and Example 4.7 from [Rouche, *et al.*, 1977]. Figure 4.1 is adapted from A.S.M.E. Journal of Basic Engineering, 1960. In section 4.2.2, the result on perturbed linear systems is from [Vidyasagar, 1978], while the result on sufficient smoothness conditions on the $\mathbf{A}(t)$ matrix is from [Middleton, 1988]. Sections 4.2.3 and 4.3 are largely adapted from [Vidyasagar, 1978], where proofs of the main results can be found. The statement of Theorem 4.9 follows that in [Bodson, 1986]. Lemma 4.2 and its proof are from [Popov, 1973]. An extensive study of absolute stability problems from a frequency-domain perspective is contained in [Narendra and Taylor, 1973], from which the definitions and theorems on positive real functions are adapted. A more recent description of positive real functions and their applications in adaptive control can be found in [Narendra and Annasswamy, 1989]. The Bellman-Gronwall lemma and its proof are adapted from [Hsu and Meyer, 1968]. The definition and theorem on total stability are based on [Hahn, 1965]. Example 4.23 is adapted from [Desoer, *et al.*, 1965].

Passivity theory (see [Popov, 1973; Desoer and Vidyasagar, 1975]) is presented in a slightly unconventional form. Passivity interpretations of adaptive control laws are discussed in [Landau, 1979]. The reader is referred to [Vidyasagar, 1978] for a detailed discussion of absolute stability. The circle criterion and its extensions to non-autonomous systems were derived by [Narendra and Goldwyn, 1964; Sandberg, 1964; Tsypkin, 1964; Zames, 1966].

Other important robustness analysis tools include singular perturbations (see, *e.g.*, [Kokotovic, *et al.*, 1986]) and averaging (see, *e.g*, [Hale, 1980; Meerkov, 1980]).

Relations between the existence of Lyapunov functions and the existence and unicity of solutions of nonlinear differential equations are discussed in [Yoshizawa, 1966, 1975].

## 4.13  Exercises

**4.1**   Show that, for a non-autonomous system, a system trajectory is generally not an invariant set.

**4.2**   Analyze the stability of the dynamics (corresponding to a mass sinking in a viscous liquid)

$$\dot{v} + 2a|v|v + bv = c \qquad a > 0, b > 0$$

**4.3**   Show that a function $V(\mathbf{x}, t)$ is radially unbounded if, and only if, there exists a class-K function $\phi$ such that

$$V(\mathbf{x}, t) \geq \phi(\|\mathbf{x}\|)$$

where the function $\phi$ satisfies

$$\lim_{\mathbf{x} \to \infty} \phi(\|\mathbf{x}\|) = \infty$$

**4.4**   The performance of underwater vehicles control systems is often constrained by the "unmodeled" dynamics of the thrusters. Assume that one decides to explicitly account for thruster dynamics, based on the model

$$\dot{\omega} = -\alpha_1 \omega |\omega| + \alpha_2 \tau \qquad \alpha_1 > 0, \alpha_2 > 0$$

$$u = b \omega |\omega| \qquad b > 0$$

where $\tau$ is the torque input to the propeller, $\omega$ is the propeller's angular velocity, and $u$ is the actual thrust generated.

Show that, for a *constant* torque input $\tau_o$, the steady-state thrust is proportional to $\tau_o$ (which is consistent witht the fact that thruster dynamics is often treated as "unmodeled").

Assuming that the coefficients $\alpha_i$ and $b$ in the above model are known with good accuracy, design and discuss the use of a simple "open-loop" observer for $u$ (given an arbitrary time-varying torque input $\tau$) in the absence of measurements of $\omega$. (Adapted from [Yoerger and Slotine, 1990].)

**4.5**   Discuss the similarity of the results of section 4.2.2 with Krasovskii's theorem of section 3.5.2.

**4.6**   Use the first instability theorem to show the instability of the vertical-up position of a pendulum.

**4.7**   Show explicitly why the linear time-varying system defined by (4.18) does not satisfy the sufficient condition (4.19).

**4.8**   Condition (4.19) on the eigenvalues of $\mathbf{A}(t) + \mathbf{A}^T(t)$ is only, of course, a *sufficient* condition. For instance, show that the linear time-varying system associated with the matrix

$$\mathbf{A}(t) = \begin{bmatrix} -1 & e^{t/2} \\ 0 & -1 \end{bmatrix}$$

*is* globally asymptotically stable.

**4.9**   Determine whether the following systems have a stable equilibrium. Indicate whether the stability is asymptotic, and whether it is global.

$(a)$
$$\begin{bmatrix} \dot{x}_1 \\ \dot{x}_2 \end{bmatrix} = \begin{bmatrix} -10 & e^{3t} \\ 0 & -2 \end{bmatrix} \begin{bmatrix} x_1 \\ x_2 \end{bmatrix}$$

$(b)$
$$\begin{bmatrix} \dot{x}_1 \\ \dot{x}_2 \end{bmatrix} = \begin{bmatrix} -1 & 2\sin t \\ 0 & -(t+1) \end{bmatrix} \begin{bmatrix} x_1 \\ x_2 \end{bmatrix}$$

$(c)$
$$\begin{bmatrix} \dot{x}_1 \\ \dot{x}_2 \end{bmatrix} = \begin{bmatrix} -1 & e^{2t} \\ 0 & -2 \end{bmatrix} \begin{bmatrix} x_1 \\ x_2 \end{bmatrix}$$

**4.10**   If a differentiable function $f$ is lower bounded and decreasing ($\dot{f} \leq 0$), then it converges to a limit. However, $\dot{f}$ does not necessarily converge to zero. Derive a counter-example. (*Hint:* You may use for $-\dot{f}$ a function that peaks periodically, but whose integral is finite.)

**4.11**    (a) Show that if a function $f$ is bounded and uniformly continuous, and there exists a positive definite function $F(f, t)$ such that

$$\int_{o}^{\infty} F(f(t), t)\, dt \;<\; \infty$$

then $f(t)$ tends to zero as $t \to \infty$.

(b) For a given autonomous nonlinear system, consider a Lyapunov function $V$ in a ball $\mathbf{B}_R$, and let $\phi$ be a scalar, differentiable, strictly monotonously increasing function of its scalar argument. Show that $[\phi(V) - \phi(0)]$ is also a Lyapunov function for the system (distinguish the cases of stability and of asymptotic stability). Suggest extensions to non-autonomous systems.

**4.12**    Consider a scalar, lower bounded, and twice continuously differentiable function $V(t)$ such that

$$\forall\, t \geq 0 \,,\;\; \dot{V}(t) \leq 0$$

Show that, for any $t \geq 0$,

$$\dot{V}(t) = 0 \quad \Rightarrow \quad \ddot{V}(t) = 0$$

# Chapter 5
# Describing Function Analysis

The frequency response method is a powerful tool for the analysis and design of linear control systems. It is based on describing a linear system by a complex-valued function, the frequency response, instead of a differential equation. The power of the method comes from a number of sources. First, *graphical* representations can be used to facilitate analysis and design. Second, *physical* insights can be used, because the frequency response functions have clear physical meanings. Finally, the method's complexity only increases mildly with system order. Frequency domain analysis, however, cannot be directly applied to nonlinear systems because frequency response functions cannot be defined for nonlinear systems.

Yet, for some nonlinear systems, an extended version of the frequency response method, called the *describing function method*, can be used to *approximately* analyze and predict nonlinear behavior. Even though it is only an approximation method, the desirable properties it inherits from the frequency response method, and the shortage of other systematic tools for nonlinear system analysis, make it an indispensable component of the bag of tools of practicing control engineers. The main use of describing function method is for the prediction of limit cycles in nonlinear systems, although the method has a number of other applications such as predicting subharmonics, jump phenomena, and the response of nonlinear systems to sinusoidal inputs.

This chapter presents an introduction to the describing function analysis of nonlinear systems. The basic ideas in the describing function method are presented in

**157**

section 5.1. Section 5.2 discusses typical "hard nonlinearities" in control engineering, since describing functions are particularly useful for studying control systems containing such nonlinearities. Section 5.3 evaluates the describing functions for these hard nonlinearities. Section 5.4 is devoted to the description of how to use the describing function method for the prediction of limit cycles.

# 5.1  Describing Function Fundamentals

In this section, we start by presenting describing function analysis using a simple example, adapted from [Hsu and Meyer, 1968]. We then provide the formal definition of describing functions and some techniques for evaluating the describing functions of nonlinear elements.

## 5.1.1  An Example of Describing Function Analysis

The interesting and classical Van der Pol equation

$$\ddot{x} + \alpha(x^2 - 1)\dot{x} + x = 0 \tag{5.1}$$

(where $\alpha$ is a positive constant) has been treated by phase-plane analysis and Lyapunov analysis in the previous chapters. Let us now study it using a different technique, which shall lead us to the concept of a describing function. Specifically, let us determine whether there exists a limit cycle in this system and, if so, calculate the amplitude and frequency of the limit cycle (pretending that we have not seen the phase portrait of the Van der Pol equation in Chapter 2). To this effect, we first assume the existence of a limit cycle with undetermined amplitude and frequency, and then determine whether the system equation can indeed sustain such a solution. This is quite similar to the assumed-variable method in differential equation theory, where we first assume a solution of certain form, substitute it into the differential equation, and then attempt to determine the coefficients in the solution.

Before carrying out this procedure, let us represent the system dynamics in a block diagram form, as shown in Figure 5.1. It is seen that the feedback system in 5.1 contains a linear block and a nonlinear block, where the linear block, although unstable, has *low-pass* properties.

Now let us assume that there is a limit cycle in the system and the oscillation signal $x$ is in the form of

$$x(t) = A\sin(\omega t)$$

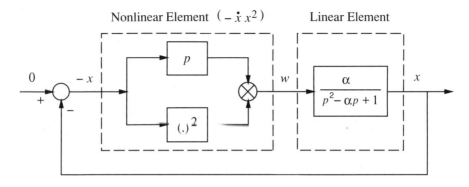

Nonlinear Element $(-\dot{x}\,x^2)$      Linear Element

**Figure 5.1 :** Feedback interpretation of the Van der Pol oscillator

with $A$ being the limit cycle amplitude and $\omega$ being the frequency. Thus,

$$\dot{x}(t) = A\,\omega\cos(\omega t)$$

Therefore, the output of the nonlinear block is

$$w = -x^2\,\dot{x} = -A^2\sin^2(\omega t)\ A\,\omega\cos(\omega t)$$

$$= -\frac{A^3\omega}{2}\,(1-\cos(2\omega t))\cos(\omega t) = -\frac{A^3\omega}{4}\,(\cos(\omega t)-\cos(3\omega t))$$

It is seen that $w$ contains a third harmonic term. Since the linear block has low-pass properties, we can reasonably assume that this third harmonic term is sufficiently attenuated by the linear block and its effect is not present in the signal flow after the linear block. This means that we can approximate $w$ by

$$w \approx -\frac{A^3}{4}\,\omega\cos\omega t = \frac{A^2}{4}\frac{\mathrm{d}}{\mathrm{d}t}[-A\sin(\omega t)]$$

so that the nonlinear block in Figure 5.1 can be approximated by the equivalent "quasi-linear" block in Figure 5.2. The "transfer function" of the quasi-linear block *depends on the signal amplitude A*, unlike a linear system transfer function (which is independent of the input magnitude).

In the frequency domain, this corresponds to

$$w = N(A,\omega)\,(-x) \tag{5.2}$$

where

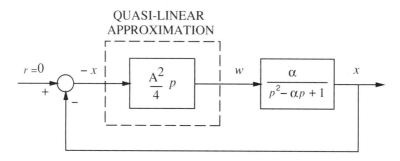

**Figure 5.2 :** Quasi-linear approximation of the Van der Pol oscillator

$$N(A, \omega) = \frac{A^2}{4} (j\omega)$$

That is, the nonlinear block can be approximated by the frequency response function $N(A, \omega)$. Since the system is assumed to contain a sinusoidal oscillation, we have

$$x = A \sin(\omega t) = G(j\omega)w = G(j\omega) N(A, \omega) (-x)$$

where $G(j\omega)$ is the linear component transfer function. This implies that

$$1 + \frac{A^2 (j\omega)}{4} \frac{\alpha}{(j\omega)^2 - \alpha(j\omega) + 1} = 0$$

Solving this equation, we obtain

$$A = 2 \quad \omega = 1$$

Note that in terms of the Laplace variable $p$, the closed-loop characteristic equation of this system is

$$1 + \frac{A^2 p}{4} \frac{\alpha}{p^2 - \alpha p + 1} = 0 \tag{5.3}$$

whose eigenvalues are

$$\lambda_{1,2} = -\frac{1}{8} \alpha (A^2 - 4) \pm \sqrt{\frac{1}{64} \alpha^2 (A^2 - 4)^2 - 1} \tag{5.4}$$

Corresponding to $A = 2$, we obtain the eigenvalues $\lambda_{1,2} = \pm j$. This indicates the existence of a limit cycle of amplitude 2 and frequency 1. It is interesting to note neither the amplitude nor the frequency obtained above depends on the parameter $\alpha$. In

Equation 5.1.

    In the phase plane, the above approximate analysis suggests that the limit cycle is a circle of radius 2, regardless of the value of $\alpha$. To verify the plausibility of this result, the real limit cycles corresponding to the different values of $\alpha$ are plotted (Figure 5.3). It is seen that the above approximation is reasonable for small value of $\alpha$, but that the inaccuracy grows as $\alpha$ increases. This is understandable because as $\alpha$ grows the nonlinearity becomes more significant and the quasi-linear approximation becomes less accurate.

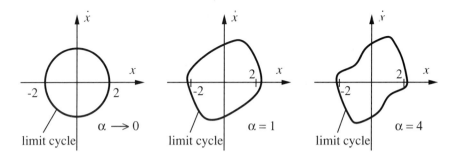

**Figure 5.3 :** Real limit cycles on the phase plane

    The stability of the limit cycle can also be studied using the above analysis. Let us assume that the limit cycle's amplitude $A$ is increased to a value larger than 2. Then, equation (5.4) shows that the closed-loop poles now have a negative real part. This indicates that the system becomes exponentially stable and thus the signal magnitude will decrease. Similar conclusions are obtained assuming that the limit cycle's amplitude $A$ is decreased to a value less than 2. Thus, we conclude that the limit cycle is stable with an amplitude of 2.

    Note that, in the above approximate analysis, the critical step is to replace the nonlinear block by the quasi-linear block which has the frequency response function $(A^2/4)\,(j\omega)$. Afterwards, the amplitude and frequency of the limit cycle can be determined from $1 + G(j\omega)\,N(A, \omega) = 0$. The function $N(A, \omega)$ is called the *describing function* of the nonlinear element. The above approximate analysis can be extended to predict limit cycles in other nonlinear systems which can be represented into the block diagram similar to Figure 5.1, as we shall do in section 5.4.

# 5.1.2 Applications Domain

Before moving on to the formal treatment of the describing function method, let us briefly discuss what kind of nonlinear systems it applies to, and what kind of information it can provide about nonlinear system behavior.

## THE SYSTEMS

Simply speaking, any system which can be transformed into the configuration in Figure 5.4 can be studied using describing functions. There are at least two important classes of systems in this category.

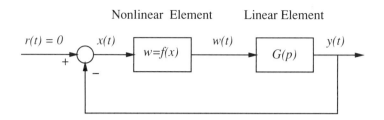

**Figure 5.4 :** A nonlinear system

The first important class consists of "almost" linear systems. By "almost" linear systems, we refer to systems which contain hard nonlinearities in the control loop but are otherwise linear. Such systems arise when a control system is designed using linear control but its implementation involves hard nonlinearities, such as motor saturation, actuator or sensor dead-zones, Coulomb friction, or hysteresis in the plant. An example is shown in Figure 5.5, which involves hard nonlinearities in the actuator.

**Example 5.1: A system containing only one nonlinearity**

Consider the control system shown in Figure 5.5. The plant is linear and the controller is also linear. However, the actuator involves a hard nonlinearity. This system can be rearranged into the form of Figure 5.4 by regarding $G_p G_1 G_2$ as the linear component $G$, and the actuator nonlinearity as the nonlinear element. $\qquad\square$

"Almost" linear systems involving sensor or plant nonlinearities can be similarly rearranged into the form of Figure 5.4.

The second class of systems consists of genuinely nonlinear systems whose dynamic equations can actually be rearranged into the form of Figure 5.4. We saw an example of such systems in the previous section.

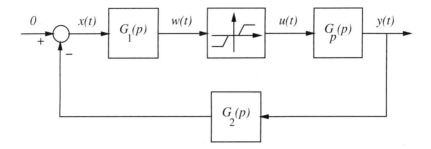

**Figure 5.5 :** A control system with hard nonlinearity

## APPLICATIONS OF DESCRIBING FUNCTIONS

For systems such as the one in Figure 5.5, limit cycles can often occur due to the nonlinearity. However, linear control cannot predict such problems. Describing functions, on the other hand, can be conveniently used to discover the existence of limit cycles and determine their stability, regardless of whether the nonlinearity is "hard" or "soft." The applicability to limit cycle analysis is due to the fact that the form of the signals in a limit-cycling system is usually approximately sinusoidal. This can be conveniently explained on the system in Figure 5.4. Indeed, asssume that the linear element in Figure 5.4 has low-pass properties (which is the case of most physical systems). If there is a limit cycle in the system, then the system signals must all be periodic. Since, as a periodic signal, the input to the linear element in Figure 5.4 can be expanded as the sum of many harmonics, and since the linear element, because of its low-pass property, filters out higher frequency signals, the output $y(t)$ must be composed mostly of the lowest harmonics. Therefore, it is appropriate to assume that the signals in the whole system are basically sinusoidal in form, thus allowing the technique in subsection 5.1.1 to be applied.

Prediction of limit cycles is very important, because limit cycles can occur frequently in physical nonlinear system. Sometimes, a limit cycle can be desirable. This is the case of limit cycles in the electronic oscillators used in laboratories. Another example is the so-called dither technique which can be used to minimize the negative effects of Coulomb friction in mechanical systems. In most control systems, however, limit cycles are undesirable. This may be due to a number of reasons:

1. limit cycle, as a way of instability, tends to cause poor control accuracy

2. the constant oscillation associated with the limit cycles can cause increasing wear or even mechanical failure of the control system hardware

3. limit cycling may also cause other undesirable effects, such as passenger

discomfort in an aircraft under autopilot

In general, although a precise knowledge of the waveform of a limit cycle is usually not mandatory, the knowledge of the limit cycle's existence, as well as that of its approximate amplitude and frequency, is critical. The describing function method can be used for this purpose. It can also guide the design of compensators so as to avoid limit cycles.

## 5.1.3  Basic Assumptions

Consider a nonlinear system in the general form of Figure 5.4. In order to develop the *basic version* of the describing function method, the system has to satisfy the following four conditions:

1. *there is only a single nonlinear component*

2. *the nonlinear component is time-invariant*

3. *corresponding to a sinusoidal input* $x = \sin(\omega t)$ *, only the fundamental component* $w_1(t)$ *in the output* $w(t)$ *has to be considered*

4. *the nonlinearity is odd*

The first assumption implies that if there are two or more nonlinear components in a system, one either has to lump them together as a single nonlinearity (as can be done with two nonlinearities in parallel), or retain only the primary nonlinearity and neglect the others.

The second assumption implies that we consider only autonomous nonlinear systems. It is satisfied by many nonlinearities in practice, such as saturation in amplifiers, backlash in gears, Coulomb friction between surfaces, and hysteresis in relays. The reason for this assumption is that the Nyquist criterion, on which the describing function method is largely based, applies only to linear time-invariant systems.

The third assumption is the *fundamental assumption* of the describing function method. It represents an *approximation*, because the output of a nonlinear element corresponding to a sinusoidal input usually contains higher harmonics besides the fundamental. This assumption implies that the higher-frequency harmonics can all be neglected in the analysis, as compared with the fundamental component. For this assumption to be valid, it is important for the linear element following the nonlinearity to have low-pass properties, *i.e.*,

$$|G(j\omega)| \gg |G(jn\omega)| \quad \text{for} \quad n = 2, 3, \ldots \tag{5.5}$$

This implies that higher harmonics in the output will be filtered out significantly. Thus, the third assumption is often referred to as the *filtering hypothesis*.

The fourth assumption means that the plot of the nonlinearity relation $f(x)$ between the input and output of the nonlinear element is symmetric about the origin. This assumption is introduced for simplicity, *i.e.*, so that the static term in the Fourier expansion of the output can be neglected. Note that the common nonlinearities discussed before all satisfy this assumption.

The relaxation of the above assumptions has been widely studied in literature, leading to describing function approaches for general situations, such as multiple nonlinearities, time-varying nonlinearities, or multiple sinusoids. However, these methods based on relaxed conditions are usually much more complicated than the basic version, which corresponds to the above four assumptions. In this chapter, we shall mostly concentrate on the basic version.

## 5.1.4  Basic Definitions

Let us now discuss how to represent a *nonlinear component* by a describing function. Let us consider a sinusoidal input to the nonlinear element, of amplitude $A$ and frequency $\omega$, *i.e.*, $x(t) = A\sin(\omega t)$, as shown in Figure 5.6. The output of the nonlinear component $w(t)$ is often a periodic, though generally non-sinusoidal, function. Note that this is always the case if the nonlinearity $f(x)$ is single-valued, because the output is $f[A\sin(\omega(t+2\pi/\omega))] = f[A\sin(\omega t)]$. Using Fourier series, the periodic function $w(t)$ can be expanded as

$$w(t) = \frac{a_o}{2} + \sum_{n=1}^{\infty} [a_n\cos(n\omega t) + b_n\sin(n\omega t)] \tag{5.6}$$

where the Fourier coefficients $a_i$'s and $b_i$'s are generally functions of $A$ and $\omega$, determined by

$$a_o = \frac{1}{\pi}\int_{-\pi}^{\pi} w(t)\,d(\omega t) \tag{5.7a}$$

$$a_n = \frac{1}{\pi}\int_{-\pi}^{\pi} w(t)\cos(n\omega t)\,d(\omega t) \tag{5.7b}$$

$$b_n = \frac{1}{\pi}\int_{-\pi}^{\pi} w(t)\sin(n\omega t)\,d(\omega t) \tag{5.7c}$$

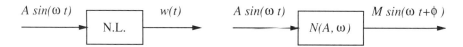

**Figure 5.6 :** A nonlinear element and its describing function representation

Due to the fourth assumption above, one has $a_0 = 0$. Furthermore, the third assumption implies that we only need to consider the fundamental component $w_1(t)$, namely

$$w(t) \approx w_1(t) = a_1 \cos(\omega t) + b_1 \sin(\omega t) = M \sin(\omega t + \phi) \tag{5.8}$$

where

$$M(A, \omega) = \sqrt{a_1^2 + b_1^2} \quad \text{and} \quad \phi(A, \omega) = \arctan(a_1/b_1).$$

Expression (5.8) indicates that the fundamental component corresponding to a sinusoidal input is a sinusoid at the same frequency. In complex representation, this sinusoid can be written as $w_1 = M e^{j(\omega t + \phi)} = (b_1 + j a_1) e^{j \omega t}$.

Similarly to the concept of frequency response function, which is the frequency-domain ratio of the sinusoidal input and the sinusoidal output of a system, we define the *describing function* of the nonlinear element to be *the complex ratio of the fundamental component of the nonlinear element by the input sinusoid*, i.e.,

$$N(A, \omega) = \frac{M e^{j(\omega t + \phi)}}{A e^{j \omega t}} = \frac{M}{A} e^{j \phi} = \frac{1}{A}(b_1 + j a_1) \tag{5.9}$$

With a describing function representing the nonlinear component, the nonlinear element, in the presence of sinusoidal input, can be treated as if it were a linear element with a frequency response function $N(A, \omega)$, as shown in Figure 5.6. The concept of a describing function can thus be regarded as an extension of the notion of frequency response. For a linear dynamic system with frequency response function $H(j\omega)$, the describing function is independent of the input gain, as can be easily shown. However, the describing function of a nonlinear element differs from the frequency response function of a linear element in that it depends on the input amplitude $A$. Therefore, representing the nonlinear element as in Figure 5.6 is also called quasi-linearization.

Generally, the describing function depends on the frequency and amplitude of the input signal. There are, however, a number of special cases. When the nonlinearity is *single-valued*, the describing function $N(A, \omega)$ is *real and independent of the input frequency* $\omega$. The realness of $N$ is due to the fact that $a_1 = 0$, which is true because $f[A\sin(\omega t)]\cos(\omega t)$, the integrand in the expression (5.7b) for $a_1$, is an odd function of $\omega t$, and the domain of integration is the symmetric interval $[-\pi, \pi]$. The frequency-independent nature is due to the fact that the integration of the single-valued function $f[A\sin(\omega t)]\sin(\omega t)$ in expression (5.7c) is done for the variable $\omega t$, which implies that $\omega$ does not explicitly appear in the integration.

Although we have implicitly assumed the nonlinear element to be a scalar nonlinear function, the definition of the describing function also applies to the case when the nonlinear element contains dynamics (*i.e.*, is described by differential equations instead of a function). The derivation of describing functions for such nonlinear elements is usually more complicated and may require experimental evaluation.

## 5.1.5  Computing Describing Functions

A number of methods are available to determine the describing functions of nonlinear elements in control systems, based on definition (5.9). We now briefly describe three such methods: analytical calculation, experimental determination, and numerical integration. Convenience and cost in each particular application determine which method should be used. One thing to remember is that precision is not critical in evaluating describing functions of nonlinear elements, because the describing function method is itself an approximate method.

### ANALYTICAL CALCULATION

When the nonlinear characteristics $w = f(x)$ (where $x$ is the input and $w$ the output) of the nonlinear element are described by an explicit function and the integration in (5.7) can be easily carried out, then analytical evaluation of the describing function based on (5.7) is desirable. The explicit function $f(x)$ of the nonlinear element may be an idealized representation of simple nonlinearities such as saturation and dead-zone, or it may be the curve-fit of an input-output relationship for the element. However, for nonlinear elements which evade convenient analytical expressions or contain dynamics, the analytical technique is difficult.

### NUMERICAL INTEGRATION

For nonlinearities whose input-output relationship $w = f(x)$ is given by graphs or tables, it is convenient to use numerical integration to evaluate the describing functions. The idea is, of course, to approximate integrals in (5.7) by discrete sums over small intervals. Various numerical integration schemes can be applied for this purpose. It is obviously important that the numerical integration be easily implementable by computer programs. The result is a plot representing the describing function, which can be used to predict limit cycles based on the method to be developed in section 5.4.

## EXPERIMENTAL EVALUATION

The experimental method is particularly suitable for complex nonlinearities and dynamic nonlinearities. When a system nonlinearity can be isolated and excited with sinusoidal inputs of known amplitude and frequency, experimental determination of the describing function can be obtained by using a harmonic analyzer on the output of the nonlinear element. This is quite similar to the experimental determination of frequency response functions for linear elements. The difference here is that not only the frequencies, but also the *amplitudes* of the input sinusoidal should be varied. The results of the experiments are a set of curves on complex planes representing the describing function $N(A, \omega)$, instead of analytical expressions. Specialized instruments are available which automatically compute the describing functions of nonlinear elements based on the measurement of nonlinear element response to harmonic excitation.

Let us illustrate on a simple nonlinearity how to evaluate describing functions using the analytical technique.

### Example 5.2: Describing function of a hardening spring

The characteristics of a hardening spring are given by

$$w = x + x^3/2$$

with $x$ being the input and $w$ being the output. Given an input $x(t) = A\sin(\omega t)$, the output $w(t) = A\sin(\omega t) + A^3 \sin^3(\omega t)/2$ can be expanded as a Fourier series, with the fundamental being

$$w_1(t) = a_1 \cos \omega t + b_1 \sin \omega t$$

Because $w(t)$ is an odd function, one has $a_1 = 0$, according to (5.7). The coefficient $b_1$ is

$$b_1 = \frac{1}{\pi}\int_{-\pi}^{\pi} [A\sin(\omega t) + A^3 \sin^3(\omega t)/2] \sin(\omega t)\, d(\omega t) \; = \; A + \frac{3}{8}A^3$$

Therefore, the fundamental is

$$w_1 = (A + \frac{3}{8}A^3)\sin(\omega t)$$

and the describing function of this nonlinear component is

$$N(A,\omega) = N(A) = 1 + \frac{3}{8}A^2$$

Note that due to the odd nature of this nonlinearity, the describing function is real, being a function only of the amplitude of the sinusoidal input.                                                 □

## 5.2  Common Nonlinearities In Control Systems

In this section, we take a closer look at the nonlinearities found in control systems. Consider the typical system block shown in Figure 5.7. It is composed of four parts: a plant to be controlled, sensors for measurement, actuators for control action, and a control law, usually implemented on a computer. Nonlinearities may occur in any part of the system, and thus make it a nonlinear control system.

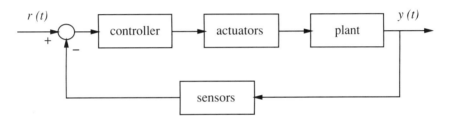

**Figure 5.7 :** Block diagram of control systems

### CONTINUOUS AND DISCONTINUOUS NONLINEARITIES

Nonlinearities can be classified as *continuous* and *discontinuous*.   Because discontinuous nonlinearities cannot be locally approximated by linear functions, they are also called "hard" nonlinearities.   Hard nonlinearities are commonly found in control systems, both in small range operation and large range operation.   Whether a system in small range operation should be regarded as nonlinear or linear depends on the magnitude of the hard nonlinearities and on the extent of their effects on the system performance.

Because of the common occurence of hard nonlinearities, let us briefly discuss the characteristics and effects of some important ones.

### Saturation

When one increases the input to a physical device, the following phenomenon is often observed: when the input is small, its increase leads to a corresponding (often proportional) increase of output; but when the input reaches a certain level, its further increase does produces little or no increase of the output. The output simply stays around its maximum value. The device is said to be in *saturation* when this happens. Simple examples are transistor amplifiers and magnetic amplifiers. A saturation nonlinearity is usually caused by limits on component size, properties of materials, and available power. A typical saturation nonlinearity is represented in Figure 5.8, where the thick line is the real nonlinearity and the thin line is an idealized saturation nonlinearity.

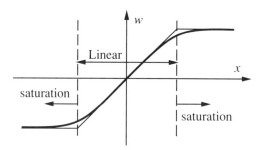

**Figure 5.8 :** A saturation nonlinearity

Most actuators display saturation characteristics. For example, the output torque of a two-phase servo motor cannot increase infinitely and tends to saturate, due to the properties of the magnetic material. Similarly, valve-controlled hydraulic servo motors are saturated by the maximum flow rate.

Saturation can have complicated effects on control system performance. Roughly speaking, the occurence of saturation amounts to reducing the gain of the device (*e.g.*, the amplifier) as the input signals are increased. As a result, if a system is unstable in its linear range, its divergent behavior may be suppressed into a self-sustained oscillation, due to the inhibition created by the saturating component on the system signals. On the other hand, in a linearly stable system, saturation tends to slow down the response of the system, because it reduces the effective gain.

### On-off nonlinearity

An extreme case of saturation is the *on-off* or relay nonlinearity. It occurs when the linearity range is shrunken to zero and the slope in the linearity range becomes vertical. Important examples of on-off nonlinearities include output torques of gas jets for spacecraft control (as in example 2.5) and, of course, electrical relays. On-off nonlinearities have effects similar to those of saturation nonlinearities. Furthermore they can lead to "chattering" in physical systems due to their discontinuous nature.

**Dead-zone**

In many physical devices, the output is zero until the magnitude of the input exceeds a certain value. Such an input-output relation is called a *dead-zone*. Consider for instance a d.c. motor. In an idealistic model, we assume that any voltage applied to the armature windings will cause the armature to rotate, with small voltage causing small motion. In reality, due to the static friction at the motor shaft, rotation will occur only if the torque provided by the motor is sufficiently large. Similarly, when transmitting motion by connected mechanical components, dead zones result from manufacturing clearances. Similar dead-zone phenomena occur in valve-controlled pneumatic actuators and in hydraulic components.

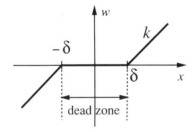

**Figure 5.9 :** A dead-zone nonlinearity

Dead-zones can have a number of possible effects on control systems. Their most common effect is to decrease static output accuracy. They may also lead to limit cycles or system instability because of the lack of response in the dead zone. In some cases, however, they may actually stabilize a system or suppress self-oscillations. For example, if a dead-zone is incorporated into an ideal relay, it may lead to the avoidance of the oscillation at the contact point of the relay, thus eliminating sparks and reducing wear at the contact point. In chapter 8, we describe a dead-zone technique to improve the robustness of adaptive control systems with respect to measurement noise.

**Backlash and hysteresis**

*Backlash* often occurs in transmission systems. It is caused by the small gaps which exist in transmission mechanisms. In gear trains, there always exist small gaps between a pair of mating gears, due to the unavoidable errors in manufacturing and assembly. Figure 5.10 illustrates a typical situation. As a result of the gaps, when the driving gear rotates a smaller angle than the gap $b$, the driven gear does not move at all, which corresponds to the dead-zone (OA segment in Figure 5.10); after contact has been established between the two gears, the driven gear follows the rotation of the driving gear in a linear fashion (AB segment). When the driving gear rotates in the reverse direction by a distance of $2b$, the driven gear again does not move, corresponding to the BC segment in Figure 5.10.  After the contact between the two gears is re-established, the driven gear follows the rotation of the driving gear in the reverse direction (CD segment).  Therefore, if the driving gear is in periodic motion, the driven gear will move in the fashion represented by the closed path EBCD. Note that the height of B, C, D, E in this figure depends on the amplitude of the input sinusoidal.

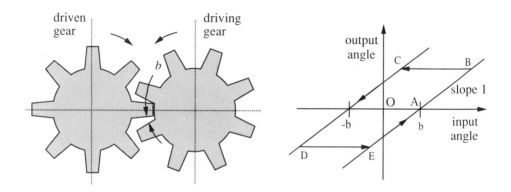

**Figure 5.10 :** A backlash nonlinearity

A critical feature of backlash is its multi-valued nature.  Corresponding to each input, two output values are possible.  Which one of the two occur depends on the history of the input.  We remark that a similar multi-valued nonlinearity is hysteresis, which is frequently observed in relay components.

Multi-valued nonlinearities like backlash and hysteresis usually lead to energy storage in the system.  Energy storage is a frequent cause of instability and self-sustained oscillation.

# 5.3   Describing Functions of Common Nonlinearities

In this section, we shall compute the describing functions for a few common nonlinearities. This will not only allow us to familiarize ourselves with the frequency domain properties of these common nonlinearities, but also will provide further examples of how to derive describing functions for nonlinear elements.

### SATURATION

The input-output relationship for a saturation nonlinearity is plotted in Figure 5.11, with $a$ and $k$ denoting the range and slope of the linearity. Since this nonlinearity is single-valued, we expect the describing function to be a real function of the input

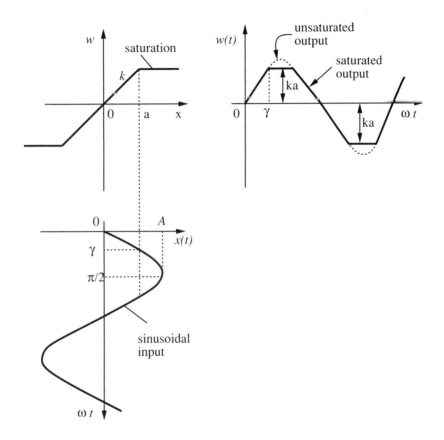

**Figure 5.11 :** Saturation nonlinearity and the corresponding input-output relationship

amplitude.

Consider the input $x(t) = A\sin(\omega t)$. If $A \le a$, then the input remains in the linear range, and therefore, the output is $w(t) = kA\sin(\omega t)$. Hence, the describing function is simply a constant $k$.

Now consider the case $A > a$. The input and the output functions are plotted in Figure 5.11. The output is seen to be symmetric over the four quarters of a period. In the first quarter, it can be expressed as

$$w(t) = \begin{cases} kA\sin(\omega t) & 0 \le \omega t \le \gamma \\ ka & \gamma < \omega t \le \pi/2 \end{cases}$$

where $\gamma = \sin^{-1}(a/A)$. The odd nature of $w(t)$ implies that $a_1 = 0$ and the symmetry over the four quarters of a period implies that

$$b_1 = \frac{4}{\pi} \int_0^{\pi/2} w(t) \sin(\omega t) \, d(\omega t)$$

$$= \frac{4}{\pi} \int_0^{\gamma} kA \sin^2(\omega t) \, d(\omega t) + \frac{4}{\pi} \int_{\gamma}^{\pi/2} ka \sin(\omega t) \, d(\omega t)$$

$$= \frac{2kA}{\pi} [ \, \gamma + \frac{a}{A} \sqrt{1 - \frac{a^2}{A^2}} \, ] \tag{5.10}$$

Therefore, the describing function is

$$N(A) = \frac{b_1}{A} = \frac{2k}{\pi} [ \, \sin^{-1} \frac{a}{A} + \frac{a}{A} \sqrt{1 - \frac{a^2}{A^2}} \, ] \tag{5.11}$$

The normalized describing function ($N(A)/k$) is plotted in Figure 5.12 as a function of $A/a$ . One can observe three features for this describing function:

1. $N(A) = k$ if the input amplitude is in the linearity range

2. $N(A)$ decreases as the input amplitude increases

3. there is no phase shift

The first feature is obvious, because for small signals the saturation is not displayed. The second is intuitively reasonable, since saturation amounts to reduce the ratio of the output to input. The third is also understandable because saturation does not cause the delay of the response to input.

As a special case, one can obtain the describing function for the relay-type (on-off) nonlinearity shown in Figure 5.13. This case corresponds to shrinking the linearity range in the saturation function to zero, *i.e.*, $a \to 0, k \to \infty$, but $ka = M$. Though $b_1$ can be obtained from (5.10) by taking the limit, it is more easily obtained directly as

$$b_1 = \frac{4}{\pi} \int_0^{\pi/2} M \sin(\omega t) \, d(\omega t) = \frac{4}{\pi} M$$

Therefore, the describing function of the relay nonlinearity is

$$N(A) = \frac{4M}{\pi A} \tag{5.12}$$

The normalized describing function ($N/M$) is plotted in Figure 5.13 as a function of

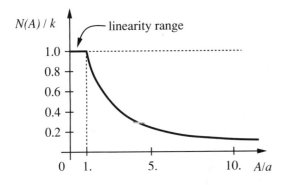

**Figure 5.12 :** Describing function of the saturation nonlinearity

input amplitude. Although the describing function again has no phase shift, the flat segment seen in Figure 5.12 is missing in this plot, due to the completely nonlinear nature of the relay. The asymptic properties of the describing function curve in Figure 5.13 are particularly interesting. When the input is infinitely small, the describing function is infinitely large. When the input is infinitely large, the describing function is infinitely small. One can gain an intuitive understanding of these properties by considering the ratio of the output to input for the on-off nonlinearity.

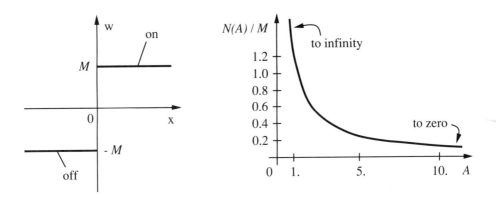

**Figure 5.13 :** Relay nonlinearity and its describing function

## DEAD-ZONE

Consider the dead-zone characteristics shown in Figure 5.9, with the dead-zone width being $2\delta$ and its slope $k$. The response corresponding to a sinusoidal input $x(t) = A\sin(\omega t)$ into a dead-zone of width $2\delta$ and slope $k$, with $A \geq \delta$, is plotted in Figure 5.14. Since the characteristics is an odd function, $a_1 = 0$. The response is also seen to be symmetric over the four quarters of a period. In one quarter of a period, *i.e.*, when $0 \leq \omega t \leq \pi/2$, one has

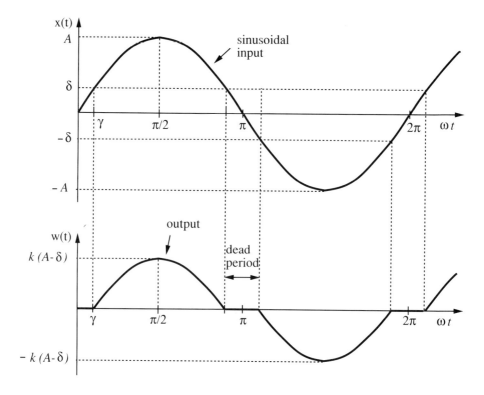

**Figure 5.14 :** Input and output functions for a dead-zone nonlinearity

$$w(t) = \begin{cases} 0 & 0 \leq \omega t \leq \gamma \\ k(A\sin(\omega t) - \delta) & \gamma \leq \omega t \leq \pi/2 \end{cases}$$

where $\gamma = \sin^{-1}(\delta/A)$. The coefficient $b_1$ can be computed as follows

$$b_1 = \frac{4}{\pi} \int_0^{\pi/2} w(t) \sin(\omega t) \, d(\omega t) = \frac{4}{\pi} \int_\gamma^{\pi/2} k(A\sin(\omega t) - \delta) \sin(\omega t) \, d(\omega t)$$

$$= \frac{2kA}{\pi} (\frac{\pi}{2} - \sin^{-1}\frac{\delta}{A} - \frac{\delta}{A} \sqrt{1 - \frac{\delta^2}{A^2}}) \tag{5.13}$$

This leads to

$$N(A) = \frac{2k}{\pi} (\frac{\pi}{2} - \sin^{-1}\frac{\delta}{A} - \frac{\delta}{A} \sqrt{1 - \frac{\delta^2}{A^2}})$$

This describing function $N(A)$ is a *real* function and, therefore, there is no phase shift (reflecting the absence of time-delay). The normalized describing function is plotted in Figure 5.15. It is seen that $N(A)/k$ is zero when $A/\delta < 1$, and increases up to 1 with $A/\delta$. This increase indicates that the effect of the dead-zone gradually diminishes as the amplitude of the input signal is increased, consistently with intuition.

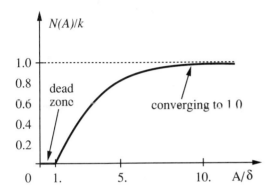

**Figure 5.15 :** Describing function of the dead-zone nonlinearity

## BACKLASH

The evaluation of the describing functions for backlash nonlinearity is more tedious. Figure 5.16 shows a backlash nonlinearity, with slope $k$ and width $2b$. If the input amplitude is smaller than $b$, there is no output. In the following, let us consider the input being $x(t) = A\sin(\omega t)$, $A \geq b$. The output $w(t)$ of the nonlinearity is as shown in the figure. In one cycle, the function $w(t)$ can be represented as

$$w(t) = (A - b)k \qquad\qquad \frac{\pi}{2} < \omega t \leq \pi - \gamma$$

$$w(t) = (A\sin(\omega t) + b)k \qquad \pi - \gamma < \omega t \leq \frac{3\pi}{2}$$

$$w(t) = -(A - b)k \qquad\qquad \frac{3\pi}{2} < \omega t \leq 2\pi - \gamma$$

$$w(t) = (A\sin(\omega t) - b)k \qquad 2\pi - \gamma < \omega t \leq \frac{5\pi}{2}$$

where $\gamma = \sin^{-1}(1 - 2b/A)$.

Unlike the previous nonlinearities, the function $w(t)$ here is neither odd nor even. Therefore, $a_1$ and $b_1$ are both nonzero. Using (5.7b) and (5.7c), we find through some tedious integrations that

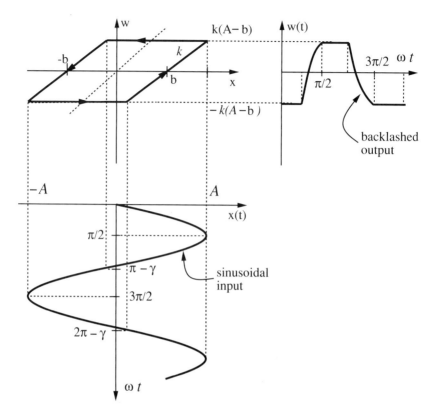

**Figure 5.16 :** Input and output functions for a backlash nonlinearity

$$a_1 = \frac{4kb}{\pi} (\frac{b}{A} - 1)$$

$$b_1 = \frac{Ak}{\pi} [\frac{\pi}{2} - \sin^{-1}(\frac{2b}{A} - 1) - (\frac{2b}{A} - 1) \sqrt{1 - (\frac{2b}{A} - 1)^2}$$

Therefore, the describing function of the backlash is given by

$$|N(A)| = \frac{1}{A} \sqrt{a_1^2 + b_1^2} \tag{5.14a}$$

$$\underline{/N(A)} = \tan^{-1}(a_1/b_1) \tag{5.14b}$$

The amplitude of the describing function for backlash is plotted in Figure 5.17.

We note a few interesting points :

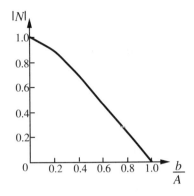

**Figure 5.17 :** Amplitude of describing function for backlash

1. $|N(A)| = 0$    *if* $A = b$.

2. $|N(A)|$  increases, when $b/A$ decreases.

3. $|N(A)| \rightarrow 1$    *as* $b/A \rightarrow 0$.

The phase angle of the describing function is plotted in Figure 5.18. Note that a phase lag (up to 90°) is introduced, unlike the previous nonlinearities. This phase lag is the reflection of the time delay of the backlash, which is due to the gap *b*. Of course, a larger *b* leads to a larger phase lag, which may create stability problems in feedback control systems.

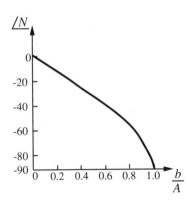

**Figure 5.18 :** Phase angle of describing function for backlash (degree)

## 5.4  Describing Function Analysis of Nonlinear Systems

For a nonlinear system containing a nonlinear element, we now know how to obtain a decribing function for the nonlinear element. The next step is to formalize the procedure in subsection 5.1.1 for the prediction of limit cycles, based on the describing function representation of the nonlinearity. The basic approach to achieve this is to apply an extended version of the famous Nyquist criterion in linear control to the equivalent system. Let us begin with a short review of the Nyquist criterion and its extension.

## 5.4.1 The Nyquist Criterion and Its Extension

Consider the linear system of Figure 5.19. The characteristic equation of this system is

$$\delta(p) = 1 + G(p)\,H(p) = 0$$

Note that $\delta(p)$, often called the *loop transfer function*, is a rational function of $p$, with its zeros being the poles of the closed-loop system, and its poles being the poles of the open-loop transfer function $G(p)\,H(p)$. Let us rewrite the characteristic equation as

$$G(p)\,H(p) = -1$$

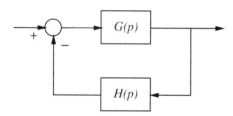

**Figure 5.19 :** Closed-loop linear system

Based on this equation, the famous Nyquist criterion can be derived straightforwardly from the Cauchy theorem in complex analysis. The criterion can be summarized (assuming that $G(p)\,H(p)$ has no poles or zeros on the $j\omega$ axis) in the following procedure (Figure 5.20):

1. draw, in the $p$ plane, a so-called Nyquist path enclosing the right-half plane

2. map this path into another complex plane through $G(p)H(p)$

3. determine $N$, the number of clockwise encirclements of the plot of $G(p)H(p)$ around the point $(-1,0)$

4. compute $Z$, the number of zeros of the loop transfer function $\delta(p)$ in the right-half $p$ plane, by

$$Z = N + P \quad, \text{ where } P \text{ is the number of unstable poles of } \delta(p)$$

Then the value of $Z$ is the number of unstable poles of the closed-loop system.

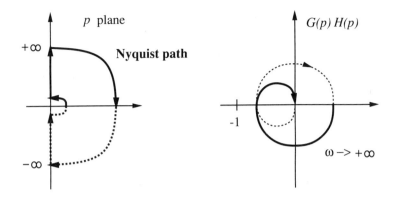

**Figure 5.20** : The Nyquist criterion

A simple formal extension of the Nyquist criterion can be made to the case when a constant gain $K$ (possibly a complex number) is included in the forward path in Figure 5.21. This modification will be useful in interpreting the stability analysis of limit cycles using the describing function method. The loop transfer function becomes

$$\delta(p) = 1 + K\,G(p)H(p)$$

with the corresponding characteristic equation

$$G(p)\,H(p) = -\,1/K$$

The same arguments as used in the derivation of Nyquist criterion suggest the same procedure for determining unstable closed-loop poles, with the minor difference that now *Z represents the number of clockwise encirclements of the $G(p)\,H(p)$ plot <u>around the point $-1/K$</u>*. Figure 5.21 shows the corresponding extended Nyquist plot.

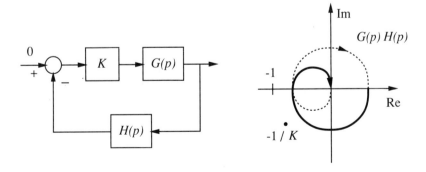

**Figure 5.21** : Extension of the Nyquist criterion

## 5.4.2 Existence of Limit Cycles

Let us now assume that there exists a self-sustained oscillation of amplitude $A$ and frequency $\omega$ in the system of Figure 5.22. Then the variables in the loop must satisfy the following relations:

$$x = -y$$

$$w = N(A,\omega)x$$

$$y = G(j\omega)w$$

Therefore, we have $y = G(j\omega)N(A,\omega)(-y)$. Because $y \neq 0$, this implies

$$G(j\omega)\,N(A,\omega) + 1 = 0 \tag{5.15}$$

which can be written as

$$G(j\omega) = -\frac{1}{N(A,\omega)} \tag{5.16}$$

Therefore, the amplitude $A$ and frequency $\omega$ of the limit cycles in the system must satisfy (5.16). If the above equation has no solutions, then the nonlinear system has no limit cycles.

Expression (5.16) represents two nonlinear equations (the real part and imaginary part each give one equation) in the two variables $A$ and $\omega$. There are usually a finite number of solutions. It is generally very difficult to solve these equations by analytical methods, particularly for high-order systems, and therefore, a graphical approach is usually taken. The idea is to plot both sides of (5.16) in the complex plane and find the intersection points of the two curves.

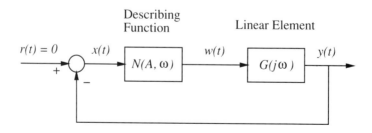

**Figure 5.22 :** A nonlinear system

## FREQUENCY-INDEPENDENT DESCRIBING FUNCTION

First, we consider the simpler case when the describing function $N$ being a function of the gain $A$ only, *i.e.*, $N(A, \omega) = N(A)$. This includes all single-valued nonlinearities and some important double-valued nonlinearities such as backlash. The equality becomes

$$G(j\omega) = -\frac{1}{N(A)} \tag{5.17}$$

We can plot both the frequency response function $G(j\omega)$ (varying $\omega$) and the negative inverse describing function ($1/N(A)$) (varying $A$) in the complex plane, as in Figure 5.23. If the two curves intersect, then there exist limit cycles, and the values of $A$ and $\omega$ corresponding to the intersection point are the solutions of Equation (5.17). If the curves intersect $n$ times, then the system has $n$ possible limit cycles. Which one is actually reached depends on the initial conditions. In Figure 5.23, the two curves intersect at one point $K$. This indicates that there is one limit cycle in the system. The amplitude of the limit cycle is $A_k$, the value of $A$ corresponding to the point $K$ on the $-1/N(A)$ curve. The frequency of the limit cycle is $\omega_k$, the value of $\omega$ corresponding to the point $K$ on the $G(j\omega)$ curve.

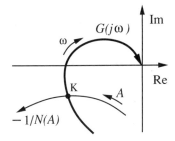

**Figure 5.23** : Detection of limit cycles

Note that for single-valued nonlinearities, $N$ is real and therefore the plot of $-1/N$ always lies on the real axis. It is also useful to point out that, as we shall discuss later, the above procedure only gives a *prediction* of the existence of limit cycles. The validity and accuracy of this prediction should be confirmed by computer simulations.

We already saw in section 5.1.1 an example of the prediction of limit cycles, for the Van der Pol equation.

## FREQUENCY-DEPENDENT DESCRIBING FUNCTION

For the general case, where the describing function depends on both input amplitude and frequency ( $N = N(A, \omega)$ ), the method can be applied, but with more complexity. Now the right-hand side of (5.15), $- 1/N(A, \omega)$ , corresponds to a family of curves on the complex plane with A as the running parameter and $\omega$ fixed for each curve, as shown in Figure 5.24. There are generally an *infinite* number of intersection points between the $G(j\omega)$ curve and the $- 1/N(A, \omega)$ curves. Only the intersection points with matched $\omega$ indicate limit cycles.

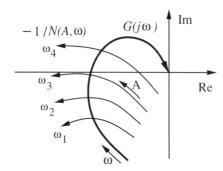

**Figure 5.24 :** Limit cycle detection for frequency-dependent describing functions

To avoid the complexity of matching frequencies at intersection points, it may be advantageous to consider the graphical solution of (5.16) directly, based on the plots of $G(j\omega)N(A, \omega)$. With *A* fixed and $\omega$ varying from 0 to $\infty$, we obtain a curve representing $G(j\omega)N(A,\omega)$. Different values of *A* correspond to a family of curves, as shown in Figure 5.25. A curve passing *through the point* $(- 1,0)$ in the complex plane indicates the existence of a limit cycle, with the value of *A* for the curve being the amplitude of the limit cycle, and the value of $\omega$ at the point $(- 1,0)$ being the frequency of the limit cycle. While this technique is much more straightforward than the previous one, it requires repetitive computation of the $G(j\omega)$ in generating the family of curves, which may be handled easily by computer.

## 5.4.3 Stability of Limit Cycles

As pointed out in chapter 2, limit cycles can be stable or unstable. In the above, we have discussed how to detect the existence of limit cycles. Let us now discuss how to determine the stability of a limit cycle, based on the extended Nyquist criterion in section 5.4.1.

Consider the plots of frequency response and inverse describing function in Figure 5.26. There are two intersection points in the figure, predicting that the system

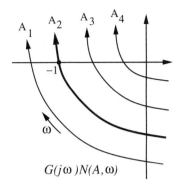

**Figure 5.25** : Solving equation (5.15) graphically

has two limit cycles. Note that the value of $A$ corresponding to point $L_1$ is smaller than the value of $A$ corresponding to $L_2$. For simplicity of discussion, we assume that the linear transfer function $G(p)$ has no unstable poles.

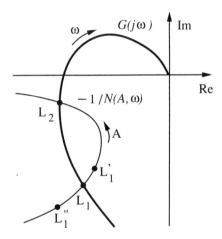

**Figure 5.26** : Limit Cycle Stability

Let us first discuss the stability of the limit cycle at point $L_1$. Assume that the system initially operates at point $L_1$, with the limit cycle amplitude being $A_1$, and its frequency being $\omega_1$. Due to a slight disturbance, the amplitude of the input to the nonlinear element is slightly increased, and the system operating point is moved from $L_1$ to $L_1'$. Since the new point $L_1'$ is encircled by the curve of $G(j\omega)$, according to the extended Nyquist criterion mentioned in section 5.4.1, the system at this operating point is unstable, and the amplitudes of the system signals will increase. Therefore, the operating point will continue to move along the curve $-1/N(A)$ toward the other limit cycle point $L_2$. On the other hand, if the system is disturbed so that the amplitude $A$ is decreased, with the operating point moved to the point $L_1''$, then $A$ will continue to decrease and the operating point moving away from $L_1$ in the other direction. This is because $L_1^{''}$ is not encircled by the curve $G(j\omega)$ and thus the extended Nyquist plot asserts the stability of the system. The above discussion indicates that a slight disturbance can destroy the oscillation at point $L_1$ and, therefore, that this limit cycle is unstable. A similar analysis for the limit cycle at point $L_2$ indicates that that limit cycle is stable.

Summarizing the above discussion and the result in the previous subsection, we obtain a criterion for existence and stability of limit cycles:

**Limit Cycle Criterion**: *Each intersection point of the curve $G(j\omega)$ and the curve $-1/N(A)$ corresponds to a limit cycle. If points near the intersection and along the increasing-A side of the curve $-1/N(A)$ are not encircled by the curve $G(j\omega)$, then the corresponding limit cycle is stable. Otherwise, the limit cycle is unstable.*

## 5.4.4  Reliability of Describing Function Analysis

Empirical evidence over the last three decades, and later theoretical justification, indicate that the describing function method can effectively solve a large number of practical control problems involving limit cycles. However, due to the approximate nature of the technique, it is not surprising that the analysis results are sometimes not very accurate. Three kinds of inaccuracies are possible:

1. The amplitude and frequency of the predicted limit cycle are not accurate

2. A predicted limit cycle does not actually exist

3. An existing limit cycle is not predicted

The first kind of inaccuracy is quite common. Generally, the predicted amplitude and frequency of a limit cycle always deviate somewhat from the true values. How much the predicted values differ from the true values depends on how well the nonlinear system satisfies the assumptions of the describing function method. In order to obtain accurate values of the predicted limit cycles, simulation of the nonlinear system is necessary.

The occurrence of the other two kinds of inaccuracy is less frequent but has more serious consequences. Usually, their occurrence can be detected by examining the linear element frequency response and the relative positions of the $G$ plot and $-1/N$ plot.

*Violation of filtering hypothesis*: The validity of the describing function method relies on the filtering hypothesis defined by (5.5). For some linear elements, this hypothesis is not satisfied and errors may result in the describing function analysis. Indeed, a number of failed cases of describing function analysis occur in systems whose linear element has resonant peaks in its frequency response $G(j\omega)$.

*Graphical Conditions*: If the $G(j\omega)$ locus is tangent or almost tangent to the $-1/N$ locus, then the conclusions from a describing function analysis might be erroneous. Such an example is shown in Figure 5.27(a). This is because the effects of neglected higher harmonics or system model uncertainty may cause the change of the intersection situations, particularly when filtering in the linear element is weak. As a result, the second and third types of errors listed above may occur. A classic case of this problem involves a second-order servo with backlash studied by Nychols. While describing function analysis predicts two limit cycles (a stable one at high frequency and an unstable one at low frequency), it can be shown that the low-frequency unstable limit cycle does not exist.

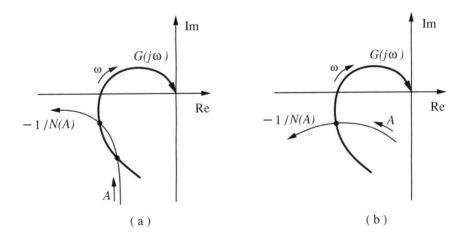

**Figure 5.27** : Reliability of limit cycle prediction

Conversely, if the $-1/N$ locus intersects the $G$ locus almost perpendicularly, then the results of the describing function are usually good. An example of this situation is shown in Figure 5.27(b).

## 5.5  Summary

The describing function method is an extension of the frequency response method of linear control. It can be used to *approximately* analyze and predict the behavior of important classes of nonlinear systems, including systems with hard nonlinearities. The desirable properties it inherits from the frequency response method, such as its graphical nature and the physically intuitive insights it can provide, make it an important tool for practicing engineers. Applications of the describing function method to the prediction of limit cycles were detailed. Other applications, such as predicting subharmonics, jump phenomena, and responses to external sinusoidal inputs, can be found in the literature.

## 5.6  Notes and References

An extensive and clear presentation of the describing function method can be found in [Gelb and VanderVelde, 1968]. A more recent treatment is contained in [Hedrick, *et al.*, 1982], which also discusses specific applications to nonlinear physical systems. The describing function method was developed and successfully used well before its mathematical justification was completely formalized [Bergen and Franks, 1971]. Figures 5.14 and 5.16 are adapted from [Shinners, 1978]. The Van der Pol oscillator example is adapted from [Hsu and Meyer, 1968].

## 5.7  Exercises

**5.1**   Determine whether the system in Figure 5.28 exhibits a self-sustained oscillation (a limit cycle). If so, determine the stability, frequency, and amplitude of the oscillation.

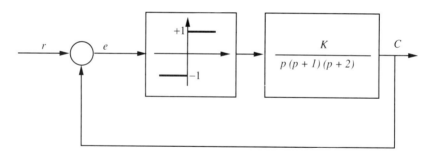

**Figure 5.28 :** A nonlinear system containing a relay

**5.2**   Determine whether the system in Figure 5.29 exhibits a self-sustained oscillation. If so, determine the stability, frequency, and amplitude of the oscillation.

**5.3**   Consider the nonlinear system of Figure 5.30. Determine the largest $K$ which preserves the stability of the system. If $K = 2 K_{max}$, find the amplitude and frequency of the self-sustained oscillation.

**5.4**   Consider the system of Figure 5.31, which is composed of a high-pass filter, a saturation function, and the inverse low-pass filter. Show that the system can be viewed as a nonlinear low-

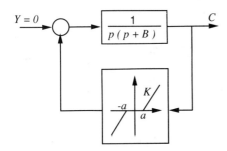

Y = 0

C

**Figure  5.29  :  A  nonlinear  system
containing a dead-zone**

pass filter, which attenuates high-frequency inputs *without introducing a phase lag.*

**5.5**    This exercise is based on a result of [Tsypkin, 1956].

Consider a nonlinear system whose output $w(t)$ is related to the input $u(t)$ by an odd function, of the form

$$w(t) \; = \; F(u(t)) \; = \; - F( - u(t) ) \tag{5.18}$$

Derive the following very simple approximate formula for the describing function $N(A)$

$$N(A) \; \approx \; \frac{2}{3A} \, [ \, F(A) + F(A/2) \, ]$$

To this effect, you may want to use the fact that

$$\frac{1}{\pi} \int_{-1}^{1} \frac{f(x)}{\sqrt{1 - x^2}} \, dx \; = \; \frac{1}{6} \, [ \, f(1) + f(-1) + 2f(1/2) + 2f(-1/2) \, ] + R$$

where  the  remainder  $R$  verifies  $R = f^6(\xi)/(2^5 6!)$  for  some  $\xi \in ( - 1, 1 )$.   Show that approximation (5.18) is quite precise (how precise?).

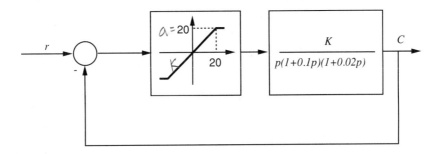

**Figure 5.30 :** A nonlinear system containing a saturation

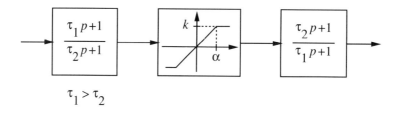

**Figure 5.31 :** A nonlinear low-pass filter

Invert (5.18) so as to obtain for the input-output relation a solution of the form

$$F(A) \approx \sum_{k=0}^{\infty} (-1)^k \frac{3A}{2^{k+1}} N(\frac{A}{2^k})$$

**5.6**   In this exercise, adapted form [Phillips and Harbor, 1988], let us consider the system of Figure 5.32, which is typical of the dynamics of electronic oscillators used in laboratories, with

$$G(p) = \frac{-5p}{p^2 + p + 25}$$

Use describing function analysis to predict whether the system exhibits a limit cycle, depending on the value of the saturation level $k$. In such cases, determine the limit cycle's frequency and amplitude.

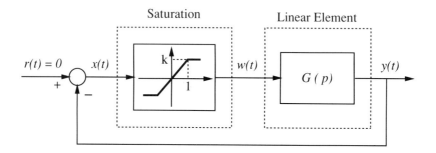

**Figure 5.32 :** Dynamics of an electronic oscillator

Interpret intuitively, by assuming that the system is started at some small initial state, and noticing that $y(t)$ can stay neither at small values (because of instability) nor at saturation values (by applying the final value theorem of linear control).

# Part II
# Nonlinear Control Systems Design

In Part I, we studied how to analyze the behavior of a nonlinear control system, assuming that the control system had been designed. Part II is devoted to the problem of designing nonlinear control systems. In this introduction, we discuss some general issues involved in nonlinear control system design, particularly emphasizing the differences of nonlinear control design problems from linear ones. In the following chapters, we will detail the specific control methods available to the designer.

As pointed out in chapter 1, the objective of control design can be stated as follows: *given a physical system to be controlled and the specifications of its desired behavior, construct a feedback control law to make the closed-loop system display the desired behavior*. In accordance with this design objective, we consider a number of key issues. First, two basic types of nonlinear control problems, nonlinear regulation and nonlinear tracking, are defined. Next, the specifications of the desired behavior of nonlinear control systems are discussed. Basic issues in constructing nonlinear controllers are then outlined. Finally, the major methods available for designing nonlinear controllers are briefly surveyed.

# II.1 Nonlinear Control Problems

If the tasks of a control system involve large range and/or high speed motions, nonlinear effects will be significant in the dynamics and nonlinear control may be necessary to achieve the desired performance. Generally, the tasks of control systems can be divided into two categories: stabilization (or regulation) and tracking (or servo). In stabilization problems, a control system, called a *stabilizer* (or a regulator), is to be designed so that the state of the closed-loop system will be stabilized around an equilibrium point. Examples of stabilization tasks are temperature control of refrigerators, altitude control of aircraft and position control of robot arms. In tracking control problems, the design objective is to construct a controller, called a *tracker*, so that the system output tracks a given time-varying trajectory. Problems such as making an aircraft fly along a specified path or making a robot hand draw straight lines or circles are typical tracking control tasks.

## STABILIZATION PROBLEMS

In order to facilitate the analytic study of stabilization and tracking design in the later chapters, let us provide some formal definitions of stabilization and tracking problems.

**Asymptotic Stabilization Problem**: *Given a nonlinear dynamic system described by*

$$\dot{\mathbf{x}} = \mathbf{f}(\mathbf{x}, \mathbf{u}, t)$$

*find a control law* $\mathbf{u}$ *such that, starting from anywhere in a region in* $\Omega$, *the state* $\mathbf{x}$ *tends to* $\mathbf{0}$ *as* $t \to \infty$ .

If the control law depends on the measurement signals directly, it is said to be a *static control law*. If it depends on the measurement through a differential equation, the control law is said to be a *dynamic control law*, *i.e.*, there is dynamics in the control law. For example, in linear control, a proportional controller is a static controller, while a lead-lag controller is a dynamic controller.

Note that, in the above definition, we allow the size of the region $\Omega$ to be large; otherwise, the stabilization problem may be adequately solved using linear control. Note also that if the objective of the control task is to drive the state to some non-zero set-point $\mathbf{x}_d$, we can simply transform the problem into a zero-point regulation problem by taking $\mathbf{x} - \mathbf{x}_d$ as the state.

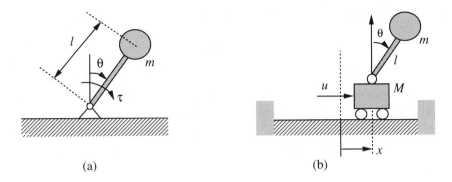

(a)                                    (b)

**Figure II.1 :** (a) a pendulum; (b) an inverted pendulum, with cart

**Example II.1: Stabilization of a pendulum**

Consider the pendulum in Figure II.1(a). Its dynamics is

$$J\ddot{\theta} - mgl\sin\theta = \tau \qquad\qquad\qquad (II.1)$$

Assume that our task is to bring the pendulum from a large initial angle, say $\theta(0) = 60^o$, to the *vertical-up* position. One choice of the stabilizer is

$$\tau = -k_d\dot{\theta} - k_p\theta - mgl\sin\theta \qquad\qquad (II.2)$$

with $k_d$ and $k_p$ denoting positive constants, This leads to the following globally stable closed-loop dynamics

$$J\ddot{\theta} + k_d\dot{\theta} + k_p\theta = 0$$

*i.e.*, the controlled pendulum behaves as a stable mass-spring-damper system. Note that the controller (II.2) is composed of a P.D. (proportional plus derivative) feedback part for stability and a feedforward part for gravity compensation. Another interesting controller is

$$\tau = -k_d\dot{\theta} - 2mgl\sin\theta \qquad\qquad\qquad (II.3)$$

which leads to the stable closed-loop dynamics

$$J\ddot{\theta} + k_d\dot{\theta} + mgl\sin\theta = 0$$

This amounts to artificially reverting the gravity field and adding viscous damping.

This example illustrates the point that feedback and feedforward control actions amount to modifying the dynamics of the plant into a desirable form.   □

However, many nonlinear stabilization problems are not so easy to solve. One such example is the inverted pendulum shown in Figure II.1(b) which can be easily shown to have the following dynamics

$$(M + m)\ddot{x} + ml\cos\theta \; \ddot{\theta} - ml\sin\theta \; \dot{\theta}^2 \; = \; u \tag{II.4a}$$

$$m\ddot{x}\cos\theta + ml \; \ddot{\theta} - ml\dot{x}\dot{q}\sin\theta + mg\sin\theta = 0 \tag{II.4b}$$

(where the mass of the cart is not assumed to be negligible). A particularly interesting task is to design a controller to bring the inverted pendulum from a vertical-down position at the middle of the lateral track to a vertical-up position at the same lateral point. This seeming simple nonlinear control problem is surprisingly difficult to solve in a systematic fashion (see Exercise II.5). This problem arises because there are two degrees of freedom and only one input.

## TRACKING PROBLEMS

The task of asymptotic tracking can be defined similarly.

**Asymptotic Tracking Problem**: *Given a nonlinear dynamics system described by*

$$\dot{\mathbf{x}} = \mathbf{f}(\mathbf{x}, \mathbf{u}, t)$$

$$\mathbf{y} = \mathbf{h}(\mathbf{x})$$

*and a desired output trajectory* $\mathbf{y}_d$, *find a control law for the input* $\mathbf{u}$ *such that, starting from any initial state in a region* $\mathbf{\Omega}$, *the tracking errors* $\mathbf{y}(t) - \mathbf{y}_d(t)$ *go to zero, while the whole state* $\mathbf{x}$ *remains bounded.*

Note that, from a practical point of view, one may require that $\mathbf{x}$ actually remain "reasonably" bounded, and, in particular, within the range of validity of the system model. This may be verified either analytically, or in simulations.

When the closed-loop system is such that proper initial states imply zero tracking error for all the time,

$$\mathbf{y}(t) \; \equiv \; \mathbf{y}_d(t) \qquad \forall \, t \geq 0$$

the control system is said to be capable of *perfect tracking*. Asymptotic tracking implies that perfect tracking is asymptotically achieved. Exponential tracking convergence can be defined similarly.

Throughout the rest of the book, unless otherwise specified, we shall make the mild assumption that the desired trajectory $y_d$ and its derivatives up to a sufficiently high order (generally equal to the system's order) are continuous and bounded. We also assume that $y_d(t)$ *and its derivatives* are available for on-line control computation. This latter assumption is satisfied by control tasks where the desired output $y_d(t)$ is planned ahead of time. For example, in robot tracking tasks, the desired position history is generally planned ahead of time and its derivatives can be easily obtained.

Actually, smooth time-histories are often generated themselves through a filtering process, thereby automatically providing higher derivatives of the desired output. In some tracking tasks, however, the assumption is not satisfied, and a so-called reference model may be used to provide the required derivative signals. For example, in designing a tracking control system for the antenna of a radar so that it will closely point toward an aircraft at all times, we only have the position of the aircraft $y_a(t)$ available at a given time instant (assuming that it is too noisy to be numerically differentiated). However, generally the tracking control law will also use the derivatives of the signals to be tracked. To solve this problem, we can generate the desired position, velocity and acceleration to be tracked by the antenna using the following second-order dynamics

$$\ddot{y}_d + k_1\dot{y}_d + k_2 y_d = k_2 y_a(t) \tag{II.5}$$

where $k_1$ and $k_2$ are chosen positive constants. Thus the problem of following the aircraft is translated into the problem of tracking the output $y_d(t)$ of the reference model. Note that the reference model serves the dual purpose of providing the desired output of the tracking system in response to the aircraft position measurements, and generating the derivatives of the desired output for tracker design. Of course, for the approach to be effective, the filtering process described by (II.5) should be fast enough for $y_d(t)$ to closely approximate $y_a(t)$.

For non-minimum phase systems (precise definitions of nonlinear non-minimum phase systems will be provided in chapter 6), perfect tracking and asymptotic tracking cannot be achieved, as seen in the following example.

**Example II.2: Tracking control of a non-minimum phase linear system**

Consider the linear system

$$\ddot{y} + 2\dot{y} + 2y = -\dot{u} + u$$

The system is non-minimum phase because it has a zero at $p = 1$. Assume that perfect tracking is achieved, *i.e.*, that $y(t) \equiv y_d(t)$, $\forall\, t \geq 0$. Then, the input $u$ satisfies

$$\dot{u} - u = -(\ddot{y}_d + 2\dot{y}_d + 2y_d)$$

Since this represents an unstable dynamics, $u$ diverges exponentially. Note that the above dynamics has a pole which exactly coincides with the unstable zero of the original system, *i.e.*, perfect tracking for non-minimum phase systems can be achieved only by infinite control inputs. By writing $u$ as

$$u = -\frac{p^2 + 2p + 2}{p - 1}\, y_d$$

we see that the perfect-tracking controller is actually inverting the plant dynamics. □

The inability of perfect tracking for a non-minimum phase linear system has its roots in its inherent tendency of "undershooting" in its step response. Thus, the control design objective for non-minimum phase systems should not be perfect tracking or asymptotic tracking. Instead, we should be satisfied with, for example, bounded-error tracking, with small tracking error being achieved for desired trajectories of particular interest.

## RELATIONS BETWEEN STABILIZATION AND TRACKING PROBLEMS

Normally, tracking problems are more difficult to solve than stabilization problems, because in tracking problems the controller should not only keep the whole state stabilized but also drive the system output toward the desired output. However, from a theoretical point of view, tracking design and stabilization design are often related. For instance, if we are to design a tracker for the plant

$$\ddot{y} + f(\dot{y}, y, u) = 0$$

so that $e(t) = y(t) - y_d(t)$ goes to zero, the problem is equivalent to the asymptotic stabilization of the system

$$\ddot{e} + f(\dot{e}, e, u, y_d, \dot{y}_d, \ddot{y}_d) = 0 \tag{II.6}$$

whose state components are $e$ and $\dot{e}$. Clearly, the tracker design problem is solved if we know how to design a stabilizer for the non-autonomous dynamics (II.6).

On the other hand, stabilization problems can often be regarded as a special case of tracking problems, with the desired trajectory being a constant. In model reference control, for instance, a set-point regulation problem is transformed into a tracking problem by incorporating a reference model to filter the supplied set-point value and generate a time-varying output as the ideal response for the tracking control system.

# II.2  Specifying the Desired Behavior

In linear control, the desired behavior of a control system can be *systematically* specified, either in the time-domain (in terms of rise time, overshoot and settling time corresponding to a step command) or in the frequency domain (in terms of regions in which the loop transfer function must lie at low frequencies and at high frequencies). In linear control design, one first lays down the quantitative specifications of the closed-loop control system, and then synthesizes a controller which meets these specifications. However, systematic specification for nonlinear systems (except those equivalent to linear systems) is much less obvious because the response of a nonlinear system to one command does not reflect its response to another command, and furthermore a frequency-domain description is not possible.

As a result, for nonlinear systems, one often looks instead for some *qualitative* specifications of the desired behavior in the operating region of interest. Computer simulation is an important complement to analytical tools in determining whether such specifications are met. Regarding the desired behavior of nonlinear control systems, a designer can consider the following characteristics:

**Stability** must be guaranteed for the nominal model (the model used for design), either in a local sense or in a global sense. The region of stability and convergence are also of interest.

**Accuracy and speed of response** may be considered for some "typical" motion trajectories in the region of operation. For some classes of systems, appropriate controller design can actually guarantee consistent tracking accuracy independently of the desired trajectory, as discussed in chapter 7.

**Robustness** is the sensitivity to effects which are not considered in the design, such as disturbances, measurement noise, unmodeled dynamics, *etc*. The system should be able to withstand these neglected effects when performing the tasks of interest.

**Cost** of a control system is determined mainly by the number and type of actuators, sensors, and computers necessary to implement it. The actuators, sensors and the controller complexity (affecting computing requirement) should be chosen consistently and suit the particular application.

A couple of remarks can be made at this point. First, stability does not imply the ability to withstand persistent disturbances of even small magnitude (as discussed in section 4.9.2). The reason is that stability of a nonlinear system is defined with respect to initial conditions, and only temporary disturbances may be translated as initial conditions. For example, a stable control system may guarantee an aircraft's ability to withstand gusts, while being inept at handling windshears even of small magnitude. This situation is different from that of linear control, where stability always implies ability to withstand bounded disturbances (assuming that the system does stay in its linear range). The effects of persistent disturbance on nonlinear system behavior are addressed by the concept of robustness. Secondly, the above qualities conflict to some extent, and a good control system can be obtained only based on effective trade-offs in terms of stability/robustness, stability/performance, cost/performance, and so on.

# II.3   Some Issues In Constructing Nonlinear Controllers

We now briefly describe a few aspects of controller design.

## A Procedure for Control Design

Given a physical system to be controlled, one typically goes through the following standard procedure, possibly with a few iterations:

1. specify the desired behavior, and select actuators and sensors;
2. model the physical plant by a set of differential equations;
3. design a control law for the system;
4. analyze and simulate the resulting control system;
5. implement the control system in hardware.

Experience, creativity, and engineering judgment are all important in this process. One should consider integrating control design with plant design if possible, similarly to the many advances in aircraft design which have been achieved using linear control techniques. Sometimes the addition or relocation of actuators and sensors may make an otherwise intractable nonlinear control problem easy.

## Modeling Nonlinear Systems

Modeling is basically the process of constructing a mathematical description (usually a set of differential equations) for the physical system to be controlled. Two points can be made about modeling. First, one should use good understanding of the system dynamics and the control tasks to obtain a tractable yet accurate model for control design. Note that more accurate models are not always better, because they may require unnecessarily complex control design and analysis and more demanding computation. The key here is to keep "essential" effects and discard insignificant effects in the system dynamics in the operating range of interest. Second, modeling is more than obtaining a nominal model for the physical system: it should also provide some characterization of the model uncertainties, which may be used for robust design, adaptive design, or merely simulation.

Model uncertainties are the differences between the model and the real physical system. Uncertainties in parameters are called parametric uncertainties while the others are called non-parametric uncertainties. For example, for the model of a controlled mass

$$m\ddot{x} = u$$

the uncertainty in *m* is parametric uncertainty, while the neglected motor dynamics, measurement noise, sensor dynamics are non-parametric uncertainties. Parametric uncertainties are often easy to characterize. For example, *m* may be known to lie somewhere between 2 kg and 5 kg. The characterization of unmodeled dynamics for nonlinear systems is often more difficult, unlike the linear control case where frequency-domain characterizations can be systematically applied.

## Feedback and Feedforward

In nonlinear control, the concept of feedback plays a fundamental role in controller design, as it does in linear control. However, the importance of feedforward is much more conspicuous than in linear control. Feedforward is used to cancel the effects of known disturbances and provide anticipative actions in tracking tasks. Very often it is impossible to control a nonlinear system stably without incorporating feedforward action in the control law. Note that a model of the plant is always required for feedforward compensation (although the model need not be very accurate).

Asymptotic tracking control always requires feedforward actions to provide the forces necessary to make the required motion. It is interesting to note that many tracking controllers can be written in the form

$$\mathbf{u} = \text{feedforward} + \text{feedback}$$

or in a similar form. The feedforward part intends to provide the necessary input for following the specified motion trajectory and canceling the effects of the known disturbances. The feedback part then stabilizes the tracking error dynamics.

As an illustration of the use of feedforward, let us consider tracking controller design in the familiar context of linear systems (as applicable to devices such as an x-y plotter, for instance). The discussion is interesting in its own right, since tracking of time-varying trajectories is not commonly emphasized in linear control texts.

### Example II.3: Tracking control of linear systems

Consider a linear (controllable and observable) minimum-phase system in the form

$$A(p)y = B(p)u \tag{II.7}$$

where

$$A(p) = a_o + a_1 p + \ldots + a_{n-1} p^{n-1} + p^n$$

$$B(p) = b_o + b_1 p + \ldots + b_m p^m$$

The control objective is to make the output $y(t)$ follow a time-varying desired trajectory $y_d(t)$. We assume that only the output $y(t)$ is measured, and that $y_d, \dot{y}_d, ..., y_d^{(r)}$ are known, with $r$ being the relative degree (the excess of poles over zeros) of the transfer function (thus, $r = n - m$).

The control design can be achieved in two steps. First, let us take the control law in the form of

$$u = v + \frac{A(p)}{B(p)} y_d \tag{II.8}$$

where $v$ is a new input to be determined. Substitution of (II.8) into (II.7) leads to

$$A(p) e = B(p) v \tag{II.9}$$

where $e(t) = y(t) - y_d(t)$ is the tracking error. The feedforward signal $(A/B)\, y_d$ can be computed as

$$\frac{A}{B} y_d = \alpha_1 y_d^{(r)} + .... + \alpha_r y_d + w$$

where the $\alpha_i$ ( $i = 1, ... , r$ ) are constants obtained from dividing $A$ by $B$, and $w(t)$ is a filtered version of $y_d(t)$.

The second step is to construct input $u$ so that the error dynamics is asymptotically stable. Since $e$ is known (by subtracting the known $y_d$ from the measured $y$), while its derivatives are not, one can stabilize $e$ by using standard linear techniques, *e.g.*, pole-placement together with a Luenberger observer. A simpler way of deriving the control law is to let

$$v = \frac{C(p)}{D(p)} e \tag{II.10}$$

with $C$ and $D$ being polynomials of order $(n - m)$. With this control law, the closed loop dynamics is

$$(AC + BD) e = 0$$

If the coefficients of $C$ and $D$ are chosen properly, the poles of the closed-loop polynomial can be placed anywhere in the complex plane (with the complex poles in conjugate pairs), as we shall see in chapter 8. Therefore, the control law

$$u = \frac{A}{B} y_d + \frac{C}{D} e \tag{II.11}$$

will guarantee that the tracking error $e(t)$ remains at zero if initial conditions satisfy $y^{(i)}(0) = y_d^{(i)}(0)$ ($i = 1, ... , r$), and exponentially converges to zero if the initial conditions do not satisfy these conditions.

The block diagram of the closed-loop system is depicted in Figure II.3. We can make the following comments about the control system:

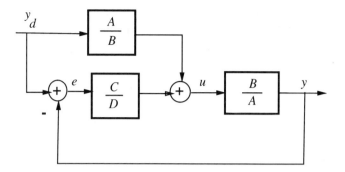

**Figure II.2 :** Linear Tracking Control system

• The feedforward part of the control law, computed by inverting the plant model, is responsible for reducing and eliminating the tracking errors, while the feedback part results in stability of the whole system. If some derivatives of the desired trajectories $y_d(t)$ are not available, one can simply omit them from the feedforward, which will only cause bounded error in tracking. Note that one may easily adapt the above controller for model reference control, with $y_d$ and its derivatives provided by the reference model.

• The control law (II.10) is equivalent to implementing a reduced-order Luenberger observer. Higher order observers can also be used, possibly for the purpose of increasing system robustness by exploiting the added flexibility.

• The above method cannot be directly used for tracking control of non-minimum phase systems (with some of the roots of $B(p)$ having positive real parts) since the inverse model $A/B$ is unstable. However, by feedforwarding low-frequency components of the desired trajectories, good tracking in the low-frequency range (lower than the left-half plane zeros of the plant) may still be achieved. For instance, by using $(A/B_1) y_d$ as the feedforward signal, the tracking error can be easily found to be

$$e(t) = \frac{C}{AC + BD} \left[ \frac{B}{B_1} - 1 \right] A y_d$$

If $B_1$ is close to $B$ at low frequencies, the control system can track a slowly varying $y_d(t)$ well. One particular choice of $B_1$ is to eliminate the right half-plane zeros of $B$, which will lead to good tracking for desired trajectories with frequencies lower than the right half-plane zeroes. ∎

## Importance of Physical Properties

In linear control, it is common practice to generate a set of differential equations for a physical system, and then forget where they came from. This presents no major problem there, at least in a theoretical sense, because linear control theory provides powerful tools for analysis and design. However, such a procedure is typically undesirable for nonlinear systems, because the number of tools available for attacking nonlinear problems is comparatively limited. In nonlinear control design, *exploitation of the physical properties* can sometimes make control design for complex nonlinear plants a simple issue, or may easily solve an otherwise intractable design problem. This point is forcibly demonstrated in the solution of the adaptive robot control problem. Adaptive control of robot manipulators was long recognized to be far out of the reach of conventional adaptive control theory, because a robot's dynamics is strongly nonlinear and has multiple inputs. However, the use of two physical facts, namely, the positive definiteness of the inertia matrix and the possibility of linearly parametrizing robot dynamics, successfully led to an adaptive controller with the desirable properties of global stability and tracking convergence, as shown in chapter 9.

## Discrete Implementations

As discussed in chapter 1, nonlinear physical systems are continuous in nature and are hard to meaningfully discretize, while digital control systems may be treated as continuous-time systems in analysis and design if high sampling rates are used (specific quantifications are discussed, *e.g.*, in section 7.3). Thus, we perform nonlinear system analysis and controller design in continuous-time form. However, of course, the control law is generally implemented digitally.

Numerical integration and differentiation are sometimes explicit parts of a controller design. Numerical differentiation may avoid the complexity of constructing the whole system state based on partial measurements (the nonlinear observer problem), while numerical integration is a standard component of most adaptive controller designs, and can also be needed more generally in dynamic controllers.

Numerical differentiation may be performed in many ways, all aimed at getting a reasonable estimate of the time-derivative, while at the same time avoiding the generation of large amounts of noise.  One can use, for instance, a filtered differentiation of the form

$$\frac{p}{p+\alpha} = \frac{p+\alpha-\alpha}{p+\alpha} = 1 - \frac{\alpha}{p+\alpha}$$

where $p$ is the Laplace variable and $\alpha \gg 1$. The discrete implementation of the above equation, assuming *e.g.*, a zero-order hold, is simply an addition:

$$\dot{x}_{new} = a_1 \dot{x}_{old} + a_2 x \qquad\qquad (\text{II}.12)$$

where the constants $a_1$ and $a_2$ are defined as

$$a_1 = -\alpha e^{-\alpha T} \qquad a_2 = 1 - \alpha(1 - e^{-\alpha T})$$

and $T$ is the sampling period. Note that this approximate procedure can actually be interpreted as building a reduced-order observer for the system. However, it does not use an explicit model, so that the system can be nonlinear and its parameters unknown. An alternative choice of filter structure is studied in Exercise II.6.

Numerical integration actually consists in *simulating in real-time* some (generally nonlinear) dynamic subcomponent required by the controller. Many methods can again be used. The simplest, which can work well at high sampling rates or for low-dimensional systems, is the so-called *Euler integration*

$$x_{new} = x_{old} + \dot{x}\, T$$

where $T$ is the sampling period. A more sophisticated approach, which is very effective in most cases but is more computationally involved than mere Euler integration, is two-step *Adams-Bashforth* integration

$$x(t) = x(t-T) + \left(\frac{3}{2}\dot{x}(t-T) - \frac{1}{2}\dot{x}(t-2T)\right) T$$

More complex techniques may also be used, depending on the desired trade-off between accuracy and on-line computational efficiency.

# II.4  Available Methods of Nonlinear Control Design

As in the analysis of nonlinear control systems, there is no general method for designing nonlinear controllers. What we have is a rich collection of alternative and complementary techniques, each best applicable to particular classes of nonlinear control problems.

### Trial-and-error

Based on the analysis methods provided in Part I, one can use trial-and-error to synthesize controllers, similarly to, *e.g.*, linear lead-lag controller design based on Bode plots. The idea is to use the analysis tools to guide the search for a controller which can then be justified by analysis and simulations. The phase plane method, the describing function method, and Lyapunov analysis can all be used for this purpose. Experience and intuition are critical in this process. However, for complex systems trial-and-error often fails.

## Feedback linearization

As discussed earlier, the first step in designing a control system for a given physical plant is to derive a meaningful *model* of the plant, *i.e.*, a model that captures the key dynamics of the plant in the operational range of interest. Models of physical systems come in various forms, depending on the modeling approach and assumptions. Some forms, however, lend themselves more easily to controller design. Feedback linearization deals with techniques for *transforming original system models into equivalent models of a simpler form.*

Feedback linearization can be used as a nonlinear design methodology. The basic idea is to first transform a nonlinear system into a (fully or partially) linear system, and then use the well-known and powerful linear design techniques to complete the control design. The approach has been used to solve a number of practical nonlinear control problems. It applies to important classes of nonlinear systems (so-called input-state linearizable or minimum-phase systems), and typically requires full state measurement. However, it does not guarantee robustness in the face of parameter uncertainty or disturbances.

Feedback linearization techniques can also be used as model-simplifying devices for robust or adaptive controllers, to be discussed next.

## Robust control

In pure model-based nonlinear control (such as the basic feedback linearization control approach), the control law is designed based on a nominal model of the physical system. How the control system will behave in the presence of model uncertainties is not clear at the design stage. In robust nonlinear control (such as, *e.g.*, sliding control), on the other hand, the controller is designed based on the consideration of both the the nominal model *and* some characterization of the model uncertainties (such as the knowledge that the load to be picked up and carried by a robot is between 2 kg and 10 kg). Robust nonlinear control techniques have proven very effective in a variety of practical control problems. They apply best to specific classes of nonlinear systems, and generally require state measurements.

## Adaptive control

Adaptive control is an approach to dealing with uncertain systems or time-varying systems. Although the term "adaptive" can have broad meanings, current adaptive control designs apply mainly to systems with known dynamic structure, but unknown constant or slowly-varying parameters. Adaptive controllers, whether developed for linear systems or for nonlinear systems, are inherently nonlinear.

Systematic theories exist for the adaptive control of linear systems. Existing adaptive control techniques can also treat important classes of nonlinear systems, with measurable states and linearly parametrizable dynamics. For these nonlinear systems, adaptive control can be viewed as an alternative and complementary approach to robust nonlinear control techniques, with which it can be combined effectively. Although most adaptive control results are for single-input single-output systems, some important nonlinear physical systems with multiple-inputs have also been studied successfully.

### Gain-scheduling

Gain scheduling (see [Rugh, 1991] for a recent discussion, and references therein) is an attempt to apply the well developed linear control methodology to the control of nonlinear systems. It was originally developed for the trajectory control of aircraft. The idea of gain-scheduling is to select a number of operating points which cover the range of the system operation. Then, at each of these points, the designer makes a linear time-invariant approximation to the plant dynamics and designs a linear controller for each linearized plant. Between operating points, the parameters of the compensators are then interpolated, or *scheduled*, thus resulting in a global compensator. Gain scheduling is conceptually simple, and, indeed, practically successful for a number of applications. The main problem with gain scheduling is that has only limited theoretical guarantees of stability in nonlinear operation, but uses some loose practical guidelines such as "the scheduling variables should change slowly" and "the scheduling variables should capture the plant's nonlinearities". Another problem is the computational burden involved in a gain-scheduling design, due to the necessity of computing many linear controllers.

## II.5   Exercises

**II.1**   Why do linear systems with a right half-plane zero exhibit the so-called "undershooting" phenomenon (the step response initially goes downward)? Is the inability of perfect tracking for non-minimum phase systems related to the undershooting phenomenon?

Consider for instance the system

$$y(p) = \frac{(1-p)}{p^2 + 2p + 2} u$$

Sketch its step response and compare it with that of the system

$$y(p) = \frac{(1+p)}{p^2 + 2p + 2} u$$

What are the differences in frequency responses?

What does the step response of a non-minimum phase linear system look like if it has two right half-plane zeros? Interpret and comment.

**II.2**    Assume that you are given a pendulum and the task of designing a control system to track the desired trajectory

$$\theta_d(t) = A \sin \omega t \qquad\qquad 0 < A \le 90^o \qquad 0 < \omega \le 10 \, Hz$$

What hardware components do you need to implement the control system? What requirements does the task impose on the the specifications of the components? Provide a detailed outline of your control system design.

**II.3**    List the model uncertainties associated with the pendulum model (II.1). Discuss how to characterize them.

**II.4**    Carry out the tracking design for the linear plants

$$y(p) \; = \; \frac{(3 + p)}{p^2 + 2p + 2} \, u \qquad\qquad y(p) \; = \; \frac{(3 - p)}{p^2 + 2p + 2} \, u$$

Simulate their responses to the desired trajectories

$$y_d(t) = \sin \omega t$$

with $\omega$ being 0.5, 1.5, and 4. rad/sec.

**II.5**    Figure out an energy-based strategy to bring the inverted pendulum in Figure II.1.b from the vertical-down position to the vertical-up position. (*Hint*: You may want first to express the system's kinetic energy in a modified coordinate system chosen such that rotation and translation are uncoupled.)

What does your controller guarantee along the $x$ direction? Does it reduce to the usual linear inverted pendulum controller when linearized?

**II.6**    An alternative to the filtered differentiation (II.12) consists in simply passing an approximate derivative through a zero-order hold discrete filter, *e.g.*,

$$\dot{x}_{new} \; = \; c_1 \dot{x}_{old} \; + \; (1 - c_1) \, \frac{x - x_{old}}{T}$$

where $c_1 = e^{-\alpha T}$. Discuss the relative merits of the two approaches.

# Chapter 6
# Feedback Linearization

Feedback linearization is an approach to nonlinear control design which has attracted a great deal of research interest in recent years. The central idea of the approach is to algebraically transform a nonlinear system dynamics into a (fully or partly) linear one, so that linear control techniques can be applied. This differs entirely from conventional linearization (*i.e.*, Jacobian linearization, as in section 3.3) in that feedback linearization is achieved by exact state transformations and feedback, rather than by linear approximations of the dynamics.

The idea of simplifying the form of a system's dynamics by choosing a different state representation is not entirely unfamiliar. In mechanics, for instance, it is well known that the form and complexity of a system model depend considerably on the choice of reference frames or coordinate systems. Feedback linearization techniques can be viewed as ways of *transforming original system models into equivalent models of a simpler form*. Thus, they can also be used in the development of robust or adaptive nonlinear controllers, as discussed in chapters 7 and 8.

Feedback linearization has been used successfully to address some practical control problems. These include the control of helicopters, high performance aircraft, industrial robots, and biomedical devices. More applications of the methodology are being developed in industry. However, there are also a number of important shortcomings and limitations associated with the feedback linearization approach. Such problems are still very much topics of current research.

This chapter provides a description of feedback linearization, including what it is, how to use it for control design and what its limitations are. In section 6.1, the basic concepts of feedback linearization are described intuitively and illustrated with simple examples. Section 6.2 introduces mathematical tools from differential geometry which are useful to generalize these concepts to a broad class of nonlinear systems. Sections 6.3 and 6.4 describe feedback linearization theory for SISO systems, and section 6.5 extends the methodology to MIMO systems.

# 6.1  Intuitive Concepts

This section describes the basic concepts of feedback linearization intuitively, using simple examples. The following sections will formalize these concepts for more general nonlinear systems.

## 6.1.1  Feedback Linearization And The Canonical Form

In its simplest form, feedback linearization amounts to canceling the nonlinearities in a nonlinear system so that the closed-loop dynamics is in a linear form. This very simple idea is demonstrated in the following example.

**Example 6.1: Controlling the fluid level in a tank**

Consider the control of the level $h$ of fluid in a tank (Figure 6.1) to a specified level $h_d$. The control input is the flow $u$ into the tank, and the initial level is $h_o$.

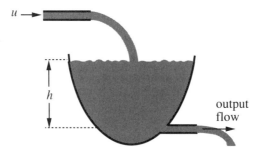

**Figure 6.1 :** Fluid level control in a tank

The dynamic model of the tank is

$$\frac{d}{dt}\left[\int_0^h A(h)\,dh\right] = u(t) \qquad u = \sqrt{2\,g\,h} \tag{6.1}$$

where $A(h)$ is the cross section of the tank and $a$ is the cross section of the outlet pipe. If the initial level $h_o$ is quite different from the desired level $h_d$, the control of $h$ involves a nonlinear regulation problem.

The dynamics (6.1) can be rewritten as

$$A(h)\,\dot{h} \;=\; u - a\sqrt{2gh}$$

If $u(t)$ is chosen as

$$u(t) = a\sqrt{2gh} + A(h)\,v \tag{6.2}$$

with $v$ being an "equivalent input" to be specified, the resulting dynamics is linear

$$\dot{h} = v$$

Choosing $v$ as

$$v = -\alpha\,\tilde{h} \tag{6.3}$$

with $\tilde{h} = h(t) - h_d$ being the level error, and $\alpha$ being a strictly positive constant, the resulting closed loop dynamics is

$$\dot{h} + \alpha\,\tilde{h} = 0 \tag{6.4}$$

This implies that $\tilde{h}(t) \to 0$ as $t \to \infty$. Based on (6.2) and (6.3), the actual input flow is determined by the nonlinear control law

$$u(t) = a\sqrt{2gh} - A(h)\,\alpha\,\tilde{h} \tag{6.5}$$

Note that, in the control law (6.5), the first part on the right-hand side is used to provide the output flow $a\sqrt{2gh}$, while the second part is used to raise the fluid level according to the the desired linear dynamics (6.4).

Similarly, if the desired level is a known time-varying function $h_d(t)$, the equivalent input $v$ can be chosen as

$$v = \dot{h}_d(t) - \alpha\,\tilde{h}$$

so as to still yield $\tilde{h}(t) \to 0$ as $t \to \infty$.                        ∎

The idea of feedback linearization, *i.e.*, of canceling the nonlinearities and imposing a desired linear dynamics, can be simply applied to a class of nonlinear systems described by the so-called *companion form*, or *controllability canonical form*. A system is said to be in companion form if its dynamics is represented by

$$x^{(n)} = f(\mathbf{x}) + b(\mathbf{x})\,u \tag{6.6}$$

where $u$ is the scalar control input, $x$ is the scalar output of interest, $\mathbf{x} = [\, x, \dot{x}, \ldots, x^{(n-1)}\,]^T$ is the state vector, and $f(\mathbf{x})$ and $b(\mathbf{x})$ are nonlinear functions of the states. This form is unique in the fact that, although derivatives of $x$ appear in this equation, no derivative of the input $u$ is present. Note that, in state-space representation, (6.6) can be written

$$\frac{d}{dt}\begin{bmatrix} x_1 \\ \cdots \\ x_{n-1} \\ x_n \end{bmatrix} = \begin{bmatrix} x_2 \\ \cdots \\ x_n \\ f(\mathbf{x}) + b(\mathbf{x})u \end{bmatrix}$$

For systems which can be expressed in the controllability canonical form, using the control input (assuming $b$ to be non-zero)

$$u = \frac{1}{b}[v - f] \tag{6.7}$$

we can cancel the nonlinearities and obtain the simple input-output relation (multiple-integrator form)

$$x^{(n)} = v$$

Thus, the control law

$$v = -k_o x - k_1 \dot{x} - \ldots - k_{n-1} x^{(n-1)}$$

with the $k_i$ chosen so that the polynomial $p^n + k_{n-1}p^{n-1} + \ldots + k_o$ has all its roots strictly in the left-half complex plane, leads to the exponentially stable dynamics

$$x^{(n)} + k_{n-1}x^{(n-1)} + \ldots + k_o x = 0$$

which implies that $x(t) \to 0$. For tasks involving the tracking of a desired output $x_d(t)$, the control law

$$v = x_d^{(n)} - k_o e - k_2 \dot{e} - \ldots - k_{n-1} e^{(n-1)} \tag{6.8}$$

(where $e(t) = x(t) - x_d(t)$ is the tracking error) leads to exponentially convergent tracking. Note that similar results would be obtained if the scalar $x$ was replaced by a vector and the scalar $b$ by an invertible square matrix.

One interesting application of the above control design idea is in robotics. The following example studies control design for a two link robot. Design for more general robots is similar and will be discussed in chapter 9.

**Example 6.2: Feedback linearization of a two-link robot**

Figure 6.2 provides the physical model of a two-link robot, with each joint equipped with a motor for providing input torque, an encoder for measuring joint position, and a tachometer for measuring joint velocity. The objective of the control design is to make the joint positions $q_1$ and $q_2$ follow desired position histories $q_{d1}(t)$ and $q_{d2}(t)$, which are specified by the motion planning system of the robot. Such tracking control problems arise when a robot hand is required to move along a specified path, *e.g.*, to draw circles.

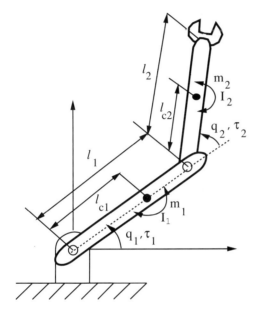

**Figure 6.2 :** A two-link robot

Using the well-known Lagrangian equations in classical dynamics, one can easily show that the dynamic equations of the robot is

$$\begin{bmatrix} H_{11} & H_{12} \\ H_{21} & H_{22} \end{bmatrix} \begin{bmatrix} \ddot{q}_1 \\ \ddot{q}_2 \end{bmatrix} + \begin{bmatrix} -h\dot{q}_2 & -h\dot{q}_1 - h\dot{q}_2 \\ h\dot{q}_1 & 0 \end{bmatrix} \begin{bmatrix} \dot{q}_1 \\ \dot{q}_2 \end{bmatrix} + \begin{bmatrix} g_1 \\ g_2 \end{bmatrix} = \begin{bmatrix} \tau_1 \\ \tau_2 \end{bmatrix} \qquad (6.9)$$

with $\mathbf{q} = [q_1 \; q_2]^T$ being the two joint angles, $\boldsymbol{\tau} = [\tau_1 \; \tau_2]^T$ being the joint inputs, and

$$H_{11} = m_1 l_{c_1}^2 + I_1 + m_2 [l_1^2 + l_{c_2}^2 + 2 l_1 l_{c_2} \cos q_2] + I_2$$

$$H_{22} = m_2 l_{c_2}^2 + I_2$$

$$H_{12} = H_{21} = m_2 l_1 l_{c_2} \cos q_2 + m_2 l_{c_2}^2 + I_2$$

$$h = m_2 l_1 l_{c_2} \sin q_2$$

$$g_1 = m_1 l_{c_1} g \cos q_1 + m_2 g [l_{c_2} \cos(q_1 + q_2) + l_1 \cos q_1]$$

$$g_2 = m_2 l_{c_2} g \cos(q_1 + q_2)$$

Equation (6.9) can be compactly expressed as

$$\mathbf{H(q)\ddot{q} + C(q, \dot{q})\dot{q} + g(q) = \tau}$$

with $\mathbf{H}$, $\mathbf{C}$ and $\mathbf{g}$ defined obviously. Thus, by multiplying both sides by $\mathbf{H}^{-1}$ (the invertibility of $\mathbf{H}$ is a physical property of the system, as discussed in Chapter 9), the above vector equation can be put easily in the form of (6.6), with $n = 2$, although this dynamics now involves multiple inputs and multiple outputs.

To achieve tracking control tasks, one can use the following control law

$$\begin{bmatrix} \tau_1 \\ \tau_2 \end{bmatrix} = \begin{bmatrix} H_{11} \, H_{12} \\ H_{21} \, H_{22} \end{bmatrix} \begin{bmatrix} v_1 \\ v_2 \end{bmatrix} + \begin{bmatrix} -h\dot{q}_2 & -h\dot{q}_1 - h\dot{q}_2 \\ h\dot{q}_1 & 0 \end{bmatrix} \begin{bmatrix} \dot{q}_1 \\ \dot{q}_2 \end{bmatrix} + \begin{bmatrix} g_1 \\ g_2 \end{bmatrix} \tag{6.10}$$

where

$$\mathbf{v} = \ddot{\mathbf{q}}_d - 2\lambda \dot{\tilde{\mathbf{q}}} - \lambda^2 \tilde{\mathbf{q}}$$

with $\mathbf{v} = [v_1 \ v_2]^T$ being the equivalent input, $\tilde{\mathbf{q}} = \mathbf{q} - \mathbf{q}_d$ being the position tracking error and $\lambda$ a positive number. The tracking error $\tilde{\mathbf{q}}$ then satisfies the equation

$$\ddot{\tilde{\mathbf{q}}} + 2\lambda \dot{\tilde{\mathbf{q}}} + \lambda^2 \tilde{\mathbf{q}} = \mathbf{0}$$

and therefore converges to zero exponentially. The control law (6.10) is commonly referred to as "computed torque" control in robotics. It can be applied to robots with arbitrary numbers of joints, as discussed in chapter 9.          □

Note that in (6.6) we have assumed that the dynamics is linear in terms of the control input $u$ (although nonlinear in the states). However, the approach can be easily extended to the case when $u$ is replaced by an invertible function $g(u)$. For example, in systems involving flow control by a valve, the dynamics may be dependent on $u^4$ rather than directly on $u$, with $u$ being the valve opening diameter. Then, by defining $w = u^4$, one can first design $w$ similarly to the previous procedure and then compute the input $u$ by $u = w^{1/4}$. This means that the nonlinearity is simply undone in the control computation.

When the nonlinear dynamics is not in a controllability canonical form, one may have to use algebraic *transformations* to first put the dynamics into the controllability form before using the above feedback linearization design, or to rely on partial linearization of the original dynamics, instead of full linearization. These are the topics of the next subsections. Conceptually, such transformations are not totally unfamiliar: even in the case of *linear* systems, pole placement is often most easily achieved by first putting the system in the controllability canonical form.

## 6.1.2 Input-State Linearization

Consider the problem of designing the control input $u$ for a single-input nonlinear system of the form

$$\dot{\mathbf{x}} = \mathbf{f}(\mathbf{x}, u)$$

The technique of input-state linearization solves this problem in two steps. First, one finds a state transformation $\mathbf{z} = \mathbf{z}(\mathbf{x})$ and an input transformation $u = u(\mathbf{x}, v)$ so that the nonlinear system dynamics is transformed into an equivalent *linear time-invariant* dynamics, in the familiar form $\dot{\mathbf{z}} = \mathbf{A}\mathbf{z} + \mathbf{b}v$. Second, one uses standard linear techniques (such as pole placement) to design $v$.

Let us illustrate the approach on a simple second-order example. Consider the system

$$\dot{x}_1 = -2x_1 + a x_2 + \sin x_1 \tag{6.11a}$$

$$\dot{x}_2 = -x_2 \cos x_1 + u \cos(2x_1) \tag{6.11b}$$

Even though linear control design can stabilize the system in a small region around the equilibrium point $(0, 0)$, it is not obvious at all what controller can stabilize it in a larger region. A specific difficulty is the nonlinearity in the first equation, which cannot be directly canceled by the control input $u$.

However, if we consider the new set of state variables

$$z_1 = x_1 \tag{6.12a}$$

$$z_2 = a x_2 + \sin x_1 \tag{6.12b}$$

then, the new state equations are

$$\dot{z}_1 = -2z_1 + z_2 \tag{6.13a}$$

$$\dot{z}_2 = -2z_1 \cos z_1 + \cos z_1 \sin z_1 + a u \cos(2z_1) \tag{6.13b}$$

Note that the new state equations also have an equilibrium point at $(0, 0)$. Now we see that the nonlinearities can be canceled by the control law of the form

$$u = \frac{1}{a \cos(2z_1)} (v - \cos z_1 \sin z_1 + 2 z_1 \cos z_1) \tag{6.14}$$

where $v$ is an equivalent input to be designed (equivalent in the sense that determining $v$ amounts to determining $u$, and vice versa), leading to a linear input-state relation

$$\dot{z}_1 = -2z_1 + z_2 \tag{6.15a}$$

$$\dot{z}_2 = v \tag{6.15b}$$

Thus, through the state transformation (6.12) and input transformation (6.14), the problem of stabilizing the original nonlinear dynamics (6.11) using the original control input $u$ has been transformed into the problem of stabilizing the new dynamics (6.15) using the new input $v$.

Since the new dynamics is linear and controllable, it is well known that the linear state feedback control law

$$v = -k_1 z_1 - k_2 z_2$$

can place the poles anywhere with proper choices of feedback gains. For example, we may choose

$$v = -2z_2 \tag{6.16}$$

resulting in the stable closed-loop dynamics

$$\dot{z}_1 = -2z_1 + z_2$$

$$\dot{z}_2 = -2z_2$$

whose poles are both placed at $-2$. In terms of the original state $x_1$ and $x_2$, this control law corresponds to the original input

$$u = \frac{1}{\cos(2x_1)} (-2ax_2 - 2\sin x_1 - \cos x_1 \sin x_1 + 2x_1 \cos x_1) \tag{6.17}$$

The original state $\mathbf{x}$ is given from $\mathbf{z}$ by

$$x_1 = z_1 \tag{6.18a}$$

$$x_2 = (z_2 - \sin z_1)/a \tag{6.18b}$$

Since both $z_1$ and $z_2$ converge to zero, the original state $\mathbf{x}$ converges to zero.

The closed loop system under the above control law is represented in the block diagram in Figure 6.3. We can detect two loops in this control system, with the inner loop achieving the linearization of the input-state relation, and the outer loop achieving the stabilization of the closed-loop dynamics. This is consistent with (6.14), where the control input $u$ is seen to be composed of a nonlinearity cancellation part and a linear compensation part.

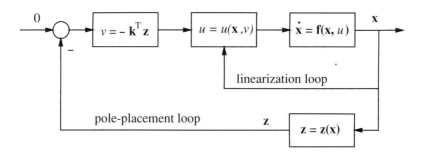

**Figure 6.3** : Input-State Linearization

A number of remarks can be made about the above control law:

• The result, though valid in a large region of the state space, is not global. The control law is not well defined when $x_1 = (\pi/4 \pm k\pi/2)$, $k = 1, 2, ...$ Obviously, when the initial state is at such singularity points, the controller cannot bring the system to the equilibrium point.

• The input-state linearization is achieved by a combination of a state transformation and an input transformation, with state feedback used in both. Thus, it is a linearization by feedback, or feedback linearization. This is fundamentally different from a Jacobian linearization for small range operation, on which linear control is based.

• In order to implement the control law, the new state components $(z_1, z_2)$ must be available. If they are not physically meaningful or cannot be measured directly, the original state $\mathbf{x}$ must be measured and used to compute them from (6.12).

• Thus, in general, we rely on the system model both for the controller design and for the computation of $\mathbf{z}$. If there is uncertainty in the model, *e.g.*, uncertainty on the parameter $a$, this uncertainty will cause error in the computation of both the new state $\mathbf{z}$ and of the control input $u$, as seen in (6.12) and (6.14).

• Tracking control can also be considered. However, the desired motion then needs to be expressed in terms of the full new state vector. Complex computations may be needed to translate the desired motion specification (in terms of physical output variables) into specifications in terms of the new states.

With the above successful design in mind, it is interesting to extend the input-state linearization idea to general nonlinear systems. Two questions arise when one speculates such generalizations:

- What classes of nonlinear systems can be transformed into linear systems?

- How to find the proper transformations for those which can?

These questions are systematically addressed in section 6.3.

## 6.1.3 Input-Output Linearization

Let us now consider a tracking control problem. Consider the system

$$\dot{\mathbf{x}} = \mathbf{f}(\mathbf{x}, u) \tag{6.19a}$$

$$y = h(\mathbf{x}) \tag{6.19b}$$

and assume that our objective is to make the output $y(t)$ track a desired trajectory $y_d(t)$ while keeping the whole state bounded, where $y_d(t)$ and its time derivatives up to a sufficiently high order are assumed to be known and bounded. An apparent difficulty with this model is that the output $y$ is only indirectly related to the input $u$, through the state variable $\mathbf{x}$ and the nonlinear state equations (6.19). Therefore, it is not easy to see how the input $u$ can be designed to control the tracking behavior of the output $y$. However, inspired by the results of section 6.1.1, one might guess that the difficulty of the tracking control design can be reduced if we can *find a direct and simple relation between the system output y and the control input u*. Indeed, this idea constitutes the intuitive basis for the so-called input-output linearization approach to nonlinear control design. Let us again use an example to demonstrate this approach.

Consider the third-order system

$$\dot{x}_1 = \sin x_2 + (x_2 + 1) x_3 \tag{6.20a}$$

$$\dot{x}_2 = x_1^5 + x_3 \tag{6.20b}$$

$$\dot{x}_3 = x_1^2 + u \tag{6.20c}$$

$$y = x_1 \tag{6.20d}$$

To generate a direct relationship between the output $y$ and the input $u$, let us differentiate the output $y$

$$\dot{y} = \dot{x}_1 = \sin x_2 + (x_2 + 1)\, x_3$$

Since $\dot{y}$ is still not directly related to the input $u$, let us differentiate again. We now obtain

$$\ddot{y} = (x_2 + 1)\ u + f_1(\mathbf{x}) \tag{6.21}$$

where $f_1(\mathbf{x})$ is a function of the state defined by

$$f_1(\mathbf{x}) = (x_1^5 + x_3)(x_3 + \cos x_2) + (x_2 + 1)x_1^2 \tag{6.22}$$

Clearly, (6.21) represents an explicit relationship between $y$ and $u$. If we choose the control input to be in the form

$$u = \frac{1}{x_2 + 1}(v - f_1) \tag{6.23}$$

where $v$ is a new input to be determined, the nonlinearity in (6.21) is canceled, and we obtain a simple linear double-integrator relationship between the output and the new input $v$,

$$\ddot{y} = v$$

The design of a tracking controller for this double-integrator relation is simple, because of the availability of linear control techniques. For instance, letting $e = y(t) - y_d(t)$ be the tracking error, and choosing the new input $v$ as

$$v = \ddot{y}_d - k_1\, e - k_2\, \dot{e} \tag{6.24}$$

with $k_1$ and $k_2$ being positive constants, the tracking error of the closed loop system is given by

$$\ddot{e} + k_2\, \dot{e} + k_1\, e = 0 \tag{6.25}$$

which represents an exponentially stable error dynamics. Therefore, if initially $e(0) = \dot{e}(0) = 0$, then $e(t) \equiv 0, \forall\, t \geq 0$, *i.e.*, prefect tracking is achieved; otherwise, $e(t)$ converges to zero exponentially.

Note that

- The control law is defined everywhere, except at the singularity points such that $x_2 = -1$.

- Full state measurement is necessary in implementing the control law, because the computations of both the derivative $\dot{y}$ and the input transformation (6.23) require the value of $\mathbf{x}$.

The above control design strategy of first generating a linear input-output relation and then formulating a controller based on linear control is referred to as the *input-output linearization* approach, and it can be applied to many systems, as will be seen in section 6.4 for SISO systems and in section 6.5 for MIMO systems. If we need to differentiate the output of a system $r$ times to generate an explicit relationship between the output $y$ and input $u$, the system is said to have *relative degree r*. Thus, the system in the above example has relative degree 2. As will be shown soon, this terminology is consistent with the notion of relative degree in linear systems (excess of poles over zeros). As we shall see later, it can also be shown formally that for any controllable system of order $n$, it will take *at most n* differentiations of any output for the control input to appear, *i.e.*, $r \leq n$. This can be understood intuitively: if it took more than $n$ differentiations, the system would be of order higher than $n$; if the control input never appeared, the system would not be controllable.

At this point, one might feel that the tracking control design problem posed at the beginning has been elegantly solved with the control law (6.23) and (6.24). However, one must remember that (6.25) only accounts for part of the closed-loop dynamics, because it has only order 2, while the whole dynamics has order 3 (the same as that of the plant, because the controller (6.23) introduces no extra dynamics). Therefore, a part of the system dynamics (described by one state component) has been rendered "unobservable" in the input-output linearization. This part of the dynamics will be called the *internal dynamics*, because it cannot be seen from the external input-output relationship (6.21). For the above example, the internal state can be chosen to be $x_3$ (because $x_3$, $y$, and $\dot{y}$ constitute a new set of states), and the internal dynamics is represented by the equation

$$\dot{x}_3 = x_1^2 + \frac{1}{x_2 + 1} ( \ddot{y}_d(t) - k_1 e - k_2 \dot{e} + f_1 ) \tag{6.26}$$

If this internal dynamics is stable (by which we actually mean that the states remain bounded during tracking, *i.e.*, stability in the BIBO sense), our tracking control design problem has indeed been solved. Otherwise, the above tracking controller is practically meaningless, because the instability of the internal dynamics would imply undesirable phenomena such as the burning-up of fuses or the violent vibration of mechanical members. Therefore, *the effectiveness of the above control design, based on the reduced-order model (6.21), hinges upon the stability of the internal dynamics.*

Let us now use some simpler examples to show that internal dynamics are stable for some systems (implying that the previous design approach is applicable), and unstable for others (implying the need for a different control design).

**Example 6.3: Internal dynamics**

Consider the nonlinear system

$$\begin{bmatrix} \dot{x}_1 \\ \dot{x}_2 \end{bmatrix} = \begin{bmatrix} x_2^3 + u \\ u \end{bmatrix} \tag{6.27a}$$

$$y = x_1 \tag{6.27b}$$

Assume that the control objective is to make $y$ track $y_d(t)$. Differentiation of $y$ simply leads to the first state equation. Thus, choosing the control law

$$u = -x_2^3 - e(t) + \dot{y}_d(t) \tag{6.28}$$

yields exponential convergence of $e$ to zero

$$\dot{e} + e = 0 \tag{6.29}$$

The same control input is also applied to the second dynamic equation, leading to the internal dynamics

$$\dot{x}_2 + x_2^3 = \dot{y}_d - e \tag{6.30}$$

which is, characteristically, non-autonomous and nonlinear. However, in view of the facts that $e$ is guaranteed to be bounded by (6.29) and $\dot{y}_d$ is assumed to be bounded, we have

$$|\dot{y}_d(t) - e| \leq D$$

where $D$ is a positive constant. Thus, we can conclude from (6.30) that $|x_2| \leq D^{1/3}$ (perhaps after a transient), since $\dot{x}_2 < 0$ when $x_2 > D^{1/3}$, and $\dot{x}_2 > 0$ when $x_2 < -D^{1/3}$.

Therefore, (6.28) does represent a satisfactory tracking control law for the system (6.27), given any trajectory $y_d(t)$ whose derivative $\dot{y}_d(t)$ is bounded. ☐

Conversely, one can easily show (Exercise 6.2) that if the second state equation in (6.27) is replaced by $\dot{x}_2 = -u$, then the resulting internal dynamics is unstable.

Finally, let us remark that, although the input-output linearization is motivated in the context of output tracking, it can also be applied to stabilization problems. For example, if $y_d(t) \equiv 0$ is the desired trajectory for the above system, the two states $y$ and $\dot{y}$ of the closed-loop system will be driven to zero by the control law (6.28), implying the stabilization of the whole system provided that the internal dynamics is stable. In addition, two useful remarks can be made about using input-output linearization for stabilization design. First, in stabilization problems, there is no reason to restrict the choice of output $y = h(x)$ to be a physically meaningful quantity (while in tracking problems the choice of output is determined by the physical task). Any function of $x$ may be used to serve as an artificial output (a designer output) to generate a linear input-output relation for the purpose of stabilization design. Second, different choices of output function leads to different internal dynamics. It is possible for one choice of output to yield a stable internal dynamics (or no internal dynamics) while another choice of output would lead to a unstable one. Therefore, one should choose, if possible, the output function to be such that the associated internal dynamics is stable.

A special case occurs when the relative degree of a system is the same as its order, *i.e.*, when the output $y$ has to be differentiated $n$ times (with $n$ being the system order) to obtain a linear input-output relation. In this case, the variables $y$, $\dot{y}$, ... , $y^{(n-1)}$ may be used as a new set of state variables for the system, and there is no internal dynamics associated with this input-output linearization. Thus, in this case, input-output linearization leads to input-state linearization, and both state regulation and output tracking (for the particular output) can be achieved easily.

## THE INTERNAL DYNAMICS OF LINEAR SYSTEMS

We must admit that it is only due to the simplicity of the system that the internal dynamics in Example 6.3 has been shown to be stable. In general, it is very difficult to directly determine the stability of the internal dynamics because it is nonlinear, non-autonomous, and coupled to the "external" closed-loop dynamics, as seen in (6.26). Although a Lyapunov or Lyapunov-like analysis may be useful for some systems, its general applicability is limited by the difficulty of finding a Lyapunov function, as discussed in chapters 3 and 4. Therefore, we naturally want to seek simpler ways of determining the stability of the internal dynamics. An examination of how the concept of internal dynamics translates in the more familiar context of *linear* systems proves helpful to this purpose.

Let us start by considering the internal dynamics of some simple linear systems.

**Example 6.4: Internal dynamics in two linear systems**

Consider the simple controllable and observable linear system

$$\begin{bmatrix} \dot{x}_1 \\ \dot{x}_2 \end{bmatrix} = \begin{bmatrix} x_2 + u \\ u \end{bmatrix} \tag{6.31a}$$

$$y = x_1 \tag{6.31b}$$

where $y(t)$ is required to track a desired output $y_d(t)$. With one differentiation of the output, we simply obtain the first state equation

$$\dot{y} = x_2 + u$$

which explicitly contains $u$. Thus, the control law

$$u = -x_2 + \dot{y}_d - (y - y_d) \tag{6.32}$$

yields the tracking error equation

$$\dot{e} + e = 0$$

(where $e = y - y_d$) and the internal dynamics

$$\dot{x}_2 + x_2 = \dot{y}_d - e(t)$$

We see from these equations that while $y(t)$ tends to $y_d(t)$ (and $\dot{y}(t)$ tends to $\dot{y}_d(t)$), $x_2$ remains bounded, and so does $u$. Therefore, (6.32) is a satisfactory tracking controller for system (6.31).

Let us now consider a slightly different system:

$$\begin{bmatrix} \dot{x}_1 \\ \dot{x}_2 \end{bmatrix} = \begin{bmatrix} x_2 + u \\ -u \end{bmatrix} \tag{6.33a}$$

$$y = x_1 \tag{6.33b}$$

The same control law as above yields the same tracking error dynamics, but now leads to the internal dynamics

$$\dot{x}_2 - x_2 = e(t) - \dot{y}_d$$

This implies that $x_2$, and accordingly $u$, both go to infinity as $t \to \infty$. Therefore, (6.32) is not a suitable tracking controller for system (6.33). $\qquad \square$

We are thus left wondering why the same tracking design method is applicable to for system (6.31) but not to system (6.33). To understand this fundamental difference between the two systems, let us consider their transfer functions, namely, for system (6.31),

$$W_1(p) = \frac{p + 1}{p^2}$$

and for system (6.33),

$$W_2(p) = \frac{p-1}{p^2}$$

We see that the two systems have the same poles but different zeros. Specifically, system (6.31), for which the design has succeeded, has a *left half-plane* zero at $-1$, while system (6.33), for which the design has failed, has an *right half-plane* zero at 1.

The above observation (the internal dynamics is stable if the plant zeros are in the left-half plane, *i.e.*, if the plant is "minimum-phase") can actually be shown to be true for all linear systems, as we do now. This is not surprising because, for non-minimum phase systems, perfect tracking of arbitrary trajectories requires infinite control effort, as seen in Example II.2.

To keep notations simple, let us consider a third-order linear system in state-space form

$$\dot{\mathbf{z}} = \mathbf{A}\mathbf{z} + \mathbf{b}u \qquad y = \mathbf{c}^T\mathbf{z} \tag{6.34}$$

and having one zero (and hence two more poles than zeros), although the procedure can be straightforwardly extended to systems with arbitrary numbers of poles and zeros. The system's input-output linearization can be facilitated if we first transform it into the so-called companion form. To do this, we note from linear control that the input/output behavior of this system can be expressed in the form

$$y = \mathbf{c}^T(p\mathbf{I} - \mathbf{A})^{-1}\mathbf{b}u = \frac{b_o + b_1 p}{a_o + a_1 p + a_2 p^2 + p^3}u \tag{6.35}$$

(where $p$ is the Laplace variable). Thus, if we define

$$x_1 = \frac{1}{a_o + a_1 p + a_2 p^2 + p^3}u$$

$$x_2 = \dot{x}_1$$

$$x_3 = \dot{x}_2$$

the system can be equivalently represented in the companion form

$$\frac{d}{dt}\begin{bmatrix} x_1 \\ x_2 \\ x_3 \end{bmatrix} = \begin{bmatrix} 0 & 1 & 0 \\ 0 & 0 & 1 \\ -a_o & -a_1 & -a_2 \end{bmatrix}\begin{bmatrix} x_1 \\ x_2 \\ x_3 \end{bmatrix} + \begin{bmatrix} 0 \\ 0 \\ 1 \end{bmatrix}u \tag{6.36a}$$

$$y = [b_o \quad b_1 \quad 0] \begin{bmatrix} x_1 \\ x_2 \\ x_3 \end{bmatrix} \tag{6.36b}$$

Let us now perform input-output linearization based on this form. The first differentiation of the output leads to

$$\dot{y} = b_o x_2 + b_1 x_3$$

and the second differentiation leads to

$$\ddot{y} = b_o \dot{x}_2 + b_1 \dot{x}_3 = b_o x_3 + b_1(-a_o x_1 - a_1 x_2 - a_2 x_3 + u) \tag{6.37}$$

It is seen that the input $u$ appears in the second differentiation, which means that the required number of differentiations (the relative degree) is indeed the same as the excess of poles over zeros (of course, since the input-output relation of $y$ to $u$ is independent of the choice of state variables, it would also take two differentiations for $u$ to appear if we used the original state-space equations (6.34) ).

Thus, the control law

$$u = (a_o x_1 + a_1 x_2 + a_2 x_3 - \frac{b_o}{b_1} x_3) + \frac{1}{b_1} (-k_1 e - k_2 \dot{e} + \ddot{y}_d) \tag{6.38}$$

where $e = y - y_d$, yields an exponentially stable tracking error

$$\ddot{e} + k_2 \dot{e} + k_1 e = 0$$

Since this is a second-order dynamics, the internal dynamics of our third-order system can be described by only one state equation. Specifically, we can use $x_1$ to complete the state vector, since one can easily show $x_1$, $y$, and $\dot{y}$ are related to $x_1$, $x_2$, and $x_3$ through a one-to-one transformation (and thus can serve as states for the system). We then easily find from (6.36a) and (6.36b) that the internal dynamics is

$$\dot{x}_1 = x_2 = \frac{1}{b_1} (y - b_o x_1)$$

that is,

$$\dot{x}_1 + \frac{b_o}{b_1} x_1 = \frac{1}{b_1} y \tag{6.39}$$

Since $y$ is bounded ($y = e + y_d$), we see that the stability of the internal dynamics depends on the location of the zero $-b_o/b_1$ of the transfer function in (6.35). If the system is minimum phase, then the zero is in the left-half plane, which implies that the internal dynamics (6.39) is stable, independently of the initial conditions and of the magnitudes of the desired $y_d$, ... , $y_d^{(r)}$ (where $r$ is the relative degree).

A classical example of the effect of a right half-plane zero is the problem of controlling the altitude of an aircraft using an elevator.

**Example 6.5: Aircraft altitude dynamics**

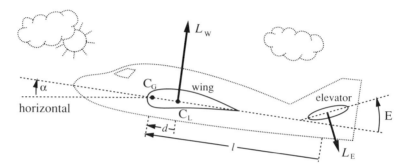

**Figure 6.4 :** Dynamic characteristics of an aircraft

A schematic diagram of the dynamics of an aircraft (in the longitudinal plane) is shown in Figure 6.4. The sum of the lift forces applied to the aircraft wings and body is equivalent to a single lift force $L_W$, applied at the "center of lift" $C_L$. The center of lift does not necessarily coincide with the center of mass $C_G$ (with a positive $d$ meaning that the center of mass is ahead of the center of lift). The mass of the aircraft is denoted by $m$ and its moment of inertia about $C_G$ is denoted by $J$. We assume that all angles are small enough to justify linear approximations, and that the forward velocity of the aircraft remains essentially constant.

The aircraft is initially cruising at a constant altitude $h = h_o$. To affect its vertical motion, the elevator (a small surface located at the aircraft tail) is rotated by an angle $E$. This generates a small aerodynamic force $L_E$ on the elevator, and thus a torque about $C_G$. This torque creates a rotation of the aircraft about $C_G$, measured by an angle $\alpha$. The lift force $L_W$ applied to the wings is proportional to $\alpha$, *i.e.*, $L_W = C_{ZW}\alpha$. Similarly, $L_E$ is proportional to the angle between the horizontal and the elevator, *i.e.*, $L_E = C_{ZE}(E-\alpha)$. Furthermore, various aerodynamic forces create friction torques proportional to $\dot{\alpha}$, of the form $b\dot{\alpha}$. In summary, a simplified model of the aircraft vertical motion can be written

$$J\ddot{\alpha}+b\dot{\alpha}+(C_{ZE}l+C_{ZW}d)\alpha = C_{ZE}l\,E \tag{6.40a}$$

$$m\ddot{h} = (C_{ZE}+C_{ZW})\alpha - C_{ZE}\,E \tag{6.40b}$$

where the first equation represents the balance of moments and the second the balance of forces.

Remark that the open-loop stability of the first equation, which defines the dynamics of the angle $\alpha$, depends on the sign of the coefficient $(C_{ZE}l + C_{ZW}d)$. In particular, the equation is open-loop stable if $d > 0$, *i.e.*, if the center of mass is ahead of the center of lift (this allows us to understand the shape of a jumbo jet, or the fact that on small aircraft passengers are first seated on the front rows).

To simplify notations, let now

$$J = 1 \quad m = 1 \quad b = 4 \quad C_{ZE} = 1 \quad C_{ZW} = 5 \quad l = 3 \quad d = 0.2.$$

The transfer functions describing the system can then be written

$$\frac{\alpha(p)}{E(p)} = \frac{3}{p^2 + 4p + 4} = \frac{3}{(p+2)^2} \tag{6.41a}$$

$$\frac{h(p)}{E(p)} = \frac{14 - 4p - p^2}{p^2(p^2 + 4p + 4)} = \frac{(6.24 + p)(2.24 - p)}{p^2(p+2)^2} \tag{6.41b}$$

where $p$ is the Laplace variable.

At time $t = 0$, a unit step input in elevator angle $E$ is applied. From the initial and final value theorems of linear control (or directly from the equations of motion), one can easily show that the corresponding initial and final vertical accelerations are

$$\ddot{h}(t=0^+) = -1 < 0 \qquad\qquad \ddot{h}(t=+\infty) = 3.5 > 0$$

The complete time responses in $\ddot{h}$ and $h$ are sketched in Figure 6.5. We see that the aircraft starts in the wrong direction, then recovers. Such behavior is typical of systems with a right half-plane zero. It can be easily understood physically, and is a direct reflection of the aircraft design itself. The initial effect of the unit step in $E$ is to create an instantaneous *downward force* on the elevator, thus creating an initial downward acceleration of the aircraft's center of mass. The unit step in elevator angle also creates a *torque* about $C_G$, which builds up the angle $\alpha$ and thus creates an increasing *upward lift force* on the wing and body. This lift force eventually takes over the downward force on the elevator. Of course, such non-minimum phase behavior is important for the pilot to know, especially when flying at low altitudes.

Let us determine the associated internal dynamics. Defining the state as $\mathbf{x} = [\alpha \ \dot{\alpha} \ h \ \dot{h}]^T$, the equations of motion can be written

$$\dot{x}_1 = x_2 \tag{6.42a}$$

$$\dot{x}_2 = -4x_2 - 4x_1 + 3E \tag{6.42b}$$

$$\dot{x}_3 = x_4 \tag{6.42c}$$

$$\dot{x}_4 = 6x_1 - E \tag{6.42d}$$

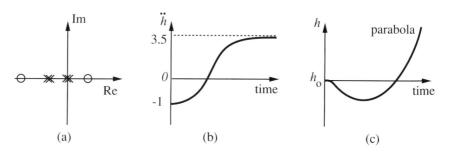

**Figure 6.5 :** Pole-zero plot and step responses in $\ddot{h}$ and $h$

The output of interest is the aircraft's altitude

$$y = x_3$$

Differentiating $y$ until the input $E$ appears yields

$$\ddot{y} = \ddot{x}_3 = \dot{x}_4 = 6x_1 - E$$

consistent with the fact that the transfer function (6.41b) has relative degree 2. Choose now a pole-placement control law for $E$

$$E = 6x_1 - \ddot{y}_d + \dot{\tilde{y}} + \tilde{y}$$

where $\tilde{y} = y - y_d$. Then

$$\ddot{\tilde{y}} + \dot{\tilde{y}} + \tilde{y} = 0 \tag{6.43}$$

The corresponding internal dynamics is

$$\dot{x}_2 = -4x_2 - 4x_1 + 3(6x_1 - \ddot{y}_d + \dot{\tilde{y}} + \tilde{y})$$

that is

$$\ddot{\alpha} + 4\dot{\alpha} - 14\alpha = 3(-\ddot{y}_d + \dot{\tilde{y}} + \tilde{y}) \tag{6.44}$$

and therefore, is unstable. Specifically, the poles on the left-hand side of (6.44) are exactly the zeros of the transfer function (6.41b). $\qquad\square$

## THE ZERO-DYNAMICS

Since for linear systems the stability of the internal dynamics is simply determined by the locations of the zeros, it is interesting to see whether this relation can be extended to nonlinear systems. To do so requires first to extend the concept of zeros to nonlinear systems, and then to determine the relation of the internal dynamics stability to this extended concept of zeros.

Extending the notion of zeros to nonlinear systems is not a trivial proposition. Transfer functions, on which linear system zeros are based, cannot be defined for nonlinear systems. Furthermore, zeros are intrinsic properties of a linear plant, while for nonlinear systems the stability of the internal dynamics may depend on the specific control input.

A way to approach these difficulties is to define a so-called *zero-dynamics* for a nonlinear system. *The zero-dynamics is defined to be the internal dynamics of the system when the system output is kept at zero by the input.* For instance, for the system (6.27), the zero-dynamics is (from (6.30))

$$\dot{x}_2 + x_2^3 = 0 \tag{6.45}$$

Noticing that the specification of maintaining the system output at zero uniquely defines the required input (namely, here, $u$ has to equal $-x_2^3$ in order to keep $x_1$ always equal to zero), we see that the zero-dynamics is an intrinsic property of a nonlinear system. The zero-dynamics (6.45) is easily seen to be asymptotically stable (by using the Lyapunov function $V = x_2^2$ ).

Similarly, for the linear system (6.34), the zero-dynamics is (from (6.39))

$$\dot{x}_1 + (b_o/b_1)\, x_1 = 0$$

Thus, in this linear system, the poles of the zero-dynamics are exactly the zeros of the system. This result is general for linear systems, and therefore, *in linear systems*, having all zeros in the left-half complex plane guarantees the global asymptotic stability of the zero-dynamics.

The reason for defining and studying the zero-dynamics is that we want to find a simpler way of determining the stability of the internal dynamics. For linear systems, the stability of the zero-dynamics implies the global stability of the internal dynamics: the left-hand side of (6.39) completely determines its stability characteristics, given that the right-hand side tends to zero or is bounded. In nonlinear systems, however, the relation is not so clear. Section 6.4 investigates this question in some detail. For stabilization problems, it can be shown that local asymptotic stability of the zero-dynamics is enough to guarantee the local asymptotic stability of the internal dynamics. Extensions can be drawn to the tracking problem. However, unlike the linear case, no results on the global stability or even large range stability can be drawn for internal dynamics of nonlinear systems, *i.e.*, only local stability is guaranteed for the internal dynamics even if the zero-dynamics is globally exponentially stable.

Similarly to the linear case, we will call a nonlinear system whose zero-dynamics is asymptotically stable an *asymptotically minimum phase* system. The concept of an exponentially minimum phase system can be defined in the same way.

Two useful remarks can be made about the zero-dynamics of nonlinear systems. First, the zero-dynamics is an intrinsic feature of a nonlinear system, which does not depend on the choice of control law or the desired trajectories. Second, examining the stability of zero-dynamics is much easier than examining the stability of internal dynamics, because the zero-dynamics only involves the internal states (while the internal dynamics is coupled to the external dynamics and desired trajectories, as seen in (6.26)).

**Example 6.6: Aircraft zero-dynamics**

Given (6.44), the zero-dynamics of the aircraft of Example 6.5 is

$$\ddot{\alpha} + 4\dot{\alpha} - 14\alpha = 0 \tag{6.46}$$

This dynamics is unstable, confirming that the system is non-minimum phase. The poles of the zero-dynamics are exactly the zeros of the transfer function (6.41b).

Now model (6.40) is actually the linearization of a more general *nonlinear* model, applicable at larger angles and angular rates. Since the zero-dynamics corresponding to the linearized model *simply is the linearization of the zero-dynamics corresponding to the nonlinear model*, thus the nonlinear system is also non-minimum phase, from Lyapunov's linearization method.  □

To summarize, control design based on input-output linearization can be made in three steps:

- *differentiate the output y until the input u appears*

- *choose u to cancel the nonlinearities and guarantee tracking convergence*

- *study the stability of the internal dynamics*

If the relative degree associated with the input-output linearization is the same as the order of the system, the nonlinear system is fully linearized and this procedure indeed leads to a satisfactory controller (assuming that the model is accurate). If the relative degree is smaller than the system order, then the nonlinear system is only partly linearized, and whether the controller can indeed be applied depends on the stability of the internal dynamics. The study of the internal dynamics stability can be simplified locally by studying that of the zero-dynamics instead. If the zero-dynamics is unstable, different control strategies should be sought, only simplified by the fact that the transformed dynamics is partly linear.

# 6.2  Mathematical Tools

The objective of the rest of this chapter is to formalize and generalize the previous intuitive concepts for a broad class of nonlinear systems. To this effect, we first introduce some mathematical tools from differential geometry and topology. To limit the conceptual and notational complexity, we discuss these tools directly in the context of nonlinear dynamic systems (instead of general topological spaces). Note that this section and the remainder of this chapter represent by far the most mathematically involved part of the book. Hurried practitioners may skip it in a *first* reading, and go directly to chapters 7-9.

In describing these mathematical tools, we shall call a vector function $\mathbf{f} : \mathbf{R}^n \to \mathbf{R}^n$ a *vector field* in $\mathbf{R}^n$, to be consistent with the terminology used in differential geometry. The intuitive reason for this term is that to every vector function $\mathbf{f}$ corresponds a field of vectors in an $n$-dimensional space (one can think of a vector $\mathbf{f}(\mathbf{x})$ emanating from every point $\mathbf{x}$). In the following, we shall only be interested in *smooth* vector fields. By smoothness of a vector field, we mean that the function $\mathbf{f}(\mathbf{x})$ has continuous partial derivatives of any required order.

Given a smooth scalar function $h(\mathbf{x})$ of the state $\mathbf{x}$, the gradient of $h$ is denoted by $\nabla h$

$$\nabla h = \frac{\partial h}{\partial \mathbf{x}}$$

The gradient is represented by a *row-vector* of elements $(\nabla h)_j = \partial h / \partial x_j$. Similarly, given a vector field $\mathbf{f}(\mathbf{x})$, the Jacobian of $\mathbf{f}$ is denoted by $\nabla \mathbf{f}$

$$\nabla \mathbf{f} = \frac{\partial \mathbf{f}}{\partial \mathbf{x}}$$

It is represented by an $n \times n$ matrix of elements $(\nabla \mathbf{f})_{ij} = \partial f_i / \partial x_j$.

### LIE DERIVATIVES AND LIE BRACKETS

Given a scalar function $h(\mathbf{x})$ and a vector field $\mathbf{f}(\mathbf{x})$, we define a new scalar function $L_{\mathbf{f}} h$, called the Lie derivative (or simply, the derivative) of $h$ with respect to $\mathbf{f}$.

**Definition 6.1**    *Let $h : \mathbf{R}^n \to \mathbf{R}$ be a smooth scalar function, and $\mathbf{f} : \mathbf{R}^n \to \mathbf{R}^n$ be a smooth vector field on $\mathbf{R}^n$, then the <u>Lie derivative of h with respect to $\mathbf{f}$</u> is a scalar function defined by $L_{\mathbf{f}} h = \nabla h \, \mathbf{f}$.*

Thus, the Lie derivative $L_f h$ is simply the directional derivative of $h$ along the direction of the vector $\mathbf{f}$.

Repeated Lie derivatives can be defined recursively

$$L_f^o h = h$$

$$L_f^i h = L_f(L_f^{i-1} h) = \nabla(L_f^{i-1} h) \ \mathbf{f} \qquad \qquad \text{for } i = 1, 2, .....$$

Similarly, if $\mathbf{g}$ is another vector field, then the scalar function $L_g L_f h(\mathbf{x})$ is

$$L_g L_f h = \nabla(L_f h) \ \mathbf{g}$$

One can easily see the relevance of Lie derivatives to dynamic systems by considering the following single-output system

$$\dot{\mathbf{x}} = \mathbf{f}(\mathbf{x})$$

$$y = h(\mathbf{x})$$

The derivatives of the output are

$$\dot{y} = \frac{\partial h}{\partial \mathbf{x}} \dot{\mathbf{x}} = L_f h$$

$$\ddot{y} = \frac{\partial [L_f h]}{\partial \mathbf{x}} \dot{\mathbf{x}} = L_f^2 h$$

and so on. Similarly, if $V$ is a Lyapunov function candidate for the system, its derivative $\dot{V}$ can be written as $L_f V$.

Let us move on to another important mathematical operator on vector fields, the Lie bracket.

**Definition 6.2**    *Let $\mathbf{f}$ and $\mathbf{g}$ be two vector fields on $\mathbf{R}^n$. The <u>Lie bracket of $\mathbf{f}$ and $\mathbf{g}$</u> is a third vector field defined by*

$$[\mathbf{f}, \mathbf{g}] = \nabla \mathbf{g} \ \mathbf{f} - \nabla \mathbf{f} \ \mathbf{g}$$

The Lie bracket $[\mathbf{f}, \mathbf{g}]$ is commonly written as $ad_f \mathbf{g}$ (where $ad$ stands for "adjoint"). Repeated Lie brackets can then be defined recursively by

$$ad_{\mathbf{f}}^{o}\,\mathbf{g} = \mathbf{g}$$

$$ad_{\mathbf{f}}^{i}\,\mathbf{g} = [\mathbf{f}, ad_{\mathbf{f}}^{i-1}\,\mathbf{g}] \qquad\qquad \text{for } i = 1, 2, .....$$

**Example 6.7:** The system (6.11) can be written in the form

$$\dot{\mathbf{x}} = \mathbf{f}(\mathbf{x}) + \mathbf{g}(\mathbf{x})\,u$$

with the two vector fields **f** and **g** defined by

$$\mathbf{f} = \begin{bmatrix} -2x_1 + ax_2 + \sin x_1 \\ -x_2 \cos x_1 \end{bmatrix} \qquad \mathbf{g}(\mathbf{x}) = \begin{bmatrix} 0 \\ \cos(2x_1) \end{bmatrix}$$

Their Lie bracket can be computed as

$$[\mathbf{f}, \mathbf{g}] = \begin{bmatrix} 0 & 0 \\ -2\sin(2x_1) & 0 \end{bmatrix}\begin{bmatrix} -2x_1 + ax_2 + \sin x_1 \\ -x_2\cos x_1 \end{bmatrix} - \begin{bmatrix} -2 + \cos x_1 & a \\ x_2 \sin x_1 & -\cos x_1 \end{bmatrix}\begin{bmatrix} 0 \\ \cos(2x_1) \end{bmatrix}$$

$$= \begin{bmatrix} a\cos(2x_1) \\ \cos x_1 \cos(2x_1) - 2\sin(2x_1)(-2x_1 + ax_2 + \sin x_1) \end{bmatrix} \qquad \square$$

The following lemma on Lie bracket manipulation will be useful later.

**Lemma 6.1**   *Lie brackets have the following properties*

*(i) bilinearity:*

$$[\alpha_1\mathbf{f}_1 + \alpha_2\mathbf{f}_2, \mathbf{g}] = \alpha_1[\mathbf{f}_1, \mathbf{g}] + \alpha_2[\mathbf{f}_2, \mathbf{g}]$$

$$[\mathbf{f}, \alpha_1\mathbf{g}_1 + \alpha_2\mathbf{g}_2] = \alpha_1[\mathbf{f}, \mathbf{g}_1] + \alpha_2[\mathbf{f}, \mathbf{g}_2]$$

*where* $\mathbf{f}, \mathbf{f}_1, \mathbf{f}_2, \mathbf{g}, \mathbf{g}_1$ *and* $\mathbf{g}_2$ *are smooth vector fields, and* $\alpha_1$ *and* $\alpha_2$ *are constant scalars.*

*(ii) skew-commutativity:*

$$[\mathbf{f}, \mathbf{g}] = -[\mathbf{g}, \mathbf{f}]$$

*(iii) Jacobi identity:*

$$L_{ad_{\mathbf{f}}\mathbf{g}}\,h = L_{\mathbf{f}}L_{\mathbf{g}}\,h - L_{\mathbf{g}}L_{\mathbf{f}}\,h$$

*where* $h(\mathbf{x})$ *is a smooth scalar function of* **x**.

**Proof**: The proofs of the first two properties are straightforward (Exercise 6.6). Let us derive the third property, which can be rewritten as

$$\nabla h \, [\mathbf{f}, \mathbf{g}] = \nabla(L_{\mathbf{g}} h) \, \mathbf{f} - \nabla(L_{\mathbf{f}} h) \, \mathbf{g}$$

The left-hand side of the above equation can be expanded as

$$\nabla h \, [\mathbf{f}, \mathbf{g}] = \frac{\partial h}{\partial \mathbf{x}} \left( \frac{\partial \mathbf{g}}{\partial \mathbf{x}} \mathbf{f} - \frac{\partial \mathbf{f}}{\partial \mathbf{x}} \mathbf{g} \right)$$

while the right-hand side can be expanded as

$$\nabla(L_{\mathbf{g}} h) \, \mathbf{f} - \nabla(L_{\mathbf{f}} h) \, \mathbf{g} = \nabla\left( \frac{\partial h}{\partial \mathbf{x}} \mathbf{g} \right) \mathbf{f} - \nabla\left( \frac{\partial h}{\partial \mathbf{x}} \mathbf{f} \right) \mathbf{g}$$

$$= \left( \frac{\partial h}{\partial \mathbf{x}} \frac{\partial \mathbf{g}}{\partial \mathbf{x}} + \mathbf{g}^T \frac{\partial^2 h}{\partial \mathbf{x}^2} \right) \mathbf{f} - \left( \frac{\partial h}{\partial \mathbf{x}} \frac{\partial \mathbf{f}}{\partial \mathbf{x}} + \mathbf{f}^T \frac{\partial^2 h}{\partial \mathbf{x}^2} \right) \mathbf{g} = \frac{\partial h}{\partial \mathbf{x}} \left( \frac{\partial \mathbf{g}}{\partial \mathbf{x}} \mathbf{f} - \frac{\partial \mathbf{f}}{\partial \mathbf{x}} \mathbf{g} \right)$$

where $\partial^2 h / \partial \mathbf{x}^2$ is the Hessian of $h$, a symmetric matrix.    □

The Jacobi identity can be used recursively to obtain useful technical identities. Using it twice yields

$$L_{ad_{\mathbf{f}}^2 \mathbf{g}} \, h = L_{ad_{\mathbf{f}}(ad_{\mathbf{f}} \mathbf{g})} \, h = L_{\mathbf{f}} L_{ad_{\mathbf{f}} \mathbf{g}} \, h - L_{ad_{\mathbf{f}} \mathbf{g}} \, L_{\mathbf{f}} \, h$$

$$= L_{\mathbf{f}}[L_{\mathbf{f}} L_{\mathbf{g}} \, h - L_{\mathbf{g}} L_{\mathbf{f}} \, h] - [L_{\mathbf{f}} L_{\mathbf{g}} - L_{\mathbf{g}} L_{\mathbf{f}}] \, L_{\mathbf{f}} \, h$$

$$= L_{\mathbf{f}}^2 L_{\mathbf{g}} \, h - 2 L_{\mathbf{f}} L_{\mathbf{g}} L_{\mathbf{f}} \, h + L_{\mathbf{g}} L_{\mathbf{f}}^2 \, h \tag{6.47}$$

Similar identities can be obtained for higher-order Lie brackets.

## DIFFEOMORPHISMS AND STATE TRANSFORMATIONS

The concept of diffeomorphism can be viewed as a generalization of the familiar concept of coordinate transformation. It is formally defined as follows:

**Definition 6.3**    *A function* $\phi : \mathbf{R}^n \to \mathbf{R}^n$, *defined in a region* $\Omega$, *is called a* _diffeomorphism_ *if it is smooth, and if its inverse* $\phi^{-1}$ *exists and is smooth.*

If the region $\Omega$ is the whole space $\mathbf{R}^n$, then $\phi(\mathbf{x})$ is called a *global* diffeomorphism. Global diffeomorphisms are rare, and therefore one often looks for *local* diffeomorphisms, *i.e.*, for transformations defined only in a finite neighborhood of a given point. Given a nonlinear function $\phi(\mathbf{x})$, it is easy to check whether it is a local diffeomorphism by using the following lemma, which is a straightforward consequence of the well-known implicit function theorem.

**Lemma 6.2**   *Let $\phi(\mathbf{x})$ be a smooth function defined in a region $\Omega$ in $\mathbf{R}^n$. If the Jacobian matrix $\nabla\phi$ is non-singular at a point $\mathbf{x} = \mathbf{x}_o$ of $\Omega$, then $\phi(\mathbf{x})$ defines a local diffeomorphism in a subregion of $\Omega$.*

A diffeomorphism can be used to transform a nonlinear system into another nonlinear system in terms of a new set of states, similarly to what is commonly done in the analysis of linear systems. Consider the dynamic system described by

$$\dot{\mathbf{x}} = \mathbf{f}(\mathbf{x}) + \mathbf{g}(\mathbf{x})\,u$$

$$y = h(\mathbf{x})$$

and let a new set of states be defined by

$$\mathbf{z} = \phi(\mathbf{x})$$

Differentiation of $\mathbf{z}$ yields

$$\dot{\mathbf{z}} = \frac{\partial\phi}{\partial\mathbf{x}}\dot{\mathbf{x}} = \frac{\partial\phi}{\partial\mathbf{x}}(\mathbf{f}(\mathbf{x}) + \mathbf{g}(\mathbf{x})u)$$

One can easily write the new state-space representation as

$$\dot{\mathbf{z}} = \mathbf{f}^*(\mathbf{z}) + \mathbf{g}^*(\mathbf{z})\,u$$

$$y = h^*(\mathbf{z})$$

where $\mathbf{x} = \phi^{-1}(\mathbf{z})$ has been used, and the functions $\mathbf{f}^*$, $\mathbf{g}^*$ and $h^*$ are defined obviously.

**Example 6.8: A non-global diffeomorphism**

The nonlinear vector function

$$\begin{bmatrix} z_1 \\ z_2 \end{bmatrix} = \phi(\mathbf{x}) = \begin{bmatrix} 2x_1 + 5x_1 x_2^2 \\ 3\sin x_2 \end{bmatrix} \tag{6.48}$$

is well defined for all $x_1$ and $x_2$. Its Jacobian matrix is

$$\frac{\partial\phi}{\partial\mathbf{x}} = \begin{bmatrix} 2 + 5x_2^2 & 10\,x_1 x_2 \\ 0 & 3\cos x_2 \end{bmatrix}$$

which has rank 2 at $\mathbf{x} = (0, 0)$. Therefore, Lemma 6.2 indicates that the function (6.48) defines a local diffeomorphism around the origin. In fact, the diffeomorphism is valid in the region

$$\Omega = \{(x_1, x_2), \ |x_2| < \pi/2\}$$

because the inverse exists and is smooth for **x** in this region. However, outside this region, $\phi$ does not define a diffeomorphism, because the inverse does not uniquely exist.  □

## THE FROBENIUS THEOREM

The Frobenius theorem is an important tool in the formal treatment of feedback linearization for $n^{\text{th}}$-order nonlinear systems. It provides a necessary and sufficient condition for the solvability of a special class of partial differential equations. Before presenting the precise statement of the theorem, let us first gain a basic understanding by discussing the case $n = 3$.

Consider the set of first-order partial differential equations

$$\frac{\partial h}{\partial x_1} f_1 + \frac{\partial h}{\partial x_2} f_2 + \frac{\partial h}{\partial x_3} f_3 = 0 \tag{6.49a}$$

$$\frac{\partial h}{\partial x_1} g_1 + \frac{\partial h}{\partial x_2} g_2 + \frac{\partial h}{\partial x_3} g_3 = 0 \tag{6.49b}$$

where $f_i(x_1, x_2, x_3)$ and $g_i(x_1, x_2, x_3)$ $(i = 1, 2, 3)$ are known scalar functions of $x_1, x_2, x_3$, and $h(x_1, x_2, x_3)$ is an unknown function. Clearly, this set of partial differential equations is uniquely defined by the two vectors $\mathbf{f} = [f_1 \ f_2 \ f_3]^T$, $\mathbf{g} = [g_1 \ g_2 \ g_3]^T$. If a solution $h(x_1, x_2, x_3)$ exists for the above partial differential equations, we shall say the set of vector fields $\{\mathbf{f}, \mathbf{g}\}$ is *completely integrable*.

The question now is to determine when these equations are solvable. This is not obvious at all, *a priori*. The Frobenius theorem provides a relatively simple condition: Equation (6.49) has a solution $h(x_1, x_2, x_3)$ if, and only if, there exists scalar functions $\alpha_1(x_1, x_2, x_3)$ and $\alpha_2(x_1, x_2, x_3)$ such that

$$[\mathbf{f}, \mathbf{g}] = \alpha_1 \mathbf{f} + \alpha_2 \mathbf{g}$$

*i.e.*, if the Lie bracket of **f** and **g** can be expressed as a linear combination of **f** and **g**. This condition is called the *involutivity condition* on the vector fields $\{\mathbf{f}, \mathbf{g}\}$. Geometrically it means that the vector $[\mathbf{f}, \mathbf{g}]$ is in the plane formed by the two vectors **f** and **g**. Thus, the Frobenius theorem states that the set of vector fields $\{\mathbf{f}, \mathbf{g}\}$ is completely integrable if, and only if, it is involutive. Note that the involutivity condition can be relatively easily checked, and therefore, the solvability of (6.49) can be determined accordingly.

Let us now discuss the Frobenius theorem in the general case, after giving formal definitions of complete integrability and involutivity.

**Definition 6.4** *A linearly independent set of vector fields* $\{\mathbf{f}_1, \mathbf{f}_2, \dots, \mathbf{f}_m\}$ *on* $\mathbf{R}^n$ *is said to be* <u>*completely integrable*</u> *if, and only if, there exist* $n-m$ *scalar functions* $h_1(\mathbf{x})$, $h_2(\mathbf{x})$, ...., $h_{n-m}(\mathbf{x})$ *satisfying the system of partial differential equations*

$$\nabla h_i \, \mathbf{f}_j = 0 \tag{6.50}$$

*where* $1 \le i \le n-m$, $1 \le j \le m$, *and the gradients* $\nabla h_i$ *are linearly independent.*

Note that with the number of vectors being $m$ and the dimension of the associated space being $n$, the number of unknown scalar functions $h_i$ involved is $(n - m)$ and the number of partial differential equations is $m(n-m)$.

**Definition 6.5** *A linearly independent set of vector fields* $\{\mathbf{f}_1, \mathbf{f}_2, \dots, \mathbf{f}_m\}$ *is said to be* <u>*involutive*</u> *if, and only if, there are scalar functions* $\alpha_{ijk} : \mathbf{R}^n \to \mathbf{R}$ *such that*

$$[\mathbf{f}_i, \mathbf{f}_j](\mathbf{x}) = \sum_{k=1}^{m} \alpha_{ijk}(\mathbf{x}) \, \mathbf{f}_k(\mathbf{x}) \qquad \forall \, i, j \tag{6.51}$$

Involutivity means that if one forms the Lie bracket of any pairs of vector fields from the set $\{\mathbf{f}_1, \mathbf{f}_2, \dots, \mathbf{f}_m\}$, then the resulting vector field can be expressed as a linear combination of the original set of vector fields. Note that

- Constant vector fields are always involutive. Indeed, the Lie bracket of two constant vectors is simply the zero vector, which can be trivially expressed as linear combination of the vector fields.

- A set composed of a single vector $\mathbf{f}$ is involutive. Indeed,

$$[\mathbf{f}, \mathbf{f}] = (\nabla \mathbf{f}) \, \mathbf{f} - (\nabla \mathbf{f}) \, \mathbf{f} = 0$$

- From Definition 6.5, checking whether a set of vector fields $\{\mathbf{f}_1, \dots, \mathbf{f}_m\}$ is involutive amounts to checking whether

$$\mathrm{rank}\Big(\mathbf{f}_1(\mathbf{x}) \;\dots\; \mathbf{f}_m(\mathbf{x})\Big) = \mathrm{rank}\Big(\mathbf{f}_1(\mathbf{x}) \;\dots\; \mathbf{f}_m(\mathbf{x}) \;\; [\mathbf{f}_i, \mathbf{f}_j](\mathbf{x})\Big)$$

for all $\mathbf{x}$ and all $i, j$.

We can now state the Frobenius theorem formally.

**Theorem 6.1 (Frobenius)** *Let* $\mathbf{f}_1, \mathbf{f}_2, \dots, \mathbf{f}_m$ *be a set of linearly independent vector fields. The set is completely integrable if, and only if, it is involutive.*

**Example 6.9:** Consider the set of partial differential equations

$$4x_3 \frac{\partial h}{\partial x_1} - \frac{\partial h}{\partial x_2} = 0$$

$$-x_1 \frac{\partial h}{\partial x_1} + (x_3^2 - 3x_2) \frac{\partial h}{\partial x_2} + 2x_3 \frac{\partial h}{\partial x_3} = 0$$

The associated vector fields are $\{\mathbf{f}_1 , \mathbf{f}_2\}$ with

$$\mathbf{f}_1 = [4x_3 \quad -1 \quad 0]^T \qquad\qquad \mathbf{f}_2 = [-x_1 \quad (x_3^2-3x_2) \quad 2x_3]^T$$

In order to determine whether this set of partial differential equations is solvable (or whether $\{\mathbf{f}_1 \ \mathbf{f}_2\}$ is completely integrable), let us check the involutivity of the set of vector fields $\{\mathbf{f}_1 \ \mathbf{f}_2\}$. One easily finds that

$$[\mathbf{f}_1, \mathbf{f}_2] = [-12x_3 \quad 3 \quad 0]^T$$

Since $[\mathbf{f}_1, \mathbf{f}_2] = -3\mathbf{f}_1 + 0\mathbf{f}_2$, this set of vector fields is involutive. Therefore, the two partial differential equations are solvable. $\qquad\qquad\qquad\qquad\qquad\qquad\qquad\square$

# 6.3 Input-State Linearization of SISO Systems

In this section, we discuss input-state linearization for single-input nonlinear systems represented by the state equations

$$\dot{\mathbf{x}} = \mathbf{f}(\mathbf{x}) + \mathbf{g}(\mathbf{x})u \qquad\qquad (6.52)$$

with $\mathbf{f}$ and $\mathbf{g}$ being smooth vector fields. We study when such systems can be linearized by state and input transformations, how to find such transformations, and how to design controllers based on such feedback linearizations.

Note that systems in the form (6.52) are said to be *linear in control* or *affine*. It is useful to point out that if a nonlinear system has the form

$$\dot{\mathbf{x}} = \mathbf{f}(\mathbf{x}) + \mathbf{g}(\mathbf{x}) \, w[u + \phi(\mathbf{x})]$$

with $w$ being an *invertible* scalar function and $\phi$ being an arbitrary functional, a simple variable substitution $v = w[u + \phi(\mathbf{x})]$ puts the dynamics into the form of (6.52). One can design a control law for $v$ and then compute $u$ by inverting $w$, *i.e.*, $u = w^{-1}(v) - \phi(\mathbf{x})$.

## DEFINITION OF INPUT-STATE LINEARIZATION

In order to proceed with a detailed study of input-state linearization, a formal definition of this concept is necessary.

**Definition 6.6** *A single-input nonlinear system in the form (6.52), with* $\mathbf{f}(\mathbf{x})$ *and* $\mathbf{g}(\mathbf{x})$ *being smooth vector fields on* $\mathbf{R}^n$, *is said to be* *input-state linearizable* *if there exists a region* $\Omega$ *in* $\mathbf{R}^n$, *a diffeomorphism* $\phi : \Omega \rightarrow \mathbf{R}^n$, *and a nonlinear feedback control law*

$$u = \alpha(\mathbf{x}) + \beta(\mathbf{x})v \tag{6.53}$$

*such that the new state variables* $\mathbf{z} = \phi(\mathbf{x})$ *and the new input* $v$ *satisfy a linear time-invariant relation*

$$\dot{\mathbf{z}} = \mathbf{A}\,\mathbf{z} + \mathbf{b}\,v \tag{6.54}$$

*where*

$$
\mathbf{A} = \begin{bmatrix}
0 & 1 & 0 & . & . & 0 \\
0 & 0 & 1 & . & . & . \\
. & . & . & . & . & . \\
. & . & . & . & . & . \\
0 & 0 & 0 & . & . & 1 \\
0 & 0 & 0 & . & . & 0
\end{bmatrix}
\qquad
\mathbf{b} = \begin{bmatrix}
0 \\
0 \\
. \\
. \\
0 \\
1
\end{bmatrix}
$$

The new state $\mathbf{z}$ is called the *linearizing state*, and the control law (6.53) is called the *linearizing control law*. To simplify notations, we will often use $\mathbf{z}$ to denote not only the transformed state, but the diffeomorphism $\phi$ itself, *i.e.* write

$$\mathbf{z} = \mathbf{z}(\mathbf{x})$$

This slight abuse of notations should not create any confusion.

Note that the transformed linear dynamics has its $\mathbf{A}$ matrix and $\mathbf{b}$ vector of a special form, corresponding to a linear companion form. However, generality is not lost by restricting ourselves to this special linear equivalent dynamics, because any representation of a linear controllable system is equivalent to the companion form (6.54) through a linear state transformation and pole placement. Therefore, if (6.53) can be transformed into a linear system, it can be transformed into the form prescribed by (6.54) by using additional linear transformations in state and input.

We easily see from the canonical form (6.54) that feedback linearization is a special case of input-output linearization, where the output function leads to a relative degree $n$. This means that if a system is input-output linearizable with relative degree $n$, it must be input-state linearizable. On the other hand, if a system is input-state linearizable, with the first new state $z_1$ representing the output, the system is input-output linearizable with relative degree $n$. Therefore, we can summarize the relationship between input-output linearization and input-state linearization as follows:

**Lemma 6.3**     *An $n^{th}$-order nonlinear system is input-state linearizable if, and only if, there exists a scalar function $z_1(\mathbf{x})$ such that the system's input-output linearization with $z_1(\mathbf{x})$ as output function has relative degree n.*

Note, however, that the above lemma provides no guidance about how to find the desirable output function $z_1(\mathbf{x})$.

## CONDITIONS FOR INPUT-STATE LINEARIZATION

At this point, a natural question is: can all nonlinear state equations in the form of (6.52) be input-state linearized? If not, when do such linearizations exist? The following theorem provides a definitive answer to that question, and constitutes one of the most fundamental results of feedback linearization theory.

**Theorem 6.2**     *The nonlinear system (6.52), with $\mathbf{f}(\mathbf{x})$ and $\mathbf{g}(\mathbf{x})$ being smooth vector fields, is input-state linearizable if, and only if, there exists a region $\Omega$ such that the following conditions hold:*

- *the vector fields $\{\mathbf{g}, ad_{\mathbf{f}}\,\mathbf{g}\,,\,....\,,\;\, ad_{\mathbf{f}}{}^{n-1}\,\mathbf{g}\}$ are linearly independent in $\Omega$*

- *the set $\{\mathbf{g}, ad_{\mathbf{f}}\,\mathbf{g}\,,\,...\,,\;\, ad_{\mathbf{f}}{}^{n-2}\,\mathbf{g}\}$ is involutive in $\Omega$*

Before proving this result, let us make a few remarks about the above conditions:

- The first condition can be interpreted as simply representing a *controllability condition* for the nonlinear system (6.52). For linear systems, the vector fields $\{\mathbf{g}, ad_{\mathbf{f}}\mathbf{g}\,,\,....\,, ad_{\mathbf{f}}{}^{n-1}\mathbf{g}\}$ become $\{\mathbf{b}\,, \mathbf{A}\,\mathbf{b}\,,\,....\,, \mathbf{A}^{n-1}\mathbf{b}\}$, and therefore their independence is equivalent to the invertibility of the familiar linear controllability matrix. It is also easy to show that if a system's linear approximations in a closed connected region $\Omega$ in $\mathbf{R}^n$ are all controllable, then, under some mild smoothness assumptions, the system can be driven from any point in $\Omega$ to any point in $\Omega$. However, as mentioned in chapter 3, a nonlinear system can be controllable while its linear approximation is not. The first condition above can be shown to represent a generalized controllability condition which also accounts for such cases.

- The involutivity condition is less intuitive. It is trivially satisfied for linear systems (which have constant vector fields), but not generically satisfied in the nonlinear case.

Let us now prove the above theorem. We first state a technical lemma.

**Lemma 6.4**    *Let $z(\mathbf{x})$ be a smooth function in a region $\Omega$. Then, in $\Omega$, the set of equations*

$$L_\mathbf{g}\,z = L_\mathbf{g}\,L_\mathbf{f}\,z = \dots = L_\mathbf{g}\,L_\mathbf{f}^{\,k}\,z = 0 \tag{6.55a}$$

*is equivalent to*

$$L_\mathbf{g}\,z = L_{ad_\mathbf{f}\mathbf{g}}\,z = \dots = L_{ad_\mathbf{f}^{\,k}\mathbf{g}}\,z = 0 \tag{6.55b}$$

*for any positive integer $k$.*

**Proof:** Let us show that (6.55a) implies (6.55b).

When $k = 0$, the result is obvious. When $k = 1$, we have from Jacobi's identity (Lemma 6.1)

$$L_{ad_\mathbf{f}\mathbf{g}}\,z = L_\mathbf{f}\,L_\mathbf{g}\,z - L_\mathbf{g}\,L_\mathbf{f}\,z = 0 - 0 = 0$$

When $k = 2$, we further have, using Jacobi's identity twice as in (6.47)

$$L_{ad_\mathbf{f}^2\mathbf{g}}\,z = L_\mathbf{f}^2\,L_\mathbf{g}\,z - 2L_\mathbf{f}\,L_\mathbf{g}\,L_\mathbf{f}\,z + L_\mathbf{g}\,L_\mathbf{f}^2\,z = 0 - 0 + 0 = 0$$

Repeating this procedure, we can show by induction that (6.55a) implies (6.55b) for any $k$.

One proceeds similarly to show that (6.55b) implies (6.55a) (by using Jacobi's identity the other way around).             $\square$

We are now ready for the proof of Theorem 6.2 itself.

**Proof of Theorem 6.2:** Let us first prove the *necessity* of the conditions. Assume that there exist a state transformation $\mathbf{z} = \mathbf{z}(\mathbf{x})$ and an input transformation $u = \alpha(\mathbf{x}) + \beta(\mathbf{x})v$ such that $\mathbf{z}$ and $v$ satisfy (6.54). Expanding the first line of (6.54), we obtain

$$\dot{z}_1 = \frac{\partial z_1}{\partial \mathbf{x}}(\mathbf{f} + \mathbf{g}u) = z_2$$

Proceeding similarly with the other components of $\mathbf{z}$ leads to a set of partial differential equations

$$\frac{\partial z_1}{\partial \mathbf{x}}\mathbf{f} + \frac{\partial z_1}{\partial \mathbf{x}}\,\mathbf{g}\,u = z_2$$

$$\frac{\partial z_2}{\partial \mathbf{x}}\mathbf{f} + \frac{\partial z_2}{\partial \mathbf{x}}\,\mathbf{g}\,u = z_3$$

$$\dots$$

$$\frac{\partial z_n}{\partial \mathbf{x}}\mathbf{f} + \frac{\partial z_n}{\partial \mathbf{x}}\,\mathbf{g}\,u = v$$

Since $z_1, \dots, z_n$ are independent of $u$, while $v$ is not, we conclude from the above equations that

$$L_\mathbf{g} z_1 = L_\mathbf{g} z_2 = \;.....\; = L_\mathbf{g} z_{n-1} = 0 \qquad\qquad L_\mathbf{g} z_n \neq 0 \qquad\qquad\qquad (6.56a)$$

$$L_\mathbf{f} z_i = z_{i+1} \qquad i = 1, 2, \dots, n-1 \qquad\qquad\qquad (6.56b)$$

The above equations on the $z_i$ can be compressed into a set of constraint equations on $z_1$ alone. Indeed, using Lemma 6.4, equation (6.56a) implies that

$$\nabla z_1 \, ad_\mathbf{f}^k \, \mathbf{g} = 0 \quad k = 0, 1, 2, \dots, n-2 \qquad\qquad\qquad (6.57a)$$

Furthermore by proceeding as in the proof of Lemma 6.4, we can show

$$\nabla z_1 \, ad_\mathbf{f}^{n-1} \mathbf{g} = (-1)^{n-1} \, L_\mathbf{g} \, z_n$$

This implies that

$$\nabla z_1 \, ad_\mathbf{f}^{n-1} \, \mathbf{g} \neq 0 \qquad\qquad\qquad (6.57b)$$

The first property we can now infer from (6.57) is that the vector fields $\mathbf{g}, ad_\mathbf{f} \, \mathbf{g}, \dots, ad_\mathbf{f}^{n-1} \, \mathbf{g}$ must be linearly independent. Indeed, if for some number $i$ ($i \leq n-1$), there existed scalar functions $\alpha_1(\mathbf{x}), \dots, \alpha_{i-1}(\mathbf{x})$ such that

$$ad_\mathbf{f}^i \, \mathbf{g} = \sum_{k=0}^{i-1} \alpha_k \, ad_\mathbf{f}^k \, \mathbf{g}$$

we would have

$$ad_\mathbf{f}^{n-1} \, \mathbf{g} = \sum_{k=n-i-1}^{n-2} \alpha_k \, ad_\mathbf{f}^k \, \mathbf{g}$$

This, together with (6.57a), would imply that

$$\nabla z_1 \, ad_\mathbf{f}^{n-1} \, \mathbf{g} = \sum_{k=n-i-1}^{n-2} \alpha_k \, \nabla z_1 \, ad_\mathbf{f}^k \, \mathbf{g} = 0$$

a contradiction to (6.57b).

The second property we can infer from (6.57) is that the vector fields are involutive. This follows from the existence of a scalar function $z_1$ satisfying the $n-1$ partial differential equations in (6.57a), and from the necessity part of the Frobenius theorem. Thus, we have completed the necessity part of the proof of Theorem 6.2.

Let us now prove that the two conditions in Theorem 6.2 are also *sufficient* for the input-state linearizability of the nonlinear system in (6.52), *i.e.*, that we can find a state transformation and an input transformation such that (6.54) is satisfied. The reasoning is as follows. Since the involutivity condition is satisfied, from Frobenius theorem there exists a non zero scalar function $z_1(\mathbf{x})$ satisfying

$$L_{\mathbf{g}} z_1 = L_{ad_{\mathbf{f}}\mathbf{g}} z_1 = \dots = L_{ad_{\mathbf{f}}^{n-2}\mathbf{g}} z_1 = 0 \tag{6.58}$$

From Lemma 6.4, the above equations can be written

$$L_{\mathbf{g}} z_1 = L_{\mathbf{g}} L_{\mathbf{f}} z_1 = \dots = L_{\mathbf{g}} L_{\mathbf{f}}^{n-2} z_1 = 0 \tag{6.59}$$

This means that if we use $\mathbf{z} = [\, z_1 \quad L_{\mathbf{f}} z_1 \quad \dots \quad L_{\mathbf{f}}^{n-1} z_1 \,]^T$ as a new set of state variables, the first $n-1$ state equations verify

$$\dot{z}_k = z_{k+1} \qquad k = 1, \dots, n-1$$

while the last state equation is

$$\dot{z}_n = L_{\mathbf{f}}^n z_1 + L_{\mathbf{g}} L_{\mathbf{f}}^{n-1} z_1 \, u \tag{6.60}$$

Now the question is whether $L_{\mathbf{g}} L_{\mathbf{f}}^{n-1} z_1$ can be equal to zero. Since the vector fields $\{\, \mathbf{g}, ad_{\mathbf{f}}\mathbf{g}, \dots, ad_{\mathbf{f}}^{n-1}\mathbf{g} \,\}$ are linearly independent in $\Omega$, and noticing, as in the proof of Lemma 6.4, that (6.58) also leads to

$$L_{\mathbf{g}} L_{\mathbf{f}}^{n-1} z_1 = (-1)^{n-1} L_{ad_{\mathbf{f}}^{n-1}\mathbf{g}} z_1$$

we must have

$$L_{ad_{\mathbf{f}}^{n-1}\mathbf{g}} z_1(\mathbf{x}) \neq 0 \qquad\qquad \forall\, \mathbf{x} \in \Omega \tag{6.61}$$

Otherwise, the non-zero vector $\nabla z_1$ would satisfy

$$\nabla z_1 \,[\, \mathbf{g} \quad ad_{\mathbf{f}}\mathbf{g} \quad \dots \quad ad_{\mathbf{f}}^{n-1}\mathbf{g} \,] = 0$$

and thus would be orthogonal to $n$ linearly independent vectors, a contradiction.

Therefore, by taking the control law to be

$$u = (-L_{\mathbf{f}}^n z_1 + v)/(L_{\mathbf{g}} L_{\mathbf{f}}^{n-1} z_1)$$

equation (6.60) simply becomes

$$\dot{z}_n = v$$

which shows that the input-output linearization of the nonlinear system has been achieved.    □

## HOW TO PERFORM INPUT-STATE LINEARIZATION

Based on the previous discussion, the input-state linearization of a nonlinear system can be performed through the following steps:

- Construct the vector fields $\mathbf{g}, ad_{\mathbf{f}}\, \mathbf{g}, \dots, ad_{\mathbf{f}}^{n-1}\mathbf{g}$ for the given system

- Check whether the controllability and involutivity conditions are satisfied

- If both are satisfied, find the first state $z_1$ (the output function leading to input-output linearization of relative degree $n$) from equations (6.58), *i.e.*,

$$\nabla z_1 \, ad_\mathbf{f}^i \mathbf{g} = 0 \qquad i = 0, \dots, n-2 \tag{6.62a}$$

$$\nabla z_1 \, ad_\mathbf{f}^{n-1} \mathbf{g} \neq 0 \tag{6.62b}$$

- Compute the state transformation $\mathbf{z}(\mathbf{x}) = [\, z_1 \quad L_\mathbf{f} z_1 \quad \dots \quad L_\mathbf{f}^{n-1} z_1 \,]^T$ and the input transformation (6.53), with

$$\alpha(\mathbf{x}) = -\frac{L_\mathbf{f}^n z_1}{L_\mathbf{g} L_\mathbf{f}^{n-1} z_1} \tag{6.63a}$$

$$\beta(\mathbf{x}) = \frac{1}{L_\mathbf{g} L_\mathbf{f}^{n-1} z_1} \tag{6.63b}$$

Let us now demonstrate the above procedure on a simple physical example [Marino and Spong, 1986; Spong and Vidyasagar, 1989].

**Example 6.10:** Consider the control of the mechanism in Figure 6.6, which represents a link driven by a motor through a torsional spring (a single-link flexible-joint robot), in the vertical plane. Its equations of motion can be easily derived as

$$I\ddot{q}_1 + MgL\sin q_1 + k(q_1 - q_2) = 0 \tag{6.64a}$$

$$J\ddot{q}_2 - k(q_1 - q_2) = u \tag{6.64b}$$

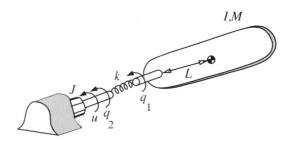

**Figure 6.6 :** A flexible-joint mechanism

Because nonlinearities (due to gravitational torques) appear in the first equation, while the control input $u$ enters only in the second equation, there is no obvious way of designing a large range controller. Let us now consider whether input-state linearization is possible.

First, let us put the system's dynamics in a state-space representation. Choosing the state vector as

$$\mathbf{x} = [q_1 \ \dot{q}_1 \ q_2 \ \dot{q}_2]^T$$

the corresponding vector fields **f** and **g** can be written

$$\mathbf{f} = [ \ x_2 \quad -\frac{MgL}{I}\sin x_1 - \frac{k}{I}(x_1 - x_3) \quad x_4 \quad \frac{k}{J}(x_1 - x_3) \ ]^T$$

$$\mathbf{g} = [ \ 0 \ \ 0 \ \ 0 \ \ \frac{1}{J} \ ]^T$$

Second, let us check the controllability and involutivity conditions   The controllability matrix is obtained by simple computation

$$[ \ \mathbf{g} \quad ad_{\mathbf{f}}\,\mathbf{g} \quad ad^2_{\mathbf{f}}\,\mathbf{g} \quad ad^3_{\mathbf{f}}\,\mathbf{g} \ ] = \begin{bmatrix} 0 & 0 & 0 & -\dfrac{k}{IJ} \\[2mm] 0 & 0 & \dfrac{k}{IJ} & 0 \\[2mm] 0 & -\dfrac{1}{J} & 0 & \dfrac{k}{J^2} \\[2mm] \dfrac{1}{J} & 0 & -\dfrac{k}{J^2} & 0 \end{bmatrix}$$

It has rank 4 for $k > 0, IJ < \infty$. Furthermore, since the vector fields $\{\mathbf{g}, ad_{\mathbf{f}}\,\mathbf{g}, ad_{\mathbf{f}}{}^2\mathbf{g}\}$ are constant, they form an involutive set. Therefore, the system in (6.64) is input-state linearizable.

Third, let us find out the state transformation $\mathbf{z} = \mathbf{z}(\mathbf{x})$ and the input transformation $u = \alpha(\mathbf{x}) + \beta(\mathbf{x})v$ so that input-state linearization is achieved. From (6.62), and given the above expression of the controllability matrix, the first component $z_1$ of the new state vector $\mathbf{z}$ should satisfy

$$\frac{\partial z_1}{\partial x_2} = 0 \qquad \frac{\partial z_1}{\partial x_3} = 0 \qquad \frac{\partial z_1}{\partial x_4} = 0 \qquad \frac{\partial z_1}{\partial x_1} \neq 0$$

Thus, $z_1$ must be a function of $x_1$ only. The simplest solution to the above equations is

$$z_1 = x_1 \tag{6.65a}$$

The other states can be obtained from $z_1$

$$z_2 = \nabla z_1\,\mathbf{f} = x_2 \tag{6.65b}$$

$$z_3 = \nabla z_2\,\mathbf{f} = -\frac{MgL}{I}\sin x_1 - \frac{k}{I}(x_1 - x_3) \tag{6.65c}$$

$$z_4 = \nabla z_3\, \mathbf{f} = -\frac{MgL}{I} x_2 \cos x_1 - \frac{k}{I}(x_2 - x_4) \tag{6.65d}$$

Accordingly, the input transformation is

$$u = (v - \nabla z_4\, \mathbf{f})/(\nabla z_4\, \mathbf{g})$$

which can be written explicitly as

$$u = \frac{IJ}{k}(v - a(\mathbf{x})) \tag{6.66}$$

where

$$a(\mathbf{x}) = \frac{MgL}{I}\sin x_1 \,(x_2^{\,2} + \frac{MgL}{I}\cos x_1 + \frac{k}{I}) + \frac{k}{I}(x_1 - x_3)\,(\frac{k}{I} + \frac{k}{J} + \frac{MgL}{I}\cos x_1)$$

As a result of the above state and input transformations, we end up with the following set of linear equations

$$\dot{z}_1 = z_2$$

$$\dot{z}_2 = z_3$$

$$\dot{z}_3 = z_4$$

$$\dot{z}_4 = v$$

thus completing the input-state linearization.

Finally, note that

• The above input-state linearization is actually global, because the diffeomorphism $\mathbf{z}(\mathbf{x})$ and the input transformation are well defined everywhere. Specifically, the inverse of the state transformation (6.65) is

$$x_1 = z_1$$

$$x_2 = z_2$$

$$x_3 = z_1 + \frac{I}{k}\left(z_3 + \frac{MgL}{I}\sin z_1\right)$$

$$x_4 = z_2 + \frac{I}{k}\left(z_4 + \frac{MgL}{I} z_2 \cos z_1\right)$$

which is well defined and differentiable everywhere. The input transformation (6.66) is also well defined everywhere, of course.

• In this particular example, the transformed variables have physical meanings. We see that

$z_1$ is the link position, $z_2$ the link velocity, $z_3$ the link acceleration, and $z_4$ the link jerk. This further illustrates our earlier remark that the complexity of a nonlinear physical model is strongly dependent on the choice of state variables.

• In hindsight, of course, we also see that the same result could have been derived simply by differentiating equation (6.64a) twice, *i.e.*, from the input-output linearization perspective of Lemma 6.3.                                                                ☐

Note that inequality (6.62b) can be replaced by the normalization equation

$$\nabla z_1 \, ad_{\mathbf{f}}^{\,n-1}\mathbf{g} \;=\; 1$$

without affecting the input-state linearization. This equation and (6.62a) constitute a total of $n$ linear equations,

$$
[\, ad_{\mathbf{f}}^{o}\mathbf{g} \quad ad_{\mathbf{f}}^{1}\mathbf{g} \quad \ldots \quad ad_{\mathbf{f}}^{n-2}\mathbf{g} \quad ad_{\mathbf{f}}^{n-1}\mathbf{g} \,]
\begin{bmatrix}
\partial z_1/\partial x_1 \\
\partial z_1/\partial x_2 \\
\ldots \\
\partial z_1/\partial x_{n-1} \\
\partial z_1/\partial x_n
\end{bmatrix}
=
\begin{bmatrix}
0 \\
0 \\
\ldots \\
0 \\
1
\end{bmatrix}
$$

Given the independence condition on the vector fields, the partial derivatives $\partial z_1/\partial x_1$ , ..... , $\partial z_1/\partial x_n$ can be computed uniquely from the above equations. The state variable $z_1$ can then be found, in principle, by sequentially integrating these partial derivatives. Note that analytically solving this set of partial differential equations for $z_1$ may be a nontrivial step (although numerical solutions may be relatively easy due to the recursive nature of the equations).

## CONTROLLER DESIGN BASED ON INPUT-STATE LINEARIZATION

With the state equation transformed into a linear form, one can easily design controllers for either stabilization or tracking purposes. A stabilization example has already been provided in the intuitive section 6.1, where $v$ is designed to place the poles of the equivalent linear dynamics, and the physical input $u$ is then computed using the corresponding input transformation. One can also design tracking controllers based on the equivalent linear system, provided that the desired trajectory can be expressed in terms of the first linearizing state component $z_1$.

Consider again the flexible link example. Its equivalent linear dynamics can be expressed as

$$z_1^{(4)} = v$$

Assume that it is desired to have the link position $z_1$ track a prespecified trajectory $z_{d1}(t)$. The control law

$$v = z_{d1}^{(4)} - a_3 \tilde{z}_1^{(3)} - a_2 \tilde{z}_1^{(2)} - a_1 \dot{\tilde{z}}_1 - a_o \tilde{z}_1$$

(where $\tilde{z}_1 = z_1 - z_{d1}$ ) leads to the tracking error dynamics

$$\tilde{z}_1^{(4)} + a_3 \tilde{z}_1^{(3)} + a_2 \tilde{z}_1^{(2)} + a_1 \dot{\tilde{z}}_1 + a_o \tilde{z}_1 = 0$$

The above dynamics is exponentially stable if the positive constants $a_i$ are chosen properly. To find the physical input $u$, one then simply uses (6.66).

# 6.4 Input-Output Linearization of SISO Systems

In this section, we discuss input-output linearization of single-input nonlinear systems described by the state space representation

$$\dot{\mathbf{x}} = \mathbf{f}(\mathbf{x}) + \mathbf{g}(\mathbf{x})u \tag{6.67a}$$

$$y = h(\mathbf{x}) \tag{6.67b}$$

where $y$ is the system output. By input-output linearization we mean the generation of a *linear* differential relation between the output $y$ and a new input $v$ ($v$ here is similar to the equivalent input $v$ in input-state linearization). Specifically, we shall discuss the following issues:

- How to generate a linear input-output relation for a nonlinear system?

- What are the internal dynamics and zero-dynamics associated with the input-output linearization?

- How to design stable controllers based on input-output linearizations?

### GENERATING A LINEAR INPUT-OUTPUT RELATION

As discussed in section 6.1.3, the basic approach of input-output linearization is simply to differentiate the output function $y$ repeatedly until the input $u$ appears, and then design $u$ to cancel the nonlinearity. However, in some cases, the second part of the approach may not be carried out, because the system's relative degree is undefined.

### The Case of Well Defined Relative Degree

Let us place ourselves in a region (an open connected set) $\Omega_{\mathbf{x}}$ in the state space. Using the notations of differential geometry, the process of repeated differentiation means that we start with

$$\dot{y} = \nabla h \, (\mathbf{f} + \mathbf{g} u) = L_{\mathbf{f}} \, h(\mathbf{x}) + L_{\mathbf{g}} \, h(\mathbf{x}) u$$

If $L_{\mathbf{g}} \, h(\mathbf{x}) \neq 0$ for *some* $\mathbf{x} = \mathbf{x}_0$ in $\Omega_{\mathbf{x}}$, then, by continuity, that relation is verified in a finite neighborhood $\Omega$ of $\mathbf{x}_0$. In $\Omega$, the input transformation

$$u = \frac{1}{L_{\mathbf{g}} \, h} \, (-L_{\mathbf{f}} \, h + v)$$

results in a linear relation between $y$ and $v$, namely $\dot{y} = v$.

If $L_{\mathbf{g}} \, h(\mathbf{x}) = 0$ for *all* $\mathbf{x}$ in $\Omega_{\mathbf{x}}$, we can differentiate $\dot{y}$ to obtain

$$\ddot{y} = L_{\mathbf{f}}^2 \, h(\mathbf{x}) + L_{\mathbf{g}} \, L_{\mathbf{f}} \, h(\mathbf{x}) u$$

If $L_{\mathbf{g}} \, L_{\mathbf{f}} \, h(\mathbf{x})$ is again zero for *all* $\mathbf{x}$ in $\Omega_{\mathbf{x}}$, we shall differentiate again and again,

$$y^{(i)} = L_{\mathbf{f}}^i \, h(\mathbf{x}) + L_{\mathbf{g}} \, L_{\mathbf{f}}^{i-1} \, h(\mathbf{x}) u$$

until for some integer $r$

$$L_{\mathbf{g}} \, L_{\mathbf{f}}^{r-1} \, h(\mathbf{x}) \neq 0$$

for *some* $\mathbf{x} = \mathbf{x}_0$ in $\Omega_{\mathbf{x}}$. Then, by continuity, the above relation is verified in a finite neighborhood $\Omega$ of $\mathbf{x}_0$. In $\Omega$, the control law

$$u = \frac{1}{L_{\mathbf{g}} \, L_{\mathbf{f}}^{r-1} h} \, (-L_{\mathbf{f}}^r h + v) \tag{6.68}$$

applied to

$$y^{(r)} = L_{\mathbf{f}}^r \, h(\mathbf{x}) + L_{\mathbf{g}} \, L_{\mathbf{f}}^{r-1} \, h(\mathbf{x}) u \tag{6.69}$$

yields the simple linear relation

$$y^{(r)} = v \tag{6.70}$$

As discussed in section 6.1.3, the number $r$ of differentiations required for the input $u$ to appear is called the *relative degree* of the system, an extension of the usual definition of relative degree for linear systems. As also noticed earlier, and as we shall soon formalize, one necessarily has $r \leq n$ (where $n$ is the system order). If $r = n$, then input-output linearization actually yields input-state linearization in $\Omega$, as stated in Lemma 6.3.

Based on the above procedure, we can give the following formal definition.

**Definition 6.7**    *The SISO system is said to have <u>relative degree</u> r in a region $\Omega$ if,*
$\forall \, \mathbf{x} \in \Omega$

$$L_{\mathbf{g}} L_{\mathbf{f}}^{\,i} \, h(\mathbf{x}) = 0 \qquad 0 \le i < r-1 \tag{6.71a}$$

$$L_{\mathbf{g}} L_{\mathbf{f}}^{\,r-1} h(\mathbf{x}) \ne 0 \tag{6.71b}$$

### The Case of Undefined Relative Degree

It is often the case that one is *a priori* interested in the properties of the system about a specific operating *point* $\mathbf{x}_0$. The definition of the relative degree then requires particular care.

As before, let us differentiate the output $y$ until the input $u$ appears (*i.e.*, until the coefficient of $u$ is not identically zero in a neighborhood of $\mathbf{x}_0$). Then, if the coefficient of $u$ is nonzero *at* $\mathbf{x}_0$

$$L_{\mathbf{g}} L_{\mathbf{f}}^{\,r-1} h(\mathbf{x}_0) \ne 0$$

this also implies, by continuity, that (6.71b) is verified over a finite neighborhood of $\mathbf{x}_0$. We shall then say that the system has relative degree $r$ <u>*at the point*</u> $\mathbf{x}_0$.

However, it is possible that when the input appears, its coefficient $L_{\mathbf{g}} L_{\mathbf{f}}^{\,r-1} h(\mathbf{x})$ is zero *at* $\mathbf{x}_0$, but nonzero at some points $\mathbf{x}$ arbitrarily close to $\mathbf{x}_0$. The relative degree of the nonlinear system is then *undefined* at $\mathbf{x}_0$. This case is illustrated by the following simple example.

**Example 6.11:** Consider the system

$$\ddot{x} = \rho(x, \dot{x}) + u$$

where $\rho$ is some smooth nonlinear function of the state $\mathbf{x} = [\,x \quad \dot{x}\,]^T$. If

$$y = x$$

is defined as the output of interest, then the system is obviously in companion form, with relative degree 2.

However, if instead one takes

$$y = x^2$$

as output, then

$$\dot{y} = 2\,x\,\dot{x}$$

$$\ddot{y} = 2\,x\,\ddot{x} + 2\,\dot{x}^2 = 2\,x\,(\rho + u) + 2\,\dot{x}^2$$

In other words,

$$L_g L_f h = 2\,x$$

and therefore, at $\mathbf{x} = \mathbf{0}$ and for this choice of output, the system has neither relative degree 1 nor relative degree 2. $\qquad\square$

In some particular cases, as above, a simple change of output will allow one to define an equivalent but easily solvable control problem. In general, however, input-output linearization at a point cannot be straightforwardly achieved when the relative degree is undefined. In the rest of this section, we consider only systems having a well defined relative degree in a region $\Omega$ (which shall often be a neighborhood of an operating point of interest $\mathbf{x}_o$ ).

## NORMAL FORMS

When the relative degree $r$ is defined and $r < n$, the nonlinear system (6.67) can be transformed, using $y, \dot{y}, \ldots , y^{(r-1)}$ as part of the new state components, into a so-called "normal form", which shall allow us to take a more formal look at the notions of internal dynamics and zero-dynamics introduced in section 6.1.3. Let

$$\boldsymbol{\mu} = [\mu_1 \ \ \mu_2 \ \ \ldots \ \ \mu_r]^T = [\,y \ \ \dot{y} \ \ \ldots \ \ y^{(r-1)}\,]^T \tag{6.72}$$

In a neighborhood $\Omega$ of a point $\mathbf{x}_o$ , the *normal form* of the system can be written as

$$\dot{\boldsymbol{\mu}} = \begin{bmatrix} \mu_2 \\ \cdots \\ \cdots \\ \mu_r \\ a(\boldsymbol{\mu}, \boldsymbol{\psi}) + b(\boldsymbol{\mu}, \boldsymbol{\psi})\,u \end{bmatrix} \tag{6.73a}$$

$$\dot{\boldsymbol{\psi}} = \mathbf{w}(\boldsymbol{\mu}, \boldsymbol{\psi}) \tag{6.73b}$$

with the output defined as

$$y = \mu_1 \tag{6.74}$$

The $\mu_i$ and $\psi_j$ are referred to as *normal coordinates* or *normal states* in $\Omega$ (or at $\mathbf{x}_o$ ). Note that the companion form subsystem (6.73a) is simply another expression of (6.69), while the subsystem (6.73b) does not contain the system input $u$.

To show that the nonlinear system (6.67) can indeed be transformed into the normal form (6.73), we have to show not only that such a coordinate transformation exists, but also that it is a true state transformation. In other words, we need to show that we can construct a (local) *diffeomorphism*

$$\phi(\mathbf{x}) = [\, \mu_1 \; \cdots \; \mu_r \; \psi_1 \; \cdots \; \psi_{n-r} \,]^T \tag{6.75}$$

such that (6.73) is verified. According to Lemma 6.2, to show that $\phi$ is a diffeomorphism, it suffices to show that its Jacobian is invertible, *i.e.*, that the *gradients* $\nabla\mu_i$ and $\nabla\psi_j$ are all linearly independent.

In the following, we first show that the gradients $\nabla\mu_i$ are linearly independent, *i.e.*, that the components of $\mu$ can serve as state variables. This result, of course, is not overly surprising, since it simply says that the system output $y$ and its first $r-1$ derivatives can serve as state variables. We then show that $n-r$ other vector fields $\psi_j$ can be found to complete the new state vector, in other words, can be found such that the gradients $\nabla\mu_i$ and $\nabla\psi_j$ are all linearly independent.

Let us first show that the gradients $\nabla\mu_i$ are linearly independent.

**Lemma 6.5**    *If the relative degree of the system (6.67) is $r$ in the region $\Omega$, then the gradients $\nabla\mu_1, \nabla\mu_2, \ldots, \nabla\mu_r$ are linearly independent in $\Omega$.*

**Proof**: Let us first remark that, in terms of $\mu$, equations (6.71) can be written simply as

$$\nabla\mu_i \, \mathbf{g} = 0 \qquad 1 \le i < r \tag{6.76a}$$

$$\nabla\mu_r \, \mathbf{g} \ne 0 \tag{6.76b}$$

We present the proof for the case $r = 3$ (the general case can be shown in a similar fashion). Assume that there exist smooth functions $\alpha_i(\mathbf{x})$ such that (everywhere in $\Omega$)

$$\alpha_1 \nabla\mu_1 + \alpha_2 \nabla\mu_2 + \alpha_3 \nabla\mu_3 = 0 \tag{6.77a}$$

Multiplying (6.77a) by $\mathbf{g}$ yields

$$[\, \alpha_1 \nabla\mu_1 + \alpha_2 \nabla\mu_2 + \alpha_3 \nabla\mu_3 \,] \mathbf{g} = 0$$

which, from (6.76), implies that $\alpha_3 = 0$ (everywhere in $\Omega$).

Replacing $\alpha_3 = 0$ in (6.77a), in turn, implies that

$$\alpha_1 \nabla\mu_1 + \alpha_2 \nabla\mu_2 = 0 \tag{6.77b}$$

Multiplying (6.77b) by the Lie bracket $ad_f\mathbf{g}$, and using Jacobi's identity and (6.71a), yields

$$0 = \alpha_1 L_{ad_f\mathbf{g}} \, \mu_1 + \alpha_2 L_{ad_f\mathbf{g}} \, \mu_2 = \alpha_1 [L_f L_\mathbf{g} h - L_\mathbf{g} L_f h] + \alpha_2 [L_f L_\mathbf{g} - L_\mathbf{g} L_f] L_f h = - \alpha_2 L_\mathbf{g} L_f^2 h$$

which, from (6.71b), implies that $\alpha_2 = 0$ (everywhere in $\Omega$).

Replacing $\alpha_2 = 0$ in (6.77b) implies that

$$\alpha_1 \nabla \mu_1 = 0 \tag{6.77c}$$

Multiplying (6.77c) by $ad_f^2 \mathbf{g}$, and using (6.47) and (6.71a), leads to

$$0 = \alpha_1 L_{ad_f^2 \mathbf{g}} \mu_1 = \alpha_1 [L_f^2 L_\mathbf{g} h - 2 L_f L_\mathbf{g} L_f h + L_\mathbf{g} L_f^2 h] = \alpha_1 L_\mathbf{g} L_f^2 h$$

which, from (6.71b), implies that $\alpha_1 = 0$ (everywhere in $\Omega$).

Thus, (6.77a) can only be verified if all the coefficients $\alpha_i$ equal zero everywhere in $\Omega$. This shows that the gradients $\nabla \mu_i$ are linearly independent. $\qquad\blacksquare$

Note that an immediate implication of this result is that, consistently with intuition, the relative degree of an $n^{th}$-order system is always smaller than $n$ (because there cannot be more than $n$ linearly independent vectors in an $n$-dimensional space).

Let us now show that there exist $n-r$ more functions $\psi_j$ to complete the coordinate transformation.

**Proof**: As noticed earlier, given Lemma 6.2, we simply need to show that gradients vectors $\nabla \psi_j$ can be found such that the $\nabla \mu_i$ and $\nabla \psi_j$ are all linearly independent. The development is illustrated in Figure 6.7.

At $\mathbf{x}_o$, equation (6.76a) indicates that the first $r-1$ vectors $\nabla \mu_i$ ($i = 1, \ldots, r-1$), which we have just shown to be linearly independent, are all within the hyperplane orthogonal to $\mathbf{g}$. Since the dimension of that hyperplane is $n-1$, one can find

$$n - r = (n - 1) - (r - 1)$$

vectors in that hyperplane that are linearly independent of the $\nabla \mu_i$ and linearly independent of each other. *Let us call these vectors* $\nabla \psi_j$ ($j = 1, \ldots, n-r$). By definition, they verify

$$\nabla \psi_j \, \mathbf{g} = 0 \qquad 1 \leq j \leq n-r \tag{6.78}$$

A subtle point at this stage is that gradient vector fields are not just any vector fields (they must satisfy curl conditions, cf. also section 3.5.3). Fortunately, Frobenius theorem (Theorem 6.1) comes to the rescue, since its application to the single vector field $\mathbf{g}$ (which is obviously involutive) guarantees that there does exist $n-1$ linearly independent *gradient* functions $\nabla h_k$ which satisfy $\nabla h_k \, \mathbf{g} = 0$.

Furthermore, from (6.76b), $\mu_r$ is *not* in the hyperplane orthogonal to $\mathbf{g}$. Thus the gradients $\nabla \mu_i$ ($i = 1, \ldots, r$) and $\nabla \psi_j$ ($j = 1, \ldots, n-r$) are all linearly independent. Thus, the Jacobian of the transformation (6.75) is invertible.

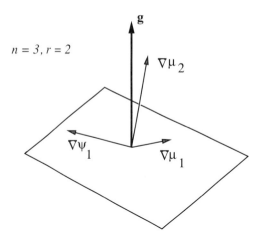

$n = 3, r = 2$

**Figure 6.7 :** Illustrating the construction of $\psi$

By continuity, the Jacobian remains invertible in a neighborhood $\Omega_1$ of $\mathbf{x}_o$ . Redefining $\Omega$ as the intersection of $\Omega_1$ and $\Omega$, the transformation $\phi$ thus defines a diffeomorphism in $\Omega$. Hence, in $\Omega$ this transformation is a true state transformation, which puts the nonlinear system into the form (6.73), with, in (6.73a)

$$a(\mathbf{\mu}, \mathbf{\psi}) = L_\mathbf{f}{}^r h(\mathbf{x}) = L_\mathbf{f}{}^r h[\phi^{-1}(\mathbf{\mu},\mathbf{\psi})]$$

$$b(\mathbf{\mu}, \mathbf{\psi}) = L_\mathbf{g} L_\mathbf{f}{}^{r-1} h(\mathbf{x}) = L_\mathbf{g} L_\mathbf{f}{}^{r-1} h[\phi^{-1}(\mathbf{\mu},\mathbf{\psi})]$$

while in (6.73b) the input $u$ indeed does not appear, since from (6.78) the $\psi_j$ verify

$$L_\mathbf{g}\, \psi_j(\mathbf{x}) = 0 \qquad \forall \mathbf{x} \in \Omega \qquad\qquad \square$$

From a practical point of view, explicitly finding a vector field $\psi$ to complete the transformation into a normal form requires the (often nontrivial) step of solving the set (6.78) of partial differential equations in the $\psi_j$. The following example, adapted from [Isidori, 1989], illustrates this procedure.

**Example 6.12:** Consider the nonlinear system

$$\dot{\mathbf{x}} = \begin{bmatrix} -x_1 \\ 2\,x_1\,x_2 + \sin x_2 \\ 2\,x_2 \end{bmatrix} + \begin{bmatrix} e^{2x_2} \\ 1/2 \\ 0 \end{bmatrix} u \qquad (6.79)$$

$$y = h(\mathbf{x}) = x_3$$

Since

$$\dot{y} = 2\,x_2$$

$$\ddot{y} = 2\,\dot{x}_2 = 2\,(2\,x_1\,x_2 + \sin x_2) + u$$

the system has relative degree 2, and

$$L_f h(\mathbf{x}) = 2 x_2 \qquad L_g h(\mathbf{x}) = 0$$

$$L_g L_f h(\mathbf{x}) = 1$$

In order to find the normal form, let us take

$$\mu_1 = h(\mathbf{x}) = x_3$$

$$\mu_2 = L_f\ h(\mathbf{x}) = 2 x_2$$

The third function $\psi(\mathbf{x})$ required to complete the transformation should satisfy

$$L_g\,\psi = \frac{\partial \psi}{\partial x_1} e^{2 x_2} + \frac{1}{2}\frac{\partial \psi}{\partial x_2} = 0$$

One solution of this equation is

$$\psi(\mathbf{x}) = 1 + x_1 - e^{2 x_2}$$

Consider now the associated state transformation $\mathbf{z} = [\ \mu_1\ \ \mu_2\ \ \psi\ ]^T$. Its Jacobian matrix is

$$\frac{\partial \mathbf{z}}{\partial \mathbf{x}} = \begin{bmatrix} 0 & 0 & 1 \\ 0 & 2 & 0 \\ 1 & -2\,e^{2 x_2} & 0 \end{bmatrix}$$

which is non-singular for any $\mathbf{x}$. The inverse transformation is given by

$$x_1 = -1 + \psi + e^{\mu_2}$$

$$x_2 = \frac{1}{2}\mu_2$$

$$x_3 = \mu_1$$

Thus, this state transformation is valid globally. With the above set of new coordinates, the system dynamics is put into the normal form

$$\dot{\mu}_1 = \mu_2$$

$$\dot{\mu}_2 = 2\,(-1 + \psi + e^{\mu_2})\,\mu_2 + 2 \sin (\mu_2/2) + u$$

$$\dot{\psi} = (1 - \psi - e^{\mu_2})(1 + 2\,\mu_2 e^{\mu_2}) - 2 \sin (\mu_2/2)\,e^{\mu_2} \qquad \square$$

## THE ZERO-DYNAMICS

By means of input-output linearization, the dynamics of a nonlinear system is decomposed into an external (input-output) part and an internal ("unobservable") part. Since the external part consists of a linear relation between $y$ and $v$ (or equivalently, the controllability canonical form between $y$ and $u$, as in (6.73a)), it is easy to design the input $v$ so that the output $y$ behaves as desired. Then, the question is whether the internal dynamics will also behave well, *i.e.*, whether the internal states will remain bounded. Since the control design must account for the whole dynamics (and therefore cannot tolerate the instability of internal dynamics), the internal behavior has to be addressed carefully.

The internal dynamics associated with the input-output linearization simply corresponds to the last $(n-r)$ equations $\dot{\boldsymbol{\psi}} = \mathbf{w}(\boldsymbol{\mu}, \boldsymbol{\psi})$ of the normal form. Generally, this dynamics depends on the output states $\boldsymbol{\mu}$. However, as we now detail, we can define an *intrinsic* property of the nonlinear system by considering the system's internal dynamics when the control input is such that the output $y$ is maintained at zero. Studying this so-called *zero-dynamics* will allow us to make some conclusions about the stability of the internal dynamics.

The constraint that the output $y$ is identically zero implies that all of its time derivatives are zero. Thus, the corresponding internal dynamics of the system, or zero-dynamics, describes motion restricted to the $(n-r)$-dimensional smooth surface (manifold) $M_O$ defined by $\boldsymbol{\mu} = \mathbf{0}$. In order for the system to operate in zero-dynamics, *i.e.*, for the state $\mathbf{x}$ to stay on the surface $M_O$, the initial state of the system $\mathbf{x}(0)$ must be on the surface, and, furthermore, the input $u$ must be such that $y$ stays at zero, *i.e.*, such that $y^{(r)}(t) = 0$. From (6.69), this means that $u$ must equal

$$u_O(\mathbf{x}) = - \frac{L_{\mathbf{f}}^r h(\mathbf{x})}{L_{\mathbf{g}} L_{\mathbf{f}}^{r-1} h(\mathbf{x})}$$

Corresponding to this input, and assuming that indeed the system's initial state is on the surface, *i.e.*, that $\boldsymbol{\mu}(0) = \mathbf{0}$, the system dynamics can be simply written in normal form as

$$\dot{\boldsymbol{\mu}} = \mathbf{0} \tag{6.80a}$$

$$\dot{\boldsymbol{\psi}} = \mathbf{w}(\mathbf{0}, \boldsymbol{\psi}) \tag{6.80b}$$

By definition, (6.80b) is the <u>*zero-dynamics*</u> of the nonlinear system (6.67).

The evolution of the system state when operating in zero-dynamics is illustrated in Figure 6.8. Note that, in normal coordinates, the control input $u_O$ can be written as a function only of the internal states $\boldsymbol{\psi}$

$$u_O(\boldsymbol{\psi}) = - \frac{a(\mathbf{0}, \boldsymbol{\psi})}{b(\mathbf{0}, \boldsymbol{\psi})}$$

**Example 6.13: Zero-dynamics of a nonlinear system**

Consider again the nonlinear system of Example 6.12. Its internal dynamics is represented by the equation

$$\dot{\psi} = (1 - \psi - e^{\mu_2})(1 + 2\,\mu_2\,e^{\mu_2}) - 2\sin(\mu_2/2)\,e^{\mu_2}$$

Its zero-dynamics is obtained by letting $\mu_1 = 0$ and $\mu_2 = 0$

$$\dot{\psi} = -\psi$$

The input $u_o = 0$ maintains the output always equal to zero.     ◻

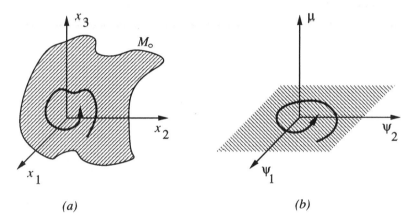

*(a)*     *(b)*

**Figure 6.8 :** Evolution of the system state on the zero-dynamics manifold, for $n = 3, r = 1$.
(a) in original state coordinates, (b) in normal coordinates.

Note that computing a system's internal dynamics (or zero-dynamics) does not necessarily require to first put the system explicitly in normal form, *i.e.* to solve the p.d.e.'s defining $\psi$, which were carefully devised such that $u$ does not appear in $\dot{\psi}$. Indeed, since $u$ (or $u_o$) is *known* as a function of the state $\mathbf{x}$, it is often easier, given $\mu$, to simply find $n-r$ vector fields $\rho_j$ to complete the state transformation, and then replace $u$ (or $u_o$) by its expression, as we did in section 6.1.3. One can verify that the corresponding state transformation is one-to-one either directly, or by checking that the Jacobian of the transformation is invertible.

Recall from section 6.1.3 that linear systems whose zero-dynamics are stable are called minimum phase. For convenience of reference, we shall extend the terminology to nonlinear systems.

**Definition 6.8**    *The nonlinear system (6.67) is said to be <u>asymptotically minimum phase</u> if its zero-dynamics is asymptotically stable.*

The concept of exponentially minimum phase can be defined similarly. If the zero dynamics is asymptotically stable for any $\boldsymbol{\psi}(0)$, we shall say the system is globally asymptotically minimum phase. Otherwise, we shall say the system is locally minimum phase. For instance, the system (6.79) is globally asymptotically minimum phase.

Let us now consider controller designs based on input-output linearization. Essentially, the idea is to design a controller based on the linear input-output relation and then check whether the internal dynamics is stable or not. In the following, we first state a result for local stabilization, and then offer some discussions of global stabilization and of tracking control.

### LOCAL ASYMPTOTIC STABILIZATION

Consider again the nonlinear system (6.67). It is natural to wonder whether choosing the artificial input $v$ in (6.70) as a simple *linear pole-placement* controller provides any guarantee about the stability of the overall system. In other words, assume that in (6.70) we let

$$v = -k_{r-1}y^{(r-1)} - ... - k_1\dot{y} - k_o y$$

where the coefficients $k_i$ are chosen such that the polynomial

$$K(p) = p^r + k_{r-1}p^{r-1} + ..... + k_1 p + k_o \tag{6.81}$$

has all its roots strictly in the left-half plane. The actual control input $u$ can then be written, from (6.68)

$$u(\mathbf{x}) = \frac{1}{L_{\mathbf{g}} L_{\mathbf{f}}^{r-1} y}[-L_{\mathbf{f}}^r y - k_{r-1}y^{(r-1)} - ... - k_1\dot{y} - k_o y] \tag{6.82}$$

The following result indicates that, *provided that the zero-dynamics is asymptotically stable*, this control law indeed stabilizes the whole system *locally*.

**Theorem 6.3**    *Assume that the system (6.67) has relative degree r, and that its zero-dynamics is locally asymptotically stable. Choose constants $k_i$ such that the polynomial (6.81) has all its roots strictly in the left-half plane. Then, the control law (6.82) yields a locally asymptotically stable closed-loop system.*

> **Proof:** The result can be easily understood by writing the closed-loop dynamics in normal coordinates, where by design

$$\dot{\boldsymbol{\mu}} = \begin{bmatrix} 0 & 1 & 0 & . & . & 0 \\ . & . & . & . & . & . \\ 0 & 0 & 0 & . & . & 1 \\ -k_o & -k_1 & -k_2 & . & . & -k_{r-1} \end{bmatrix} \boldsymbol{\mu} = \mathbf{A}\boldsymbol{\mu}$$

$$\dot{\boldsymbol{\psi}} = \mathbf{w}(\boldsymbol{\mu}, \boldsymbol{\psi}) = \mathbf{A}_1\boldsymbol{\mu} + \mathbf{A}_2\boldsymbol{\psi} + h.o.t.$$

where *h.o.t.* denotes higher order terms in the Taylor expansion about $\mathbf{x}_o = 0$, and the matrices $\mathbf{A}, \mathbf{A}_1,$ and $\mathbf{A}_2$ are defined obviously. The above equations can be written

$$\frac{\mathrm{d}}{\mathrm{d}t}\begin{bmatrix} \boldsymbol{\mu} \\ \boldsymbol{\psi} \end{bmatrix} = \begin{bmatrix} \mathbf{A} & \mathbf{O} \\ \mathbf{A}_1 & \mathbf{A}_2 \end{bmatrix}\begin{bmatrix} \boldsymbol{\mu} \\ \boldsymbol{\psi} \end{bmatrix} + h.o.t.$$

Now since the zero-dynamics is asymptotically stable, its linearization $\dot{\boldsymbol{\psi}} = \mathbf{A}_2\boldsymbol{\psi}$ is either asymptotically stable or marginally stable, according to Lyapunov's linearization theorem (Theorem 3.1).

If $\mathbf{A}_2$ is asymptotically stable (*i.e.*, has all its eigenvalues strictly in the left-half plane), then the above linearized approximation of the whole dynamics is asymptotically stable, and therefore, using again Lyapunov's linearization theorem, the nonlinear system is locally asymptotically stable.

If $\mathbf{A}_2$ is only marginally stable, stability of the closed loop system can still be derived by using the so-called center-manifold theory [Byrnes and Isidori, 1988].  ☐

In view of the fact that local stabilization problems may also be solved by linear control methods (through linearization and pole placement), one may wonder about the usefulness of the above theorem. The answer to that question is that the above stabilization method can treat systems whose linearizations contain uncontrollable but *marginally stable* modes, while linear control methods requires the linearized system to be strictly stabilizable.

For stabilization problems, where state convergence is the objective, one is usually free to choose the output function $y = h(\mathbf{x})$, and thus affect the zero-dynamics. Therefore, it might be possible to choose an output function such that the corresponding zero-dynamics is asymptotically stable. As a result, the controller specified in the above theorem will be able to stabilize the nonlinear system.

**Example 6.14:** Consider the nonlinear system

$$\dot{x}_1 = x_1^2 x_2 \tag{6.83a}$$

$$\dot{x}_2 = 3x_2 + u \tag{6.83b}$$

The system's linearization at $\mathbf{x} = \mathbf{0}$ (where $\mathbf{x} = [x_1 \quad x_2]^T$) is

$$\dot{x}_1 = 0$$

$$\dot{x}_2 = 3x_2 + u$$

and thus has an uncontrollable mode corresponding to a pure integrator.

Let us define the output function

$$y = -2x_1 - x_2 \tag{6.84}$$

Corresponding to this output, the relative degree of the system is 1, because

$$\frac{dy}{dt} = -2\dot{x}_1 - \dot{x}_2 = -2x_1{}^2x_2 - 3x_2 - u$$

The associated zero-dynamics (obtained by setting $y = 0$) is simply

$$\dot{x}_1 = -2x_1{}^3$$

and thus is asymptotically stable (cf. Example 3.9). Therefore, according to Theorem 6.3, the control law

$$u = -2x_1{}^2x_2 - 4x_2 - 2x_1$$

locally stabilizes the nonlinear system.                                                   ◻

## GLOBAL ASYMPTOTIC STABILIZATION

As we saw earlier, the stability of the zero-dynamics only guarantees the *local* stability of a control system based on input-output linearization (unless the relative degree equals the system order, in which case there is no internal dynamics), while the stability of the internal dynamics itself is generally untractable (except in very simple cases, as in section 6.1.3).   Thus, it is natural to wonder whether input-output linearization ideas can be of any use in (often most practically important) *global* stabilization problems.   Similarly, one may wonder how to achieve system stabilization using a given input-output linearization in case the zero-dynamics is unstable.

An approach to global asymptotic stabilization based on partial feedback linearization is to simply consider the control problem as a standard Lyapunov controller design problem, but simplified by the fact that putting the system in normal form makes part of the dynamics linear.   As in most Lyapunov designs based on mathematical rather than physical insights, the approach has a trial-and-error flavor, but may actually lead to satisfactory solutions. Let us study it on examples, where the control law based on partial linearization has actually to be modified in order to guarantee the global stability of the whole system.

The basic idea, after putting the system in normal form, is to view $\mu$ as the "input" to the internal dynamics, and $\psi$ as its "output". The first step is then to find a "control law" $\mu_o = \mu_o(\psi)$ which stabilizes the internal dynamics, and an associated Lyapunov function $V_o$ demonstrating that stabilizing property. This is generally easier than finding a stabilizing control law for the original system, because the internal dynamics is of lower order. The second step is then to get back to the original global control problem, define a Lyapunov function candidate $V$ appropriately *as a modified version of* $V_o$, and choose the control input $v$ so that $V$ be a Lyapunov function for the whole closed-loop dynamics.

Consider, for instance, the problem of stabilizing a nonlinear system whose normal form is

$$\dot{y} = v \tag{6.85a}$$

$$\ddot{z} + \dot{z}^3 + yz = 0 \tag{6.85b}$$

where $v$ is the control input and $\psi = [z \ \dot{z}]^T$. Considering the internal dynamics (6.85b), we see that if, we had, for instance, $y = z^2$, then the internal dynamics would be asymptotically stable (see Example 3.14). Moreover, this expression of $y$ would then also be compatible with the requirement that $y$ tends to zero. This remark suggests the following design.

Let $V_o$ be a Lyapunov function indicating the stability of (6.85b) *when y is formally replaced by* $y_o = z^2$. Based on Example 3.14, we can choose

$$V_o = \frac{1}{2}\dot{z}^2 + \frac{1}{4}z^4$$

Differentiating $V_o$ using the *actual* dynamics (6.85) leads to

$$\dot{V}_o = -\dot{z}^4 - z\dot{z}(y - z^2)$$

Consider now the Lyapunov function candidate

$$V = V_o + \frac{1}{2}(y - z^2)^2$$

obtained by adding to $V_o$ a quadratic "error" term in $(y - y_o)$. We get

$$\dot{V} = -\dot{z}^4 + (y - z^2)(v - 3z\dot{z})$$

This suggest the choice of control law

$$v = -y + z^2 + 3z\dot{z}$$

which yields

$$\dot{V} = -\dot{z}^4 - (y - z^2)^2$$

Application of the invariant set theorem (Theorem 3.5) shows that this choice of $v$ brings the whole system state to zero.

Interestingly, the same procedure may also be applicable to non-minimum phase systems. Consider, for instance, the problem of stabilizing a nonlinear system whose normal form is

$$\dot{y} = v \tag{6.86a}$$

$$\ddot{z} + \dot{z}^3 - z^5 + yz = 0 \tag{6.86b}$$

where again $v$ is the control input and $\boldsymbol{\psi} = [z \ \dot{z}]^T$. Note that the system is non-minimum phase, because its zero dynamics is unstable. Proceeding as before, and noticing that the internal dynamics would be asymptotically stable if we had $y = 2z^4$, we let

$$V_o = \frac{1}{2}\dot{z}^2 + \frac{1}{6}z^6$$

Differentiating $V_o$ using the *actual* dynamics leads to

$$\dot{V}_o = -\dot{z}^4 - z\dot{z}(y - 2z^4)$$

Considering now the Lyapunov function candidate

$$V = V_o + \frac{1}{2}(y - 2z^4)^2$$

we get

$$\dot{V} = -\dot{z}^4 + (y - 2z^4)(v - 8z^3\dot{z} - z\dot{z})$$

This suggest the choice of control law

$$v = -y + 2z^4 + 8z^3\dot{z} + z\dot{z}$$

which yields

$$\dot{V} = -\dot{z}^4 - (y - 2z^4)^2$$

Application of the invariant set theorem shows that this choice of $v$ brings the whole system state to zero.

Another interesting feature of the method is that it can be applied *recursively* to systems of increasing complexity, as the following example illustrates.

**Example 6.15:** Consider the system

$$\dot{x} + x^2 y^5 z\, e^{xy} = (x^4 + 2)\, u$$

$$\dot{y} + y^3 z^2 - x = 0$$

$$\ddot{z} + \dot{z}^3 - z^5 + yz = 0$$

We first define, of course, a new input $v$ such that

$$v = (x^4 + 2)\, u - x^2 y^5 z\, e^{xy}$$

so that the system is in the simpler form

$$\dot{x} = v$$

$$\dot{y} + y^3 z^2 = x$$

$$\ddot{z} + \dot{z}^3 - z^5 + yz = 0$$

Consider now only the last two equations and view for a moment $x$ as a control input. Then we know from our earlier study of the system (6.86) that the "control law" $x = x_o$ where

$$x_o = y^3 z^2 - y + 2z^4 + 8z^3\dot{z} + z\dot{z}$$

would globally asymptotically stabilize the variables $y$ and $z$, as can be shown using the Lyapunov function

$$V_o = \frac{1}{2}\dot{z}^2 + \frac{1}{6}z^6 + \frac{1}{2}(y - 2z^4)^2$$

Consider now the Lyapunov function

$$V = V_o + \frac{1}{2}(x - x_o)^2$$

We have

$$\dot{V} = \dot{V}_o + (x - x_o)(\dot{x} - \dot{x}_o) = \dot{V}_o + (x - x_o)(v - \dot{x}_o)$$

$$= -\dot{z}^4 - (y - 2z^4)^2 + (x - x_o)(y - 2z^4 + v - \dot{x}_o)$$

Define the control law $v$ as

$$v = \dot{x}_o - y + 2z^4 - x + x_o$$

where $\dot{x}_o$ is computed formally from the system dynamics. This yields

$$\dot{V} = -z^4 - (y - 2z^4)^2 - (x - x_o)^2$$

Application of the invariant set theorem then shows that this choice of $v$ brings the whole system state to zero. The original control input $u$ can then be computed from $v$

$$u = \frac{v + x^2 y^5 z\, e^{xy}}{x^4 + 2} \qquad\qquad \square$$

## TRACKING CONTROL

The simple pole-placement controller of Theorem 6.3 can be extended to asymptotic tracking control tasks. Consider, for system (6.67), the problem of tracking a given desired trajectory $y_d(t)$. Let

$$\mu_d = [\, y_d \quad \dot{y}_d \quad \cdots \quad y_d^{(r-1)}\, ]^T$$

and define the tracking error vector by

$$\tilde{\mu}(t) = \mu(t) - \mu_d(t)$$

We then have the following result:

**Theorem 6.4**   *Assume that the system (6.67) has relative degree r (defined and constant over the region of interest), that $\mu_d$ is smooth and bounded, and that the solution $\psi_d$ of the equation*

$$\dot{\psi}_d = \mathbf{w}(\mu_d, \psi_d) \qquad\qquad \psi_d(0) = \mathbf{0}$$

*exists, is bounded, and is uniformly asymptotically stable. Choose constants $k_i$ such that the polynomial (6.81) has all its roots strictly in the left-half plane. Then, by using the control law*

$$u = \frac{1}{L_{\mathbf{g}} L_{\mathbf{f}}^{r-1} \mu_1} [\, -L_{\mathbf{f}}^{r} \mu_1 + y_d^{(r)} - k_{r-1}\tilde{\mu}_r - \dots - k_o\tilde{\mu}_1\,] \qquad (6.87)$$

*the whole state remains bounded and the tracking error $\tilde{\mu}$ converges to zero exponentially.*

The proof of this theorem can be found in [Isidori, 1989]. Note that for tracking control to be exact from time $t = 0$ on, using *any* control law, requires that

$\mu(0) = \mu_d(0).$

## INVERSE DYNAMICS

In the following, we describe the concept of inverse dynamics, which provides an interesting interpretation of the previous tracking control design based on feedback linearization, and also yields insights about the tracking control of non-minimum phase systems.

For systems described by (6.67), let us find out what the initial conditions $\mathbf{x}(0)$ and control input $u$ should be in order for the plant output to track a reference output $y_r(t)$ perfectly. To do this, let us assume that the system output $y(t)$ is identical to the reference output $y_r(t)$, *i.e.*, $y(t) \equiv y_r(t)$, $\forall\, t \geq 0$. This implies that the time derivatives of all orders should be the same as those of the desired output, particularly,

$$y^{(k)}(t) = y_r^{(k)}(t) \qquad k = 0, 1, \dots, r-1 \qquad \forall\, t \geq 0 \tag{6.88}$$

In terms of normal coordinates, (6.88) can be written

$$\mu(t) = \mu_r(t) = [\, y_r(t) \quad \dot{y}_r(t) \quad \dots \quad y_r^{(r-1)}(t)\,]^T \qquad \forall\, t \geq 0$$

Thus, the control input $u(t)$ must satisfy

$$y_r^{(r)}(t) = a(\mu_r, \psi) + b(\mu_r, \psi) u(t)$$

that is,

$$u(t) = \frac{y_r^{(r)} - a(\mu_r, \psi)}{b(\mu_r, \psi)} \tag{6.89}$$

where $\psi(t)$ is the solution of the differential equation

$$\dot{\psi}(t) = \mathbf{w}[\mu_r(t), \psi(t)] \tag{6.90}$$

Given a reference trajectory $y_r(t)$, we can obtain from (6.89) the required control input for output $y(t)$ to be identically equal to $y_r(t)$. Note that this output depends on the internal states $\psi(t)$ and thus, in particular, on the initial $\psi(0)$.

Normally, by the term "dynamics", we mean the mathematical equations for computing the output $y(t)$ of a system corresponding to a given input history $u(t)$. Equations (6.89) and (6.90), on the other hand, allow us to compute the input $u(t)$ corresponding to reference output history $y_r(t)$. Therefore, (6.89) and (6.90) are often called the *inverse dynamics* of the system (6.67). Formally, $\psi$ is the state of the inverse dynamics, $\mu_r$ its input, and $u$ its output.

Note from (6.90) that a system's internal dynamics corresponds to inverting the system for the reference output $y_r$ such that

$$y_r^{(r)} = y_d^{(r)} - k_{r-1} \tilde{y}^{(r-1)} - \ldots - k_o \tilde{y}$$

while its zero-dynamics corresponds to inverting the system for a zero reference output.

## TRACKING CONTROL FOR NON-MINIMUM PHASE SYSTEMS

The control law (6.87) cannot be applied to non-minimum phase nonlinear systems, because they cannot be inverted. This is a generalization of the linear result that the inverse of the transfer function of a non-minimum phase linear system is unstable. Therefore, for such systems, we should not look for control laws which achieve perfect or asymptotic convergent tracking errors. Instead, we should find controllers which lead to small tracking errors for the desired trajectories of interest.

The control of non-minimum phase systems is a topic of active current research. One interesting approach [Hedrick and Gopalswamy, 1989] is the output-redefinition method, whose principle is to redefine the output function $y_1 = h_1(\mathbf{x})$ so that the resulting zero-dynamics is stable. Provided that the new output function $y_1$ is defined in such a way that it is essentially the same as the original output function in the frequency range of interest, exact tracking of the new output function $y_1$ then also implies good tracking of the original output $y$.

**Example 6.16: Tracking control of a non-minimum phase system**

To illustrate the idea of redefining the output, let us consider the tracking control of the linear system

$$y = \frac{(1 - \frac{p}{b}) B_o(p)}{A(p)} u \tag{6.91}$$

where $b$ is a strictly positive constant and the zeros of $B_o(p)$ are all in the left-half plane (with $p$ being the Laplace variable). This system has a right half-plane zero at $p = b$ and, therefore, perfect tracking and exponentially convergent tracking of an arbitrary desired output $y_d(t)$ by $y(t)$ is impossible. To avoid the problem of unstable zero-dynamics associated with the output $y$, let us consider the control of a "nominal output" $y_1$ instead, with $y_1$ defined by

$$y_1 = \frac{B_o(p)}{A(p)} u \tag{6.92}$$

with the desired output for $y_1$ being simply $y_d(t)$. Based on Theorem 6.4, one can easily find a control input $u$ that achieves convergent tracking of $y_1$.

With proper initial conditions, this control law leads to $y_1(t) = y_d(t)$. What is the tracking error for the real output? Since one easily sees that the true output $y(t)$ is

$$y(t) = (1 - \frac{p}{b})y_1 = (1 - \frac{p}{b})y_d \tag{6.93}$$

thus, the tracking error is proportional to the desired velocity $\dot{y}_d$,

$$y(t) - y_d(t) = -\frac{\dot{y}_d(t)}{b}$$

and, in particular, is bounded as long as $\dot{y}_d$ is bounded. Interpreting the above in the frequency domain, we also see that good tracking is achieved when the frequency content of $y_d$ is well below $b$.

An alternative choice of nominal output is

$$y_2 = \frac{B_o(p)}{A(p)(1 + \frac{p}{b})} u$$

wich is motivated by the approximation $(1 - p/b) \approx 1/(1 + p/b)$ for small $|p|/b$. With a feedback linearization design and proper initial conditions, we have $y_2(t) = y_d(t)$. This implies that the true output is

$$y(t) = (1 - \frac{p}{b})(1 + \frac{p}{b})y_d = (1 - \frac{p^2}{b^2})y_d$$

Thus, the tracking error is proportional to the desired acceleration $\ddot{y}_d$,

$$y(t) - y_d(t) = -\frac{\ddot{y}_d(t)}{b^2}$$

As compared with the previous nominal output $y_1$, this redefinition allows better tracking if $|\ddot{y}_d| < b|\dot{y}_d|$ (or, in the frequency domain, if the frequency content of $y_d$ is below $b$).  $\square$

Another practical approximation [Hauser, 1989] may be, when performing input-output linearization using successive differentiations of the output, to simply *neglect* the terms containing the input and keep differentiating the selected output a number of times equal to the system order, so that there is "approximately" no zero-dynamics. Of course, this approach can only be meaningful if the coefficients of $u$ at the intermediate steps are "small", *i.e.*, if the systems are "weakly" non-minimum phase. The approach is conceptually similar to neglecting "fast" right-half plane zeros in linear systems (in the frequency domain, $1 - \tau p \approx 1/(1 + \tau p)$ if $|\tau p| \ll 1$, *i.e.*, if the zero $(1/\tau)$ is much faster than the frequency range of interest).

Finally, an approach to dealing with non-minimum phase systems is to modify the plant itself. In linear systems, while poles can be placed using feedback, zeros are intrinsic properties of the plant and the selected output, and can be changed only by modifying the plant or the choice of output. Similarly, in nonlinear systems, the zero-dynamics is a property of the plant, the output, and the desired trajectory. It can made stable by changing the output, as discussed earlier. It can also, in principle, be modified by changing the desired trajectory directly, although this is rarely practical if the system is supposed to perform a variety of pre-specified tasks. Finally, it can be made stable by changing the plant design itself. This may involve relocation or addition of actuators and sensors, or modifying the physical construction of the plant (*e.g.*, the distribution of control surfaces on an aircraft, or the mass and stiffness distributions in a flexible-link robot).

# 6.5 * Multi-Input Systems

The concepts used in the above sections for SISO systems, such as input-state linearization, input-output linearization, zero-dynamics, and so on, can be extended to MIMO systems.

In the MIMO case, we consider, in a neighborhood of a point $\mathbf{x}_o$, square systems (*i.e.*, systems with the same number of inputs and outputs) of the form

$$\dot{\mathbf{x}} = \mathbf{f}(\mathbf{x}) + \mathbf{G}(\mathbf{x})\mathbf{u} \qquad \mathbf{y} = \mathbf{h}(\mathbf{x}) \qquad (6.94)$$

where $\mathbf{x}$ is $n\times1$ the state vector, $\mathbf{u}$ is the $m\times1$ control input vector (of components $u_i$ ), $\mathbf{y}$ is the $m\times1$ vector of system outputs (of components $y_i$ ), $\mathbf{f}$ and $\mathbf{h}$ are smooth vector fields, and $\mathbf{G}$ is a $n\times m$ matrix whose columns are smooth vector fields $\mathbf{g}_i$ .

**FEEDBACK LINEARIZATION OF MIMO SYSTEMS**

Input-output linearization of MIMO systems is obtained similarly to the SISO case, by differentiating the outputs $y_i$ until the inputs appear. Assume that $r_i$ is the smallest integer such that at least one of the inputs appears in $y_i^{(r_i)}$, then

$$y_i^{(r_i)} = L_{\mathbf{f}}^{r_i} h_i + \sum_{j=1}^{m} L_{\mathbf{g}_j} L_{\mathbf{f}}^{r_i-1} h_i \, u_j$$

with $L_{\mathbf{g}_j} L_{\mathbf{f}}^{r_i-1} h_i(\mathbf{x}) \neq 0$ for at least one $j$, in a neighborhood $\Omega_i$ of the point $\mathbf{x}_o$ . Performing the above procedure for each output $y_i$ yields

$$
\begin{bmatrix} y_1^{(r_1)} \\ ... \\ ... \\ y_m^{(r_m)} \end{bmatrix} = \begin{bmatrix} L_{\mathbf{f}}^{r_1} h_1(\mathbf{x}) \\ ... \\ ... \\ L_{\mathbf{f}}^{r_m} h_m(\mathbf{x}) \end{bmatrix} + \mathbf{E(x)\,u} \tag{6.95}
$$

where the $m \times m$ matrix $\mathbf{E(x)}$ is defined obviously.

Define then $\Omega$ as the intersection of the $\Omega_i$. If, as assumed above, the partial "relative degrees" $r_i$ are all well defined, then $\Omega$ is itself a finite neighborhood of $\mathbf{x}_o$. Furthermore, if $\mathbf{E(x)}$ is invertible over the region $\Omega$, then, similarly to the SISO case, the input transformation

$$
\mathbf{u} = \mathbf{E}^{-1} \begin{bmatrix} v_1 - L_{\mathbf{f}}^{r_1} h_1 \\ ... \\ ... \\ v_m - L_{\mathbf{f}}^{r_m} h_m \end{bmatrix} \tag{6.96}
$$

yields $m$ equations of the simple form

$$
y_i^{(r_i)} = v_i \tag{6.97}
$$

Since the input $v_i$ only affects the output $y_i$, (6.96) is called a _decoupling control law_, and the invertible matrix matrix $\mathbf{E(x)}$ is called the _decoupling matrix_ of the system. The system (6.94) is then said to have _relative degree_ $(r_1, ...., r_m)$ at $\mathbf{x}_o$, and the scalar $r = r_1 + ... + r_m$ is called the _total relative degree_ of the system at $\mathbf{x}_o$.

An interesting case corresponds to the total relative degree being $n$. In this case, there is no internal dynamics. With the control law in the form of (6.96), we thus obtain an input-state linearization of the original nonlinear system. With the equivalent inputs $v_i$ designed as in the SISO case, both stabilization and tracking can then be achieved for the system without any worry about the stability of the internal dynamics. Note that the necessary and sufficient conditions for input-state linearization of multi-input nonlinear systems to be achievable are similar to and more complex than those for single input systems [Su, _et al._, 1983].

The zero-dynamics of a MIMO system can be defined similarly to the SISO case, by constraining the outputs to be zero. The notion of a minimum phase system can also be similarly defined.

## EXTENSIONS OF THE BASIC MIMO LINEARIZATION

The above input-output linearization can be achieved only when the decoupling matrix $\mathbf{E}$ is invertible in the region $\Omega$. Given the straightforward procedure used to construct $\mathbf{E}$, this condition is rather restrictive (for instance, $\mathbf{E}$ may have a column of zeros). In the following, we discuss two methods to generate input-output linearization when the invertibility condition is violated, *i.e.*, when $\mathbf{E}$ is singular. Both techniques have an iterative nature, and formal conditions on the system (6.94) can be derived for them to converge in a finite number of steps (see *e.g.*, [Isidori, 1989], and references therein). The first technique, called dynamic extension, involves choosing some *new inputs* as the *derivatives* of some of the original system inputs, in a way that the corresponding matrix $\mathbf{E}$ is invertible. The control system is designed based on the new set of inputs, and the actual system inputs are then computed by integration. The second technique, a MIMO form of system inversion, involves deriving *new outputs* so that the resulting $\mathbf{E}$ matrix is invertible. In both cases, as in the basic version, the stability of the internal dynamics (or, locally, of the zero-dynamics) has to be verified.

## REDEFINING INPUTS: DYNAMIC EXTENSION

For notational simplicity, let us consider a system with 2 inputs and 2 outputs, and let us assume that $\mathbf{E}(\mathbf{x})$ has rank one. This means that, without loss of generality, we can *redefine the input* vector $\mathbf{u}$ (through linear transformations) so that $\mathbf{E}(\mathbf{x})$ has only one non-zero column, $\mathbf{e}_1 = \mathbf{e}_1(\mathbf{x})$, *i.e.*, so that equations (6.95) can be expressed in terms of $u_1$ only,

$$\begin{bmatrix} y_1^{(r_1)} \\ y_2^{(r_2)} \end{bmatrix} = \begin{bmatrix} L_{\mathbf{f}}^{r_1} h_1 \\ L_{\mathbf{f}}^{r_2} h_2 \end{bmatrix} + \mathbf{e}_1 \, u_1 \tag{6.98}$$

Differentiating the above and replacing in the system dynamics then leads to an equation of the form

$$\begin{bmatrix} y_1^{(r_1 + 1)} \\ y_2^{(r_2 + 1)} \end{bmatrix} = \mathbf{b}(\mathbf{x}, u_1) + \mathbf{E}_1(\mathbf{x}, u_1) \begin{bmatrix} \dot{u}_1 \\ u_2 \end{bmatrix} \tag{6.99}$$

If the matrix $\mathbf{E}_1$ is invertible, then the above equation is in the standard form of (6.95), with $\dot{u}_1$ and $u_2$ regarded as control inputs, and $u_1$ considered as an extra state. Input-output linearization can then be used straightforwardly to design these inputs, *i.e.*,

$$\begin{bmatrix} \dot{u}_1 \\ u_2 \end{bmatrix} = \mathbf{E}_1^{-1} [\mathbf{v} - \mathbf{b}(\mathbf{x}, u_1)] \tag{6.100}$$

with the vector $\mathbf{v}$ chosen to place the poles of the resulting linear input-output dynamics. However, the system input $u_1$ must now be obtained from (6.100) by integration. Thus the actual control law contains an integrator, yielding a "dynamic" controller.

If, in (6.99), the matrix $\mathbf{E}_1$ is still singular, the procedure can be repeated, amounting to adding more integrators.

## REDEFINING OUTPUTS: SYSTEM INVERSION

Let us redefine the input as in the dynamic extension case, but now stop at (6.98). Instead of going further to differentiate the left hand side, consider the variable

$$z = e_{12}(\mathbf{x}) \, y_1^{(r_1)} - e_{11}(\mathbf{x}) \, y_1^{(r_2)}$$

(where $\mathbf{e}_1 = [e_{11} \ e_{12}]^T$). Using (6.98) shows that the expression of $z$ can be computed as a function of the state $\mathbf{x}$ only (and does not contain the input $u$), namely

$$z = e_{12}(\mathbf{x}) \, L_{\mathbf{f}}^{r_1} h_1(\mathbf{x}) - e_{11}(\mathbf{x}) \, L_{\mathbf{f}}^{r_2} h_2(\mathbf{x})$$

Hence, by differentiating $z$, we obtain an equation of the form

$$\dot{z} = \gamma_o(\mathbf{x}) + \gamma_1(\mathbf{x}) \, u_1 + \gamma_2(\mathbf{x}) \, u_2$$

If the matrix

$$\mathbf{E}_2(\mathbf{x}) = \begin{bmatrix} e_{11}(\mathbf{x}) & 0 \\ \gamma_1(\mathbf{x}) & \gamma_2(\mathbf{x}) \end{bmatrix}$$

is invertible, we can regard $y_1$ and $z$ as outputs, $u_1$ and $u_2$ as inputs, and use the control law

$$\begin{bmatrix} u_1 \\ u_2 \end{bmatrix} = \mathbf{E}_2^{-1} \begin{bmatrix} v_1 - L_{\mathbf{f}}^{r_1} h_1 \\ v_2 - \gamma_o \end{bmatrix}$$

to achieve input-output linearization. This leads to

$$\begin{bmatrix} y^{(r_1)} \\ \dot{z} \end{bmatrix} = \begin{bmatrix} v_1 \\ v_2 \end{bmatrix}$$

The new inputs $v_1$ and $v_2$ can be easily designed to regulate $y$ and $z$. If the matrix $\mathbf{E}_2$ is singular, we can repeat this procedure to create new outputs.

# 6.6  Summary

Feedback linearization is based on the idea of transforming nonlinear dynamics into a linear form by using state feedback, with input-state linearization corresponding to complete linearization and input-output linearization to partial linearization. The method can be used for both stabilization and tracking control problems, single-input and multiple-input systems, and has been successfully applied to a number of practical nonlinear control problems, both as an system analysis tool and as a controller design tool.

The method also has a number of important limitations:

1. it cannot be used for all nonlinear systems

2. the full state has to be measured

3. no robustness is guaranteed in the presence of parameter uncertainty or unmodeled dynamics

The applicability of input-state linearization is quantified by a set of somewhat stringent conditions, while input-output feedback linearization cannot be applied when the relative degree is not defined and lacks systematic global results. Furthermore, analytically solving the partial differential equations defining input-state linearizing transformations is generally not systematic.

The second problem is due to the difficulty of finding convergent observers for nonlinear systems and, when an observer can be found, the lack of a general separation principle (analogous to that in linear systems) which guarantees that the straightforward combination of a stable state feedback controller and a stable observer will guarantee the stability of the closed-loop system.

The third problem is due to the fact that the exact model of the nonlinear system is not available in performing feedback linearization. The sensitivity to modeling errors may be particularly severe when the linearizing transformation is poorly conditioned.

Active research is being performed to overcome the above drawbacks. For the first problem, research is aimed at extending feedback linearization to non-minimum phase or weakly non-minimum phase systems. For the second problem, many efforts are being made to construct observers for nonlinear systems and to extend the separation principle to nonlinear systems. For the third problem, robust and adaptive control are being introduced to provide feedback linearizable systems with robustness to parametric uncertainties. Chapters 7 and 8 provide some further discussions of robust and adaptive techniques for feedback linearizable systems.

## 6.7  Notes and References

Extensive theoretical developments of the material in this chapter can be found in [Isidori, 1989], from which Examples 6.12 and 6.14 are adapted, and in references therein. The writing of this chapter was also influenced by the clear presentation in [Spong and Vidyasagar, 1989], and by the clear and concise summary of input-output linearization in [Hauser, 1989]. The reader may also consult the recent book [Nijmeijer and Van der Schaft, 1990], which contains many interesting examples. A survey of aircraft control applications is contained in [Hedrick and Gopalswamy, 1989].

The discussion on global asymptotic stabilization of section 6.4 is inspired by the recent works of [Kokotovic and Sussmann, 1989; Praly, *et al.*, 1989]. See [Sastry and Kokotovic, 1988; Sussmann, 1990] for further discussions of the relation between stability of the zero-dynamics and overall system stability.

While nonlinear observer design is far from being a mature subject, a number of important theoretical results have been derived [Hermann and Krener, 1977; Vidyasagar, 1980; Krener and Respondek, 1985]. See, *e.g.*, [Misawa and Hedrick, 1989] for a recent review.

The basic idea of "undoing" the nonlinear dynamics is quite old, and can be traced back at least to the early robotics literature [Freund, 1973; Takase, 1975]. The basis of input-state transformations was suggested by [Brockett, 1978] and fully derived in [Jacubczyk and Respondek, 1980; Hunt, *et al.*, 1983]. Many results in input-output linearization are extensions of [Krener, *et al.*, 1983]. The basic principles of system inversion are due to [Hirschorn, 1981; Singh 1982], and extend to nonlinear systems the classic linear control algorithm of [Silverman, 1969]. A detailed historical perspective of the field can be found in [Isidori, 1989].

## 6.8  Exercises

**6.1**  Simulate the nonlinear controller design of section 6.1.1 for a spherical tank of radius 0.5 meter (m) and outlet opening radius 0.1 m. Let the initial height be 0.1 m and the desired final height be 0.6 m.

**6.2** Show that the internal dynamics of the system

$$\begin{bmatrix} \dot{x}_1 \\ \dot{x}_2 \end{bmatrix} = \begin{bmatrix} x_2{}^3 + u \\ -u \end{bmatrix} \qquad y = x_1$$

is unstable.

**6.3** For the system

$$\dot{x}_1 = \sin x_2$$

$$\dot{x}_2 = x_1{}^4 \cos x_2 + u$$

design a controller to track an arbitrary desired trajectory $x_{d1}(t)$. Assume that the model is very accurate, that the state $[x_1 \ x_2]^T$ is measured, and that $x_{d1}(t), \dot{x}_{d1}(t), \ddot{x}_{d1}(t)$ are all known. Write the full expression of the controller, as a function of the measured state $[x_1 \ x_2]^T$. Check your design in simple simulations.

**6.4** Consider the system

$$\dot{x}_1 = u$$

$$\dot{x}_2 + x_2 = (1 + x_2 + x_2{}^2) \, x_1$$

where the variable $x_1$ is to be controlled to zero. Note that the zero-dynamics is exponentially stable. Show that, assuming that $x_2(0) = 0$, the stability of the internal dynamics depends on the initial condition in $x_1$ and on how fast $u$ drives $x_1$ to zero. In particular, if these are such that $x_1 \geq e^{1 - 2t/\pi}$, then $x_2 \geq \tan t$ and therefore tends to infinity in finite time.

**6.5** A different interpretation can be given to the zero-dynamics of linear systems. Indeed, note the frequency-domain equation (6.35) can be written

$$y = \frac{1}{a_o + a_1 p + a_2 p^2 + p^3} [\, b_o + b_1 p \,] \, u$$

Thus, if we assume that the roots of the denominator polynomial are strictly in the left half-plane, then there are certain control inputs which, after stable transients, *do not affect the system output y*. These control inputs verify

$$\dot{u} + \frac{b_o}{b_1} u = 0 \qquad\qquad\qquad (6.101)$$

If the system is minimum-phase, these particular inputs tend to zero as $t \to \infty$. But if the system is non-minimum phase, then there are certain *diverging* control inputs which do not affect the system output (after stable transients). This reflects a "helplessness" of the system input at certain frequencies.

Show that equation (6.101) is of the same form as that defining the zero-dynamics of the system.

For the aircraft of Example 6.5, find the control inputs which, after stable transients, do not affect the output. Illustrate your result in simulation.

Can this interpretation of the zero-dynamics be extended to nonlinear systems?

**6.6** Show the bilinearity and skew-commutativity of Lie brackets.

**6.7** Check the input-state linearizability of the system

$$\frac{d}{dt}\begin{bmatrix} x_1 \\ x_2 \\ x_3 \end{bmatrix} = \begin{bmatrix} x_2 + x_2^2 + x_3^2 \\ x_3 + \sin(x_1 - x_3) \\ x_3^2 \end{bmatrix} + \begin{bmatrix} 1 \\ 0 \\ 1 \end{bmatrix} u$$

Can

$$z_1 = x_1 - x_3 \qquad z_2 = x_2 + x_2^2 \qquad z_3 = x_3 + \sin(x_1 - x_3) + 2x_2[x_3 + \sin(x_1 - x_3)]$$

serve as linearizing states? (Adapted from [Nijmeijer and Van der Schaft, 1990].)

**6.8** Globally stabilize the nonlinear system

$$\dot{y} + y^2 e^{y^4 z} = u^3$$

$$\ddot{z} + \dot{z}^3 - z^7 + yz^2 = 0$$

where $u$ is the control input. Is the system minimum-phase?

**6.9** Put in normal form the nonlinear system

$$\dot{y} + yz^2 \ln(z^4 + 1) = u$$

$$\ddot{z} + \dot{z}^5 + z^3 + yz^2 = 5u$$

where $u$ is the control input. Can the system be locally/globally stabilized? Is the system minimum-phase?

**6.10** Consider the system of Figure 6.9, which represents a link driven by an electric motor through a rigid mechanism, in the vertical plane. The (stable) motor electrical dynamics is assumed to be slow, so that it cannot be neglected.

Using a procedure similar to that of Example 6.10, perform input-state linearization and

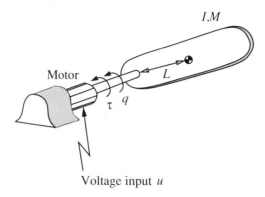

**Figure 6.9 :** A link driven by a slow motor

design a controller for this system in the following cases (where $\lambda$ is simply a scaling factor between units):

- The motor dynamics is linear and first-order:

$$\dot{\tau} + \lambda_2 \lambda \tau = \lambda u$$

- The motor dynamics is linear and second-order:

$$\ddot{\tau} + \lambda_1 \dot{\tau} + \lambda_2 \lambda \tau = \lambda u$$

- The motor dynamics is nonlinear and second-order:

$$\ddot{\tau} + \lambda_1 \dot{\tau}^3 + \lambda^5 \lambda_2 \tau^5 = \lambda^5 u^5$$

What variables do you assume to be available for measurement?

In the above calculations, the back-emf of the motor has been neglected. It actually introduces a damping torque $I \alpha_o \dot{q}$. Assuming that the motor dynamics is linear and first-order (which is often a reasonable model, simply corresponding to an equivalent "RL" circuit), perform input-state linearization and design a controller for the system. What variables do you assume to be available for measurement?

**6.11**   Globally stabilize the nonlinear system

$$\dot{y} + z y^4 = u^5$$

$$\ddot{z} + (y-1)\dot{z}^2 + z^5 = 0$$

where $u$ is the control input. Is the system minimum-phase?

**6.12**   Consider the nonlinear system

$$\dot{y} + z^3 e^{yz^2} = u$$

$$\ddot{z} - (\dot{y} + y^3)(\dot{z}^4 + 1) + z^5 + yz = 0$$

where $u$ is the control input. Can the system be locally/globally stabilized?

**6.13**    Globally stabilize the nonlinear system

$$\dot{y} + y^3 e^z = u^3$$

$$\ddot{z} - \dot{z}^4 + z^5 = y u^2 \dot{z}^2$$

where $u$ is the control input. Is the system minimum-phase?

**6.14**    Globally stabilize the nonlinear system

$$\dot{x} + x^2 y = (x^2 z^2 + 4) u$$

$$\dot{y} + y^4 e^{yz} + x = 0$$

$$\ddot{z} + \dot{z}^3 - z^7 + yz^2 = 0$$

where $u$ is the control input. Is the system minimum-phase?

**6.15**    Discuss the stabilization of the MIMO system

$$\ddot{x} + xy^2 = u_1 + 2 u_2$$

$$\dot{y} + x^3 = 2 u_1 + 4 u_2$$

using dynamic extension and inversion techniques.

Same question for the system

$$\ddot{x} + \dot{x} e^y = u_1 + 2 u_2$$

$$\dot{y} + y^2 = 2 u_1 + 4 u_2$$

**6.16**    Consider again the system of Exercise 6.10 (including motor back-emf) but assume now that the "link" actually consists of the three blades of an underwater propeller, so that the gravitational torque $- MgL \sin q$ is replaced by the hydrodynamic torque

$$\tau_{thrust} = -I \alpha_1 \dot{q} |\dot{q}|$$

(where the notation $\tau_{thrust}$ comes from the fact that the thrust generated by the propeller is proportional to $\tau_{thrust}$, cf. Exercise 4.4). Assuming that the motor dynamics is linear and first-order, perform input-state linearization and design a controller for the system. Indicate what variables you assume to be available for measurement.

# Chapter 7
# Sliding Control

In this chapter, we consider again the control of nonlinear systems of the general form studied in chapter 6, but we now allow the models to be imprecise. Model imprecision may come from actual uncertainty about the plant (*e.g.*, unknown plant parameters), or from the purposeful choice of a *simplified* representation of the system's dynamics (*e.g.*, modeling friction as linear, or neglecting structural modes in a reasonably rigid mechanical system). From a control point of view, modeling inaccuracies can be classified into two major kinds:

- *structured* (or *parametric*) uncertainties

- *unstructured* uncertainties (or *unmodeled dynamics*)

The first kind corresponds to inaccuracies on the terms actually included in the model, while the second kind corresponds to inaccuracies on (*i.e.*, underestimation of) the system order.

As discussed earlier, modeling inaccuracies can have strong adverse effects on nonlinear control systems. Therefore, any practical design must address them explicitly. Two major and complementary approaches to dealing with model uncertainty are *robust control*, which we discuss in this chapter, and *adaptive control*, which is the subject of chapter 8. The typical structure of a robust controller is composed of a nominal part, similar to a feedback linearizing or inverse control law, and of additional terms aimed at dealing with model uncertainty. The structure of an adaptive controller is similar, but in addition the model is actually updated *during operation*, based on the measured performance.

A simple approach to robust control, and the main topic of this chapter, is the so-called sliding control methodology. Intuitively, it is based on the remark that it is much easier to control 1st-order systems (*i.e.*, systems described by 1st-order differential equations), be they nonlinear or uncertain, than it is to control general $n$th-order systems (*i.e.*, systems described by $n$th-order differential equations). Accordingly, a notational simplification is introduced, which, in effect, allows $n$th-order problems to be replaced by equivalent 1st-order problems. It is then easy to show that, for the transformed problems, "perfect" performance can in principle be achieved in the presence of arbitrary parameter inaccuracies. Such performance, however, is obtained at the price of extremely high control activity. This is typically at odds with the other source of modeling uncertainty, namely the presence of neglected dynamics, which the high control activity may excite. This leads us to a modification of the control laws which, given the admissible control activity, is aimed at achieving an effective trade-off between tracking performance and parametric uncertainty. Furthermore, in some specific applications, particularly those involving the control of electric motors, the unmodified control laws can be used directly.

For the class of systems to which it applies, sliding controller design provides a systematic approach to the problem of maintaining stability and consistent performance in the face of modeling imprecisions. Furthermore, by allowing the trade-offs between modeling and performance to be quantified in a simple fashion, it can illuminate the whole design process. Sliding control has been successfully applied to robot manipulators, underwater vehicles, automotive transmissions and engines, high-performance electric motors, and power systems.

The concepts are presented first for systems with a single control input, which allows us to develop intuition about the basic aspects of nonlinear controller design. Specifically, section 7.1 introduces the main concepts and notations of sliding control, and illustrates the associated basic controller designs. Section 7.2 describes modifications of the control laws aimed at eliminating excessive control activity. Section 7.3 discusses the choice of controller design parameters. Section 7.4 then studies generalizations to multi-input systems.

# 7.1 Sliding Surfaces

Consider the single-input dynamic system

$$x^{(n)} = f(\mathbf{x}) + b(\mathbf{x}) u \tag{7.1}$$

where the scalar $x$ is the output of interest (for instance, the position of a mechanical system), the scalar $u$ is the control input (for instance, a motor torque), and $\mathbf{x} = [x \; \dot{x} \; \dots \; x^{(n-1)}]^T$ is the state vector. In equation (7.1) the function $f(\mathbf{x})$ (in general nonlinear) is not exactly known, but the *extent of the imprecision on $f(\mathbf{x})$ is upper bounded by a known continuous function of* $\mathbf{x}$ ; similarly, the control gain $b(\mathbf{x})$ is not exactly known, but is of known sign and is bounded by known, continuous functions of $\mathbf{x}$. For instance, typically, the inertia of a mechanical system is only known to a certain accuracy, and friction models only describe part of the actual friction forces. The control problem is to get the state $\mathbf{x}$ to *track a specific time-varying state* $\mathbf{x}_d = [x_d \; \dot{x}_d \; \dots \; x_d^{(n-1)}]^T$ *in the presence of model imprecision on* $f(\mathbf{x})$ and $b(\mathbf{x})$.

For the tracking task to be achievable using a finite control $u$, the initial desired state $\mathbf{x}_d(0)$ must be such that

$$\mathbf{x}_d(0) \; = \; \mathbf{x}(0) \tag{7.2}$$

In a second-order system, for instance, position or velocity cannot "jump", so that any desired trajectory feasible from time $t = 0$ necessarily starts with the same position and velocity as those of the plant. Otherwise, tracking can only be achieved after a transient.

## 7.1.1 A Notational Simplification

Let $\tilde{x} = x - x_d$ be the tracking error in the variable $x$, and let

$$\tilde{\mathbf{x}} = \mathbf{x} - \mathbf{x}_d = [\tilde{x} \; \dot{\tilde{x}} \; \dots \; \tilde{x}^{(n-1)}]^T$$

be the tracking error vector. Furthermore, let us define a time-varying surface $S(t)$ in the state-space $\mathbf{R}^{(n)}$ by the scalar equation $s(\mathbf{x};t) = 0$ , where

$$s(\mathbf{x};t) = \left(\frac{d}{dt} + \lambda\right)^{n-1} \tilde{x} \tag{7.3}$$

and $\lambda$ is a strictly positive constant, whose choice we shall interpret later. For instance, if $n = 2$ ,

$$s = \dot{\tilde{x}} + \lambda\tilde{x}$$

*i.e.*, $s$ is simply a weighted *sum* of the position error and the velocity error; if $n = 3$ ,

$$s = \ddot{\tilde{x}} + 2\lambda\dot{\tilde{x}} + \lambda^2\tilde{x}$$

Given initial condition (7.2), *the problem of tracking* $\mathbf{x} \equiv \mathbf{x}_d$ *is equivalent to that of remaining on the surface S(t) for all t > 0* ; indeed $s \equiv 0$ represents a linear differential equation whose unique solution is $\tilde{\mathbf{x}} \equiv 0$ , given initial conditions (7.2). Thus, the problem of tracking the *n*-dimensional vector $\mathbf{x}_d$ can be reduced to that of keeping the *scalar* quantity $s$ at zero.

More precisely, the problem of tracking the *n*-dimensional vector $\mathbf{x}_d$ (*i.e.*, the original $n^{\text{th}}$-order tracking problem in $x$) can in effect be replaced by a *$1^{st}$-order* stabilization problem in $s$. Indeed, since from (7.3) the expression of $s$ contains $\tilde{x}^{(n-1)}$ , we only need to differentiate $s$ *once* for the input $u$ to appear.

Furthermore, bounds on $s$ can be directly translated into bounds on the tracking error vector $\tilde{\mathbf{x}}$, and therefore the scalar $s$ represents a true measure of tracking performance. Specifically, assuming that $\tilde{\mathbf{x}}(0) = \mathbf{0}$ (the effect of non-zero initial conditions in $\tilde{\mathbf{x}}$ can be added separately), we have

$$\forall \, t \geq 0 \, , \, |s(t)| \leq \Phi \qquad => \qquad \forall \, t \geq 0 \, , \, |\tilde{x}^{(i)}(t)| \leq (2\lambda)^i \, \varepsilon \qquad (7.4)$$
$$i = 0, \ldots, n-1$$

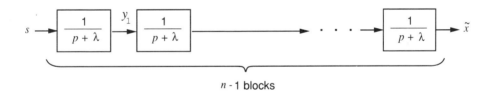

$$n - 1 \text{ blocks}$$

**Figure 7.1.a :** Computing bounds on $\tilde{x}$

where $\varepsilon = \Phi / \lambda^{n-1}$ . Indeed, by definition (7.3), the tracking error $\tilde{x}$ is obtained from $s$ through a sequence of first-order lowpass filters (Figure 7.1.a, where $p = (d/dt)$ is the Laplace operator). Let $y_1$ be the output of the first filter. We have

$$y_1(t) = \int_o^t e^{-\lambda(t-T)} s(T) \, dT$$

From $|s| \leq \Phi$ we thus get

$$|y_1(t)| \leq \Phi \int_o^t e^{-\lambda(t-T)} dT = (\Phi/\lambda)(1 - e^{-\lambda t}) \leq \Phi/\lambda$$

We can apply the same reasoning to the second filter, and so on, all the way to $y_{n-1} = \tilde{x}$ . We then get

$$| \tilde{x} | \leq \Phi/\lambda^{n-1} = \varepsilon$$

Similarly, $\tilde{x}^{(i)}$ can be thought of as obtained through the sequence of Figure 7.1.b. From the previous result, one has $| z_1 | \leq \Phi/\lambda^{n-1-i}$ , where $z_1$ is the output of the $(n - i - 1)^{\text{th}}$ filter. Furthermore, noting that

$$\frac{p}{p+\lambda} = \frac{p + \lambda - \lambda}{p + \lambda} = 1 - \frac{\lambda}{p + \lambda}$$

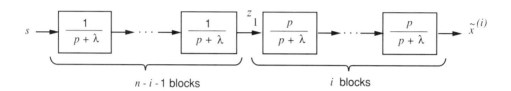

$$n - i - 1 \text{ blocks} \qquad i \text{ blocks}$$

**Figure 7.1.b :** Computing bounds on $\tilde{x}^{(i)}$

one sees that the sequence of Figure 7.1.b implies that

$$| \tilde{x}^{(i)} | \leq \left( \frac{\Phi}{\lambda^{n-1-i}} \right) \left( 1 + \frac{\lambda}{\lambda} \right)^{i} = (2\lambda)^i \, \varepsilon$$

*i.e.*, bounds (7.4). Finally, in the case that $\tilde{\mathbf{x}}(0) \neq \mathbf{0}$, bounds (7.4) are obtained asymptotically, *i.e.*, within a short time-constant $(n - 1)/\lambda$.

Thus, we have in effect replaced an $n^{\text{th}}$-order tracking problem by a $1^{\text{st}}$-order stabilization problem, and have quantified with (7.4) the corresponding transformations of performance measures.

The simplified, $1^{\text{st}}$-order problem of keeping the scalar $s$ at zero can now be achieved by choosing the control law $u$ of (7.1) such that outside of $S(t)$

$$\frac{1}{2} \frac{d}{dt} s^2 \leq - \eta \, |s| \tag{7.5}$$

where $\eta$ is a strictly positive constant. Essentially, (7.5) states that the squared "distance" to the surface, as measured by $s^2$ , decreases along all system trajectories. Thus, it constrains trajectories to point towards the surface $S(t)$, as illustrated in Figure 7.2. In particular, once on the surface, the system trajectories remain on the surface. In other words, satisfying condition (7.5), or *sliding condition*, makes the surface an *invariant set*. Furthermore, as we shall see, (7.5) also implies that some disturbances or dynamic uncertainties can be tolerated while still keeping the surface an invariant set. Graphically, this corresponds to the fact that in Figure 7.2 the trajectories off the surface can "move" while still pointing towards the surface. $S(t)$ verifying (7.5) is referred to as a *sliding surface*, and the system's behavior once on the surface is called *sliding regime* or *sliding mode*.

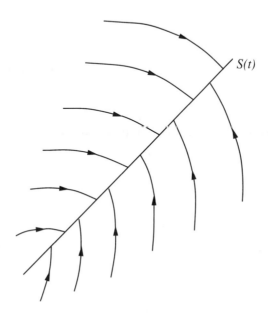

**Figure 7.2 :** The sliding condition

The other interesting aspect of the invariant set $S(t)$ is that once on it, the system trajectories are defined by the equation of the set itself, namely

$$\left(\frac{d}{dt} + \lambda\right)^{n-1} \tilde{x} = 0$$

In other words, the surface $S(t)$ is both a place and a dynamics. This fact is simply the geometric interpretation of our earlier remark that definition (7.3) allows us, in effect, to replace an $n$th-order problem by a 1st-order one.

Finally, satisfying (7.5) guarantees that if condition (7.2) is not exactly verified, *i.e.*, if $\mathbf{x}(t=0)$ is actually off $\mathbf{x}_d(t=0)$ , the surface $S(t)$ will nonetheless be reached in a *finite time* smaller than $|s(t=0)|/\eta$. Indeed, assume for instance that $s(t=0) > 0$, and let $t_{\text{reach}}$ be the time required to hit the surface $s = 0$. Integrating (7.5) between $t = 0$ and $t = t_{\text{reach}}$ leads to

$$0 - s(t=0) = s(t=t_{\text{reach}}) - s(t=0) \leq -\eta(t_{\text{reach}} - 0)$$

which implies that

$$t_{\text{reach}} \leq s(t=0)/\eta$$

One would obtain a similar result starting with $s(t=0) < 0$, and thus

$$t_{\text{reach}} \leq |s(t=0)|/\eta$$

Furthermore, definition (7.3) implies that once on the surface, the tracking error tends exponentially to zero, with a time constant $(n-1)/\lambda$ (from the sequence of $(n-1)$ filters of time constants equal to $1/\lambda$).

The typical system behavior implied by satisfying sliding condition (7.5) is illustrated in Figure 7.3 for $n = 2$. The sliding surface is a line in the phase plane, of slope $-\lambda$ and containing the (time-varying) point $\mathbf{x}_d = [x_d \; \dot{x}_d]^T$. Starting from any initial condition, the state trajectory reaches the time-varying surface in a finite time smaller than $|s(t=0)|/\eta$, and then slides along the surface towards $\mathbf{x}_d$ exponentially, with a time-constant equal to $1/\lambda$.

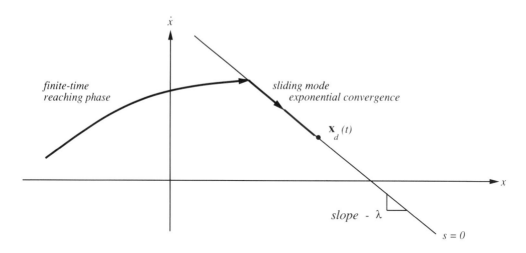

**Figure 7.3** : Graphical interpretation of equations (7.3) and (7.5) ($n = 2$)

In summary, the idea behind equations (7.3) and (7.5) is to pick-up a well-behaved function of the tracking error, $s$, according to (7.3), and then select the feedback control law $u$ in (7.1) such that $s^2$ remains a Lyapunov-like function of the closed-loop system, despite the presence of model imprecision and of disturbances. The controller design procedure then consists of two steps. First, as will be illustrated in section 7.1.3, a feedback control law $u$ is selected so as to verify sliding condition (7.5). However, in order to account for the presence of modeling imprecision and of disturbances, the control law has to be *discontinuous across S(t)*. Since the implementation of the associated control switchings is necessarily imperfect (for instance, in practice switching is not instantaneous, and the value of *s* is not known with infinite precision), this leads to *chattering* (Figure 7.4). Now (with a few important exceptions that we shall discuss in section 7.1.4), chattering is undesirable in practice, since it involves high control activity and further may excite high-frequency dynamics neglected in the course of modeling (such as unmodeled structural modes, neglected time-delays, and so on). Thus, in a second step detailed in section 7.2, the *discontinuous control law u is suitably smoothed to achieve an optimal trade-off between control bandwidth and tracking precision*: while the first step accounts for parametric uncertainty, the second step achieves robustness to high-frequency unmodeled dynamics.

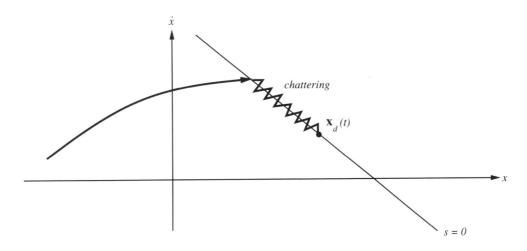

**Figure 7.4** : Chattering as a result of imperfect control switchings

## 7.1.2 * Filippov's Construction of the Equivalent Dynamics

The system's motion on the sliding surface can be given an interesting geometric interpretation, as an "average" of the system's dynamics on both sides of the surface.

The dynamics while in sliding mode can be written as

$$\dot{s} = 0 \tag{7.6}$$

By solving the above equation *formally* for the control input, we obtain an expression for *u* called the <u>*equivalent control*</u>, $u_{eq}$ , which can be interpreted as the continuous control law that would maintain $\dot{s} = 0$ if the dynamics were exactly known. For instance, for a system of the form

$$\ddot{x} = f + u$$

we have

$$u_{eq} = -f + \ddot{x}_d - \lambda\dot{\tilde{x}}$$

and the system dynamics while in sliding mode is, of course,

$$\ddot{x} = f + u_{eq} = \ddot{x}_d - \lambda\dot{\tilde{x}}$$

Geometrically, the equivalent control can be constructed as

$$u_{eq} = \alpha\, u_+ + (1 - \alpha)\, u_- \tag{7.7}$$

*i.e.*, as a *convex combination* of the values of $u$ on both sides of the surface $S(t)$. The value of $\alpha$ can again be obtained formally from (7.6), which corresponds to requiring that the system trajectories be tangent to the surface. This intuitive construction is summarized in Figure 7.5, where $\mathbf{f}_+ = [\,\dot{x} \quad f + u_+\,]^T$ , and similarly $\mathbf{f}_- = [\,\dot{x} \quad f + u_-\,]^T$ and $\mathbf{f}_{eq} = [\,\dot{x} \quad f + u_{eq}\,]$. Its formal justification was derived in the early 1960's by the Russian mathematician A. F. Filippov.

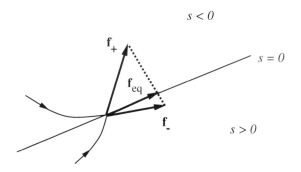

**Figure 7.5** : Filippov's construction of the equivalent dynamics in sliding mode

Recalling that the sliding motion on the surface corresponds to a limiting behavior as control switchings occur infinitely fast, the formal solution $\alpha$ of (7.6) and (7.7) can be interpreted as the average "residence time" of the trajectory on the side $s > 0$.

## 7.1.3  Perfect Performance - At a Price

Given the bounds on uncertainties on $f(\mathbf{x})$ and $b(\mathbf{x})$, constructing a control law to verify the sliding condition (7.5) is straightforward, as we now illustrate.

### A BASIC EXAMPLE

Consider the second-order system

$$\ddot{x} = f + u \tag{7.8}$$

where $u$ is the control input, $x$ is the (scalar) output of interest, and the dynamics $f$ (possibly nonlinear or time-varying) is not exactly known, but estimated as $\hat{f}$. The estimation error on $f$ is assumed to be *bounded* by some known function $F = F(x, \dot{x})$ :

$$|\hat{f} - f| \leq F \tag{7.9}$$

For instance, given the system

$$\ddot{x} + a(t)\, \dot{x}^2 \cos 3x = u \tag{7.10}$$

where $a(t)$ is unknown but verifies

$$1 \leq a(t) \leq 2$$

one has

$$\hat{f} = -1.5\, \dot{x}^2 \cos 3x \qquad\qquad F = 0.5\, \dot{x}^2 \,|\cos 3x\,|$$

In order to have the system track $x(t) \equiv x_d(t)$, we define a sliding surface $s = 0$ according to (7.3), namely:

$$s = \left(\frac{d}{dt} + \lambda\right) \tilde{x} \;=\; \dot{\tilde{x}} + \lambda \tilde{x} \tag{7.11}$$

We then have:

$$\dot{s} = \ddot{x} - \ddot{x}_d + \lambda \dot{\tilde{x}} = f + u - \ddot{x}_d + \lambda \dot{\tilde{x}} \tag{7.12}$$

The best approximation $\hat{u}$ of a continuous control law that would achieve $\dot{s} = 0$ is thus

$$\hat{u} = -\hat{f} + \ddot{x}_d - \lambda \dot{\tilde{x}} \tag{7.13}$$

Note that in terms of the discussion of section 7.1.2, $\hat{u}$ can be interpreted as our best estimate of the equivalent control. In order to satisfy sliding condition (7.5) despite uncertainty on the dynamics $f$, we add to $\hat{u}$ a term *discontinuous* across the surface

$s = 0$ :

$$u = \hat{u} - k \, \text{sgn}(s) \tag{7.14}$$

where sgn is the sign function:

$$\text{sgn}(s) = +1 \quad \text{if } s > 0$$
$$\text{sgn}(s) = -1 \quad \text{if } s < 0$$

By choosing $k = k(x, \dot{x})$ in (7.14) to be large enough, we can now guarantee that (7.5) is verified. Indeed, we have from (7.12)-(7.14)

$$\frac{1}{2}\frac{d}{dt}s^2 = \dot{s} \cdot s = [\,f - \hat{f} - k \, \text{sgn}(s)\,]\,s = (f - \hat{f})\,s \, - \, k\,|\,s\,|$$

so that, letting

$$k = F + \eta \tag{7.15}$$

we get from (7.9)

$$\frac{1}{2}\frac{d}{dt}s^2 \le -\eta\,|s|$$

as desired. Note from (7.15) that the control discontinuity $k$ across the surface $s = 0$ increases with the extent of parametric uncertainty. Also, note that $\hat{f}$ and $F$ need not depend only on $x$ or $\dot{x}$. They may more generally be functions of any *measured* variables external to system (7.8), and may also depend explicitly on time.

We can see on this basic example one of the main advantages of transforming the original tracking control problem into a simple *first-order* stabilization problem in $s$. Namely, the intuitive feedback control strategy "if the error is negative, push hard enough in the positive direction (and conversely)" *actually works for first-order systems* (recall also Example 3.9). It does not for higher-order systems.

**INTEGRAL CONTROL**

A similar result would be obtained by using integral control, *i.e.*, formally letting $(\int_o^t \tilde{x}(r)\,dr)$ be the variable of interest. The system (7.8) is now third-order relative to this variable, and (7.3) gives:

$$s = \left(\frac{d}{dt} + \lambda\right)^2 \left(\int_o^t \tilde{x}\,dr\right) = \dot{\tilde{x}} + 2\lambda\tilde{x} + \lambda^2\int_o^t \tilde{x}\,dr$$

We then obtain, instead of (7.13)

$$\hat{u} = -\hat{f} + \ddot{x}_d - 2\lambda\dot{\tilde{x}} - \lambda^2\tilde{x}$$

with (7.14) and (7.15) formally unchanged. Note that $\int_o^t \tilde{x}\, dr$ can be replaced by $\int^t \tilde{x}\, dr$, *i.e.*, the integral can be defined to within a constant. The constant can be chosen to obtain $s(t=0) = 0$ *regardless of* $\mathbf{x}_d(0)$, by letting

$$s = \dot{\tilde{x}} + 2\lambda\tilde{x} + \lambda^2\int_o^t \tilde{x}\, dr - \dot{\tilde{x}}(0) - 2\lambda\tilde{x}(0)$$

**GAIN MARGINS**

Assume now that (7.8) is replaced by

$$\ddot{x} = f + bu \tag{7.16}$$

where the (possibly time-varying or state-dependent) control gain $b$ is unknown but of known bounds (themselves possibly time-varying or state-dependent)

$$0 < b_{min} \le b \le b_{max} \tag{7.17}$$

Since the control input enters multiplicatively in the dynamics, it is natural to choose our estimate $\hat{b}$ of gain $b$ as the *geometric* mean of the above bounds:

$$\hat{b} = (b_{min}\, b_{max})^{1/2}$$

Bounds (7.17) can then be written in the form

$$\beta^{-1} \le \frac{\hat{b}}{b} \le \beta \tag{7.18}$$

where

$$\beta = (b_{max}/b_{min})^{1/2}$$

Since the control law will be designed to be robust to the bounded multiplicative uncertainty (7.18), we shall call $\beta$ the *gain margin* of our design, by analogy to the terminology used in linear control. Note that $\beta$ may be time-varying or state-dependent, and that we also have

$$\beta^{-1} \le \frac{b}{\hat{b}} \le \beta$$

Also note that the uncertainty on $b$ may come directly in the form (7.18), *e.g.*, if the control action $u$ itself is generated by a linear dynamic system.

With $s$ and $\hat{u}$ defined as before, one can then easily show that the control law

$$u = \hat{b}^{-1}[\hat{u} - k \, \text{sgn}(s)] \tag{7.19}$$

with

$$k \geq \beta(F + \eta) + (\beta - 1) \, |\hat{u}| \tag{7.20}$$

satisfies the sliding condition. Indeed, using (7.19) in the expression of $\dot{s}$ leads to

$$\dot{s} = (f - b\hat{b}^{-1}\hat{f}) + (1 - b\hat{b}^{-1})(-\ddot{x}_d + \lambda\dot{\tilde{x}}) - b\hat{b}^{-1}k \, \text{sgn}(s)$$

so that $k$ must verify

$$k \geq |\, \hat{b}b^{-1}f - \hat{f} + (\hat{b}b^{-1} - 1)(-\ddot{x}_d + \lambda\dot{\tilde{x}})\,| + \eta\hat{b}b^{-1}$$

Since $f = \hat{f} + (f - \hat{f})$, where $|f - \hat{f}| \leq F$, this in turn leads to

$$k \geq \hat{b}b^{-1}F + \eta\hat{b}b^{-1} + |\hat{b}b^{-1} - 1| \cdot |\hat{f} - \ddot{x}_d + \lambda\dot{\tilde{x}}|$$

and thus to (7.20). Note that the control discontinuity has been increased in order to account for the uncertainty on the control gain $b$.

**Example 7.1:** A simplified model of the motion of an underwater vehicle can be written

$$m\ddot{x} + c\dot{x}|\dot{x}| = u \tag{7.21}$$

where $x$ defines position, $u$ is the control input (the force provided by a propeller), $m$ is the mass of the vehicle (including the so-called added-mass, associated with motion in a fluid), and $c$ is a drag coefficient. In practice, $m$ and $c$ are not known accurately, because they only describe loosely the complex hydrodynamic effects that govern the vehicle's motion.

Defining $s$ as $s = \dot{\tilde{x}} + \lambda\tilde{x}$, computing $\dot{s}$ explicitly, and proceeding as before, a control law satisfying the sliding condition can be derived as

$$u = \hat{m}(\ddot{x}_d - \lambda\dot{\tilde{x}}) + \hat{c}\,\dot{x}|\dot{x}| - k \, \text{sgn} \, s \tag{7.22}$$

with

$$k = (F + \beta\eta) + \hat{m}(\beta - 1) |\ddot{x}_d - \lambda\dot{\tilde{x}}| \tag{7.23}$$

Note that expression (7.23) is "tighter" than the general form (7.20), reflecting the simpler structure of parametric uncertainty: intuitively, $u$ can compensate for $c\dot{x}|\dot{x}|$ directly, regardless of the uncertainty on $m$. In general, for a given problem, it is a good idea to quickly rederive a control law satisfying the sliding condition, rather than apply some pre-packed formula. $\qquad \square$

It is useful to pause at this point, and wonder whether a different control action, obtained by some other method, could achieve the same task. The answer to this question is that, given a feasible desired trajectory, there is a *unique* smooth control *time-function* that tracks it exactly, namely

$$u(t) = b(\mathbf{x}_d)^{-1} [ \ddot{x}_d - f(\mathbf{x}_d) ] \tag{7.24}$$

Thus, whatever the method, the resulting control time-function will be the same, and therefore using this particular approach simply provides a straightforward way of arriving at that time-function. Because we require perfect tracking to be achieved even in the presence of parametric uncertainty, this time-function is constructed through a process of *averaging* infinitely fast discontinuous switchings, into what we called in section 7.1.2 the equivalent control, which is precisely (7.24).

Control laws which satisfy sliding condition (7.5), and thus lead to "perfect" tracking in the face of model uncertainty, are discontinuous across the surface $S(t)$, thus leading in practice to control chattering. In general, chattering is highly undesirable, since it involves extremely high control activity, and furthermore may excite high-frequency dynamics neglected in the course of modeling. In section 7.2, we shall show how to modify the switching control laws derived above so as to eliminate chattering.

In specific (if exceptional) applications, however, control chattering is acceptable, and the pure switching control laws derived above can yield extremely high performance. We now discuss such direct applications of the previous development.

## 7.1.4  Direct Implementations of Switching Control Laws

The main direct applications of the above switching controllers include the control of electric motors, and the use of artificial dither to reduce stiction effects.

### SWITCHING CONTROL IN PLACE OF PULSE-WIDTH MODULATION

In pulse-width modulated electric motors, the control input $u$ is an electrical voltage rather than a mechanical force or acceleration. Control chattering may then be acceptable provided it is *beyond* the frequency range of the relevant unmodeled dynamics. Provided that the necessary computations (including both control law and state estimation) can be handled on-line at a high enough rate, or implemented using analog circuitry, pure sliding mode control using switched control laws can be a viable and extremely high-performance option.

### SWITCHING CONTROL WITH LINEAR OBSERVER

The difficulty in obtaining meaningful state measurements at very high sampling rates can be turned around by using state observers. For *linear* systems, the design of such observers is well known and systematic. The principle of the approach to designing a switching controller using an observer is then very simple. Instead of tracking the surface $s = 0$, the system is made to track the surface $s_e = 0$, where $s_e$ is obtained by replacing the state $\mathbf{x}$ by its estimate $\mathbf{x}_e$ in the expression of $s$. This can be achieved by computing a dynamic compensation term $\hat{u}_e$ based on the available state estimates, and using switching terms of the form $-k(\mathbf{x}_e)\ \mathrm{sgn}(s_e)$, where $k(\mathbf{x}_e)$ is chosen large enough to compensate both for parametric inaccuracies *and* for observer inaccuracies. This yields $s_e \to 0$ (as $t \to \infty$). Then, if the observer has been properly designed so that it converges despite modeling uncertainties (which, again, is easy to achieve in the linear case), we also have $s \to s_e$. Therefore, $s \to 0$, and the actual state converges towards the desired state. Furthermore, sliding mode and its robust properties are maintained on the surface $s_e = 0$, which tends towards the desired sliding surface as the observer converges.

### SWITCHING CONTROL IN PLACE OF DITHER

When uncertainty consists of effects small in magnitude but difficult to model, such as stiction or actuator ripple, switching in $s$ may be advantageously used in place of a more standard "dither" signal. Ideally, the frequency of the switching should be chosen well *beyond* that of significant structural vibration modes (in mechanical systems), while remaining *below* the actuators' bandwidth. This assumes again that meaningful state estimates can be provided at the selected switching frequency. Such an approach can particularly improve the quality of low-speed behavior, which otherwise is extremely sensitive to friction.

The examples above represent the few specific applications where chattering can be tolerated and actually exploited. In the general case, however, the question is how to derive control laws that maintain the system close to the surface $s = 0$ while avoiding chattering altogether. This is the subject of the next section.

## 7.2 Continuous Approximations of Switching Control Laws

In general, chattering must be eliminated for the controller to perform properly. This can be achieved by smoothing out the control discontinuity in a thin *boundary layer* neighboring the switching surface

$$B(t) = \{\mathbf{x}, |s(\mathbf{x};t)| \le \Phi\} \qquad \Phi > 0 \tag{7.25}$$

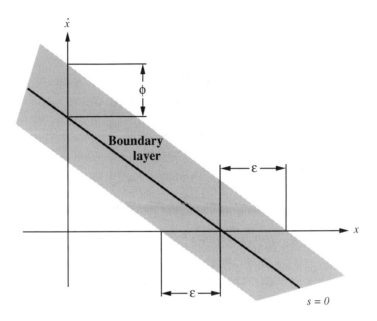

**Figure 7.6.a :** The boundary layer

where $\Phi$ is the boundary layer *thickness*, and $\varepsilon = \Phi/\lambda^{n-1}$ is the boundary layer *width*, as Figure 7.6.a illustrates for the case $n = 2$. In other words, outside of $B(t)$, we choose control law $u$ as before (*i.e.*, satisfying sliding condition (7.5)), which guarantees that the boundary layer is attractive, hence invariant: all trajectories starting inside $B(t=0)$ remain inside $B(t)$ for all $t \geq 0$ ; and we then interpolate $u$ inside $B(t)$ – for instance, replacing in the expression of $u$ the term $\mathrm{sgn}(s)$ by $s/\Phi$, inside $B(t)$, as illustrated in Figure 7.6.b.

Given the results of section 7.1.1, this leads to *tracking to within a guaranteed precision* $\varepsilon$ (rather than "perfect" tracking), and more generally guarantees that for all trajectories starting inside $B(t=0)$

$$\forall\, t \geq 0\,,\ |\tilde{x}^{(i)}(t)| \leq (2\lambda)^i\, \varepsilon \qquad i = 0\,,\ \ldots\,,\, n{-}1$$

**Example 7.2:** Consider again the system (7.10), and assume that the desired trajectory is $x_d = \sin(\pi t/2)$.

Figure 7.7 shows the tracking error and control law using the switched control law (with $\lambda = 20\,,\ \eta = 0.1$)

$$u = \hat{u} - k\,\mathrm{sgn}(s)$$
$$= 1.5\,\dot{x}^2 \cos 3x + \ddot{x}_d - 20\,\dot{\tilde{x}} - (\,0.5\,\dot{x}^2 \,|\cos 3x| + 0.1\,)\,\mathrm{sgn}[\dot{\tilde{x}} + 20\,\tilde{x}]$$

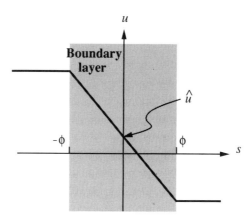

**Figure 7.6.b :** Control interpolation in the boundary layer

at a sampling rate of 1 kHz. The actual value of $a(t)$ used in the simulations is $a(t) = |\sin t| + 1$ (which verifies the assumed bound on $a(t)$). We see that tracking performance is excellent, but is obtained at the price of high control chattering.

Assume now that we interpolate the above control input in a thin boundary layer of thickness 0.1

$$u = \hat{u} - k\, \mathrm{sat}(s/\Phi)$$
$$= 1.5\, \dot{x}^2 \cos 3x + \ddot{x}_d - 20\, \dot{\tilde{x}} - (\,0.5\, \dot{x}^2\, |\cos 3x| + 0.1)\, \mathrm{sat}[\,(\dot{\tilde{x}} + 20\, \tilde{x})/0.1\,]$$

As shown in Figure 7.8, the tracking performance, while not as "perfect" as above, is still very good, and is now achieved using a smooth control law. Note that the bounds on tracking error are consistent with (7.25). ☐

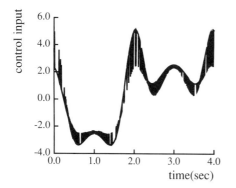
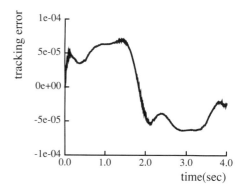

**Figure 7.7 :** Switched control input and resulting tracking performance

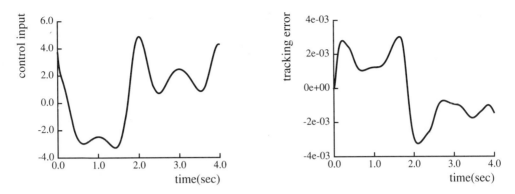

**Figure 7.8 :** Smooth control input and resulting tracking performance

The intuitive understanding of the effect of control interpolation in a boundary layer can be carried on further, and guide the selection of the design parameters $\lambda$ and $\Phi$. As we now show, *the smoothing of control discontinuity inside B(t) essentially assigns a lowpass filter structure to the local dynamics of the variable s*, thus eliminating chattering. Recognizing this filter-like structure then allows us, in essence, to tune up the control law so as to achieve a trade-off between tracking precision and robustness to unmodeled dynamics. Boundary layer thickness $\Phi$ can be made *time-varying*, and can be monitored so as to well exploit the control "bandwidth" available. The development is first detailed for the case $\beta = 1$ (no gain margin), and then generalized.

Consider again the system (7.1) with $b = \hat{b} = 1$. In order to maintain attractiveness of the boundary layer now that $\Phi$ is allowed to vary with time, we must actually modify condition (7.5). Indeed, we now need to guarantee that the distance *to the boundary layer* always decreases

$$s \geq \Phi \quad \Rightarrow \quad \frac{d}{dt}[s - \Phi] \leq -\eta$$

$$s \leq -\Phi \quad \Rightarrow \quad \frac{d}{dt}[s - (-\Phi)] \geq \eta$$

Thus, instead of simply requiring that (7.5) be satisfied outside the boundary layer, we now require that (combining the above equations)

$$|s| \geq \Phi \quad \Rightarrow \quad \frac{1}{2}\frac{d}{dt}s^2 \leq (\dot{\Phi} - \eta)|s| \tag{7.26}$$

The additional term $\dot{\Phi}|s|$ in (7.26) reflects the fact that the boundary layer attraction condition is more stringent during boundary layer contraction ($\dot{\Phi} < 0$) and less stringent during boundary layer expansion ($\dot{\Phi} > 0$). In order to satisfy (7.26), the quantity $-\dot{\Phi}$ is added to control discontinuity gain $k(\mathbf{x})$ , *i.e.*, in our smoothed implementation the term $k(\mathbf{x})\,\mathrm{sgn}(s)$ obtained from switched control law $u$ is actually replaced by $\overline{k}(\mathbf{x})\,\mathrm{sat}(s/\Phi)$ , where

$$\overline{k}(\mathbf{x}) = k(\mathbf{x}) - \dot{\Phi} \tag{7.27}$$

and sat is the saturation function:

$$\mathrm{sat}(y) = y \qquad \text{if } |y| \le 1$$

$$\mathrm{sat}(y) = \mathrm{sgn}(y) \quad \text{otherwise}$$

Accordingly, control law $u$ becomes:

$$u = \hat{u} - \overline{k}(\mathbf{x})\,\mathrm{sat}(s/\Phi)$$

Let us now consider the system trajectories *inside the boundary layer*, where they lie by construction: they can be expressed directly in terms of the variable $s$ as

$$\dot{s} = -\overline{k}(\mathbf{x})\,\frac{s}{\Phi} - \Delta f(\mathbf{x}) \tag{7.28}$$

where $\Delta f = \hat{f} - f$. Now since $\overline{k}$ and $\Delta f$ are continuous in $\mathbf{x}$, we can exploit (7.4) to rewrite (7.28) in the form

$$\dot{s} = -\overline{k}(\mathbf{x}_d)\,\frac{s}{\Phi} + \left( -\Delta f(\mathbf{x}_d) + O(\varepsilon) \right) \tag{7.29}$$

We see from (7.29) that *the variable s* (which is a measure of the algebraic distance to the surface $S(t)$) *can be viewed as the output of a first-order filter*, whose dynamics only depends on the desired state $\mathbf{x}_d(t)$, and whose inputs are, to the first order, "perturbations," *i.e.*, uncertainty $\Delta f(\mathbf{x}_d)$. Thus, chattering can indeed be eliminated, as long as high-frequency unmodeled dynamics are not excited. Conceptually, the structure of the closed-loop error dynamics can be summarized by Figure 7.9: perturbations are filtered according to (7.29) to give $s$, which in turn provides tracking error $\tilde{x}$ by further lowpass filtering, according to definition (7.3). Control action $u$ is a function of $\mathbf{x}$ and $\mathbf{x}_d$. Now, since $\lambda$ is the break-frequency of filter (7.3), it must be chosen to be "small" with respect to high-frequency unmodeled dynamics (such as unmodeled structural modes or neglected time delays). Furthermore, we can now tune the boundary layer thickness $\Phi$ so that (7.29) also represents a first-order filter of bandwidth $\lambda$. It suffices to let

$$\frac{\overline{k}(\mathbf{x}_d)}{\Phi} = \lambda \qquad (7.30)$$

which can be written from (7.27) as

$$\dot{\Phi} + \lambda\Phi = k(\mathbf{x}_d) \qquad (7.31)$$

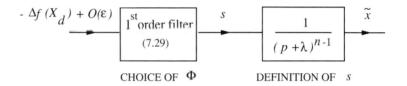

$$-\Delta f(X_d) + O(\varepsilon) \rightarrow \boxed{\begin{array}{c} 1^{\text{st}} \text{ order filter} \\ (7.29) \end{array}} \xrightarrow{\ s\ } \boxed{\dfrac{1}{(p+\lambda)^{n-1}}} \xrightarrow{\ \tilde{x}\ }$$

CHOICE OF $\Phi$        DEFINITION OF $s$

**Figure 7.9 :** Structure of the closed-loop error dynamics

Equation (7.31) defines the desired time-history of boundary layer thickness $\Phi$, and, in the light of Figure 7.9, shall be referred to as the *balance condition*. Intuitively, it amounts to tuning up the closed-loop system so that it mimics an $n^{\text{th}}$ order critically damped system. Furthermore, definition (7.27) can then be rewritten as

$$\overline{k}(\mathbf{x}) = k(\mathbf{x}) - k(\mathbf{x}_d) + \lambda\Phi \qquad (7.32)$$

The *s–trajectory*, *i.e.*, the variation of $s$ with time, is a compact descriptor of the closed-loop behavior: control activity directly depends on $s$, while by definition (7.3) tracking error $\tilde{x}$ is merely a filtered version of $s$. Furthermore, the *s–trajectory represents a time-varying measure of the validity of the assumptions on model uncertainty*. Similarly, the boundary layer thickness $\Phi$ describes the evolution of dynamic model uncertainty with time. It is thus particularly informative to plot $s(t)$, $\Phi(t)$, and $-\Phi(t)$ on a single diagram, as illustrated in Figure 7.10.

**Example 7.3:** Consider again the system described by (7.10). The complete control law is now

$$u = \ddot{x}_d - \lambda\dot{\tilde{x}} + 1.5\,\dot{x}^2\cos 3x - (0.5\,\dot{x}^2|\cos 3x| + \eta - \dot{\Phi})\,\text{sat}[(\dot{\tilde{x}} + \lambda\tilde{x})/\Phi]$$

with      $\dot{\Phi} = -\lambda\Phi + (0.5\,\dot{x}_d^2|\cos 3x_d| + \eta)$

and, assuming *e.g.*, that $\dot{x}_d(0) = 0$ initially, $\Phi(0) = \eta/\lambda$. As in Example 7.2, we let $\eta = 0.1$ and $\lambda = 20$. Typically, the arbitrary constant $\eta$ (which, formally, reflects the time to reach the boundary layer starting from the outside) is chosen to be small as compared to the average value of $k(\mathbf{x}_d)$, so as to fully exploit our knowledge of the structure of parametric uncertainty. The

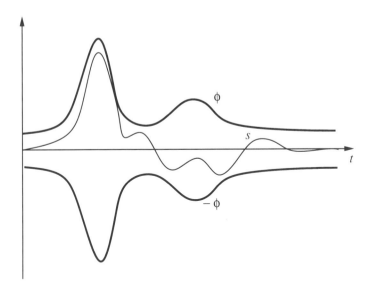

**Figure 7.10 :** The *s*–trajectories can convey much information on a single plot

value of $\lambda$ is selected based on the frequency range of unmodeled dynamics, as we shall discuss later.

The tracking error, control input, and *s*-trajectories are plotted in Figure 7.11 for the same desired trajectory $x_d = \sin(\pi t/2)$ as in Example 7.2. We see that while the maximum value of the time-varying boundary layer thickness $\Phi$ is the same as that originally chosen (purposefully) as the constant value of $\Phi$ in Example 7.2, the tracking error is consistently better (up to 4 times better) than that in Example 7.2, because varying the thickness of the boundary layer allows us to make better use of the available bandwidth.    $\square$

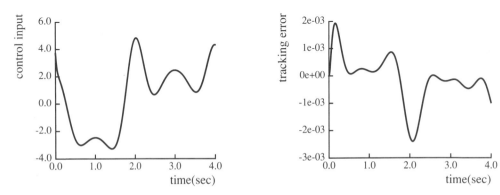

**Figure 7.11a :** Control input and resulting tracking performance

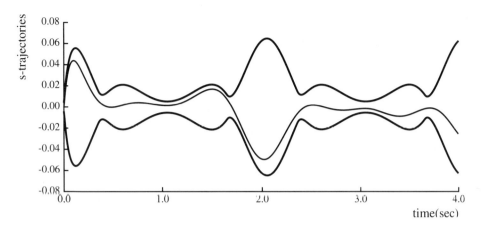

**Figure 7.11b** : *s*-trajectories with time-varying boundary layers

In the case that $\beta \neq 1$, one can easily show that (7.31) and (7.32) become (with $\beta_d = \beta(\mathbf{x}_d)$)

$$k(\mathbf{x}_d) \geq \frac{\lambda\Phi}{\beta_d} \quad => \quad \dot{\Phi} + \lambda\Phi = \beta_d k(\mathbf{x}_d) \qquad \overline{k}(\mathbf{x}) = k(\mathbf{x}) - \dot{\Phi}/\beta \qquad (7.33)$$

$$k(\mathbf{x}_d) \leq \frac{\lambda\Phi}{\beta_d} \quad => \quad \dot{\Phi} + \frac{\lambda\Phi}{\beta_d^2} = k(\mathbf{x}_d)/\beta_d \qquad \overline{k}(\mathbf{x}) = k(\mathbf{x}) - \beta\dot{\Phi} \qquad (7.34)$$

with initial condition $\Phi(0)$ defined as:

$$\Phi(0) = \beta_d\, k(\mathbf{x}_d(0))/\lambda \qquad (7.35)$$

Indeed, in order to satisfy (7.26) in the presence of uncertainty $\beta$ on the control gain we let

$$\dot{\Phi} > 0 \quad => \quad \overline{k}(\mathbf{x}) = k(\mathbf{x}) - \dot{\Phi}/\beta \qquad (7.36)$$

$$\dot{\Phi} < 0 \quad => \quad \overline{k}(\mathbf{x}) = k(\mathbf{x}) - \beta\dot{\Phi} \qquad (7.37)$$

Furthermore, the balance condition can be written, instead of (7.30), as

$$\left(\frac{\overline{k}(\mathbf{x}_d)}{\Phi}\right) \left(\frac{b(\mathbf{x}_d)}{\hat{b}(\mathbf{x}_d)}\right)_{max} = \lambda \qquad (7.38)$$

that is,

$$\overline{k}(\mathbf{x}_d) = \lambda\Phi/\beta_d$$

Applying this relation to (7.36), (7.37) leads to the desired behavior of $\dot{\Phi}$ :

$$\dot{\Phi} > 0 \quad \Rightarrow \quad \frac{\lambda\Phi}{\beta_d} = k(\mathbf{x}_d) - \dot{\Phi}/\beta_d$$

$$\dot{\Phi} < 0 \quad \Rightarrow \quad \frac{\lambda\Phi}{\beta_d} = k(\mathbf{x}_d) - \beta_d\dot{\Phi}$$

which we can combine with (7.36)-(7.37) and rewrite as (7.33)-(7.34). Finally, remark that if $\beta = \beta_d$ , one has

$$\overline{k}(\mathbf{x}) = \left(\overline{k}(\mathbf{x}) - \overline{k}(\mathbf{x}_d)\right) + \overline{k}(\mathbf{x}_d) = k(\mathbf{x}) - k(\mathbf{x}_d) + \lambda\Phi/\beta_d$$

Note that the balance conditions (7.33) and (7.34) imply that $\Phi$ and thus $\tilde{x}$ are bounded for bounded $\mathbf{x}_d$ .

The balance conditions have a simple and intuitive physical interpretation: neglecting time constants of order $1/\lambda$ , they imply that

$$\lambda^n \varepsilon \approx \beta_d\, k(\mathbf{x}_d)$$

that is

> (bandwidth)$^n$ × (tracking precision)
> $\approx$ (parametric uncertainty measured along the desired trajectory)

Such trade-off is quite similar to the situation encountered in linear time-invariant systems, but here applies to the more general nonlinear system structure (7.1), all along the desired trajectory. In particular, it shows that, in essence, *the balance conditions specify the best tracking performance attainable, given the desired control bandwidth and the extent of parameter uncertainty.* Similarly, expression (7.20) allows one to easily quantify the trade-off between accuracy and speed along a given path, for instance.

**Example 7.4:** Let us consider again the underwater vehicle of Example 7.1, and smoothen the control input using time-varying boundary layer, as described above. The *a priori* bounds on $m$ and $c$ are

$$1 \leq m \leq 5 \qquad 0.5 \leq c \leq 1.5$$

and, accordingly,

$$\hat{m} = \sqrt{5} \qquad \hat{c} = 1$$

The actual values used in the simulation are

$$m = 3 + 1.5\sin(|\dot{x}|t) \qquad\qquad c = 1.2 + .2\sin(|\dot{x}|t)$$

which are used as a metaphor of the complexity of the actual hydrodynamic effects. We let $\eta = 0.1$ and $\lambda = 10$.

The desired trajectory consists of a constant-acceleration phase at 2 m/s² for two seconds, a constant-velocity phase (at 4 m/s) for two seconds, and a constant-acceleration phase at − 2 m/s² for two seconds. The corresponding tracking error, control input, and *s* trajectories are plotted in Figure 7.12.    ☐

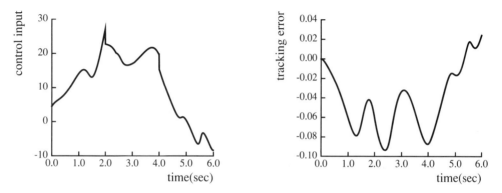

**Figure 7.12a :** Control input and resulting tracking performance

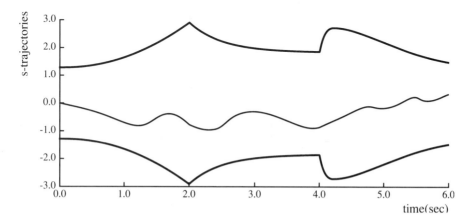

**Figure 7.12b :** *s*-trajectories with time-varying boundary layers

**REMARKS**

(i) The desired trajectory $\mathbf{x}_d$ must itself be chosen smooth enough not to excite the high-frequency unmodeled dynamics.

(ii) An argument similar to that of the above discussion shows that the choice of dynamics (7.3) used to define sliding surfaces is the "best-conditioned" among linear dynamics, in the sense that it guarantees the best tracking performance given the desired control bandwidth and the extent of parameter uncertainty.

(iii) If the model or its bounds are so imprecise that $F$ can only be chosen as a large *constant*, then $\phi$ from (7.31) is constant and large, so that the term $\overline{k}$ sat($s/\phi$) simply equals $\lambda\, s/\beta$ in the boundary layer, and therefore acts as a simple P.D. : there is no free lunch.

(iv) A well-designed controller should be capable of gracefully handling *exceptional* disturbances, *i.e.*, disturbances of intensity higher than the predicted bounds which are used in the derivation of the control law. For instance, somebody may walk into the laboratory and push violently on the system "to see how stiff it is"; an industrial robot may get jammed by the failure of some other machine; an actuator may saturate as the result of the specification of an unfeasible desired trajectory. If integral control is used in such cases, the integral term in the control action may become unreasonably large, so that once the disturbance stops, the system goes through large amplitude oscillations in order to return to the desired trajectory. This phenomenon, known as *integrator windup*, is a potential cause of instability because of saturation effects and physical limits on the motion. It can be simply avoided by *stopping integration* (i.e. maintaining the integral term constant) *as long as the system is outside the boundary layer*. Indeed, under normal circumstances the system does remain in the boundary layer; on the other hand, when the conditions return to normal after an exceptional disturbance, integration can resume as soon as the system is back in the boundary layer, since the integral term is defined to within an arbitrary constant.

(v) In the case that $\lambda$ is time-varying (as further discussed in the next section), the term

$$u' = -\dot{\lambda}\, \tilde{x}$$

should be added to the corresponding $\hat{u}$, while augmenting gain $k(\mathbf{x})$ accordingly by the quantity $|u'|(\beta - 1)$.

The degree of simplification in the system model may be varied according to the on-line computing power available: in essence, the balance conditions quantify the trade-off between model precision and tracking accuracy, as further detailed next. Furthermore, the $s$-trajectories provide a measure of the validity of the assumptions on model uncertainty and of the adequacy of bound simplifications.

# 7.3   The Modeling/Performance Trade-Offs

The balance conditions (7.33)-(7.34) have practical implications in terms of design/modeling/performance trade-offs. Neglecting time-constants of order $1/\lambda$, conditions (7.33) and (7.34) imply that

$$\lambda^n \epsilon \approx \beta_d \, k_d \tag{7.39}$$

as noticed in section 7.2. If we now consider the structure of control law (7.19), we see that the effects of parameter uncertainty on $f$ have been "dumped" in gain $k$. Conversely, better knowledge of $f$ reduces $k$ by a comparable quantity. Thus (7.39) is particularly useful in an *incremental* mode, *i.e.*, to evaluate the effects of model simplification (or conversely of increased model sophistication) on tracking performance:

$$\Delta \epsilon \approx \Delta(\beta_d \, k_d / \lambda^n) \tag{7.40}$$

In particular, marginal gains in performance are critically dependent on control bandwidth $\lambda$: if large $\lambda$'s are available, poor dynamic models may lead to respectable tracking performance, and conversely large modeling efforts produce only minor absolute improvements in tracking accuracy.

It is of course not overly surprising that system performance be very sensitive to control bandwidth $\lambda$ : (7.1) only represents *part* of the system dynamics – *e.g.*, its rigid-body component – while $\lambda$ accounts for the unmodeled part. In the right-hand side of (7.40), the effects of parametric uncertainty in (7.1) are reflected in the numerator, while the presence of dynamics neglected in the model is reflected in the denominator, since it both translates into the order $n$ of the model, and imposes upper bounds on the choice of $\lambda$.

Thus, given model (7.1), a key question is to determine how large $\lambda$ can be chosen. Although the tuning of this single scalar may in practice be done experimentally, considerable insight on the overall design can be obtained by explicitly analyzing the various factors limiting $\lambda$. In mechanical systems, for instance, given clean measurements, $\lambda$ is typically limited by three factors:

(i) *structural resonant modes*: $\lambda$ must be smaller than the frequency $\nu_R$ of the lowest unmodeled structural resonant mode; a reasonable interpretation of this constraint is, classically

$$\lambda \le \lambda_R \approx \frac{2\pi}{3} \nu_R \tag{7.41}$$

although in practice this bound may be modulated by engineering judgment, taking notably into account the natural damping of the structural modes. Furthermore, it may be worthwhile in certain cases to account for the fact that $\lambda_R$ may actually vary with the task (*e.g.*, given configuration or loads).

(ii) *neglected time delays*: along the same lines, we have a condition of the form

$$\lambda \le \lambda_A \approx \frac{1}{3T_A} \tag{7.42}$$

when $T_A$ is the largest unmodeled time-delay (for instance in the actuators).

(iii) *sampling rate*: with a full-period processing delay, one gets a condition of the form

$$\lambda \le \lambda_S \approx \frac{1}{5} \nu_{\text{sampling}} \tag{7.43}$$

where $\nu_{\text{sampling}}$ is the sampling rate.

The desired control bandwidth $\lambda$ is the minimum of the three bounds (7.41)-(7.43). Bound (7.41) essentially depends on the system's mechanical properties, while (7.42) reflects limitations on the actuators, and (7.43) accounts for the available computing power. Ideally, the most effective design corresponds to *matching* these limitations, *i.e.*, having

$$\lambda_R \approx \lambda_A \approx \lambda_S = \lambda \tag{7.44}$$

Now (7.41) and (7.42) are "hard" limitations, in the sense that they represent properties of the hardware itself, while (7.43) is "soft" as far as it reflects the performance of the computer environment *and* the complexity of the control algorithm. Assume for instance that bound (7.43) is the most stringent, which means that the system's mechanical potentials are not fully exploited. This may typically occur in modern high-performance robots (such as direct-drive arms) which feature high mechanical stiffness and high resonant frequencies. It may be worthwhile, before embarking in the development of dedicated computer architectures, to first consider *simplifying* the dynamic model used in the control algorithm. This in turn allows one to replace $\lambda = \lambda_{\text{slow}}$ by a larger $\lambda = \lambda_{\text{fast}}$ which varies inversely proportionally to the required computation time. From (7.40) and assuming that neither of bounds (7.41) or (7.42) is hit in the process, this operation is beneficial as long as

$$\frac{\Delta(\beta_d\, k_d)}{\beta_d\, k_d} \;\leq\; \left(\frac{\lambda_{fast}}{\lambda_{slow}}\right)^2 - 1 \tag{7.45}$$

Conversely, equality in (7.45) defines the threshold at which model simplification starts degrading performance despite gains in sampling rate. This threshold is rarely reached in practice: even assuming that marginal gains in model precision depend linearly on the computation time involved, $\lambda^{-2}$ still varies as the square of the required sampling period. Thus it is often advisable to reduce model complexity until computation can be achieved at a rate fully compatible with the mechanical capabilities of the arm, in other words until $\lambda_S$ is no longer the "active" limit on $\lambda$. The performance increase resulting from this simple operation may in turn be adequate and avoid major development efforts on the computer hardware.

The trade-off between modeling inaccuracy and performance can be further improved only by updating the model on-line. This can indeed be achieved when some components of the model depend linearly on unknown but constant parameters, allowing the corresponding uncertainties to be mapped in terms of a single unknown constant vector. This is the topic of chapter 8, adaptive control.

# 7.4 * Multi-Input Systems

This section discusses extensions of the previous development to multi-input systems. The point of view taken here is essentially mathematical. In chapter 9, we shall discuss how the exploitation of known physical properties of the systems, such as conservation of energy, may often make such multi-input extensions simpler and more powerful.

Consider a nonlinear multi-input system of the form

$$x_i^{(n_i)} = f_i(\mathbf{x}) + \sum_{j=1}^{m} b_{ij}(\mathbf{x})\, u_j \qquad i = 1, \ldots, m \qquad j = 1, \ldots, m$$

where the vector $\mathbf{u}$ of components $u_j$ is the control input vector, and the state $\mathbf{x}$ is composed of the $x_i$'s and their first $(n_i - 1)$ derivatives. As mentioned in chapter 6, such systems are called *square* systems, since they have as many control inputs $u_j$ as outputs to be controlled $x_i$. We are interested again in the problem of having the state $\mathbf{x}$ track a desired time-varying state $\mathbf{x}_d$, in the presence of parametric uncertainties.

We make two assumptions. First, we assume that the *matching conditions* discussed in chapter 6 are verified, *i.e.*, that parametric uncertainties are within the range space of the input matrix **B** (of components $b_{ij}$). Since **B** is a square $m \times m$ matrix, this simply means that **B** is invertible over the whole state-space, a controllability-like assumption. Second, we assume that the estimated input matrix $\hat{\textbf{B}}$ is invertible, continuously dependent on parametric uncertainty, and such that $\hat{\textbf{B}} = \textbf{B}$ in the absence of parametric uncertainty.

As in the single-input case, we shall write uncertainties on **f** in additive form, and uncertainties on the input matrix **B** in multiplicative form

$$|\hat{f}_i - f_i| \le F_i \qquad\qquad i = 1, \ldots, m \tag{7.46}$$

$$\textbf{B} = (\textbf{I} + \Delta)\,\hat{\textbf{B}} \qquad |\Delta_{ij}| \le D_{ij} \qquad i = 1, \ldots, m \qquad j = 1, \ldots, m \tag{7.47}$$

where **I** is the $n \times n$ identity matrix. Note that the structure of expression (7.46) is slightly different from that of (7.47), since the notion of a gain margin is mostly a scalar concept, while (7.47) shall prove more convenient for the purpose of matrix manipulation.

Let us define a vector **s** of components $s_i$ by

$$s_i = \left(\frac{d}{dt} + \lambda_i\right)^{n_i - 1} \tilde{x}_i$$

which, for notational compactness, we shall write

$$s_i = x_i^{(n_i - 1)} - x_{ri}^{(n_i - 1)}$$

This defines a vector $\textbf{x}_r^{(n-1)}$ of components $x_{ri}^{(n_i - 1)}$, which can be computed from **x** and $\textbf{x}_d$. As in the single-input case, the controller design can be translated in terms of finding a control law for the vector **u** that verifies individual sliding conditions of the form

$$\frac{1}{2}\frac{d}{dt} s_i^2 \le -\eta_i |s_i| \qquad (\eta_i > 0) \tag{7.48}$$

in the presence of parametric uncertainty. Letting **k** sgn(**s**) be the vector of components $k_i$ sgn($s_i$), and choosing the control law to be of the form

$$\textbf{u} = \hat{\textbf{B}}^{-1}\,(\,\textbf{x}_r^{(n-1)} - \hat{\textbf{f}} - \textbf{k}\,\text{sgn}(\textbf{s})\,) \tag{7.49}$$

similarly to the single-input case, we can write

$$\dot{s}_i = \hat{f}_i - f_i + \sum_{j=1}^{n} \Delta_{ij} (x_{ri}^{(n_i-1)} - \hat{f}_j)$$
$$- \sum_{j \neq i} \Delta_{ij} k_j \, \text{sgn}(s_j) - (1 + \Delta_{ii}) k_i \, \text{sgn}(s_i)$$

Thus, the sliding conditions are verified if

$$(1 - D_{ii}) k_i \geq F_i + \sum_{j=1}^{n} D_{ij} |x_{ri}^{(n_i-1)} - \hat{f}_j| - \sum_{j \neq i} D_{ij} k_j + \eta_i$$

$$i = 1, \ldots, n$$

and, in particular, if the vector **k** is chosen such that

$$(1 - D_{ii}) k_i + \sum_{j \neq i} D_{ij} k_j = F_i + \sum_{j=1}^{n} D_{ij} |x_{ri}^{(n_i-1)} - \hat{f}_j| + \eta_i \qquad (7.50)$$

$$i = 1, \ldots, n$$

Expression (7.50) represents a set of $m$ equations in the $m$ switching gains $k_i$. Do these equations have a solution **k** (then necessarily unique), and are the components $k_i$ all positive (or zero)? The answer to both questions is yes, thanks to an interesting result of matrix algebra, known as the Frobenius-Perron theorem.

**Theorem (Frobenius-Perron)** *Consider a square matrix* **A** *with non-negative elements. Then, the largest real eigenvalue $\rho_1$ of* **A** *of is non-negative. Furthermore, consider the equation*

$$(\mathbf{I} - \rho^{-1} \mathbf{A}) \mathbf{y} = \mathbf{z}$$

*where all components of the vector* **z** *are non-negative. If $\rho > \rho_1$, then the above equation admits a unique solution* **y**, *whose components $y_i$ are all non-negative.*

Applying the Frobenius-Perron theorem to the matrix of components $D_{ij}$, and noticing that our second assumption on the system implies that $\rho_1 < 1$, shows that equation (7.50) uniquely defines a set of non-negative $k_i$. Thus, the control law (7.49), with **k** defined by (7.50), satisfies the sliding condition in the presence of parametric uncertainties bounded by (7.46).

As in the single-input case, the switching control laws derived above can be smoothly interpolated in boundary layers, so as to eliminate chattering, thus leading to a trade-off between parametric uncertainty and performance. The reader is referred to section 7.6 for details.

Note that the point of view taken in this section is essentially mathematical.

Chapter 9 shall discuss how the exploitation of physical properties of the systems, such as conservation of energy, often makes multi-input designs simpler and more powerful. This will become particularly important in adaptive versions of the designs.

# 7.5 Summary

The aim of a sliding controller is to

(i) Design a control law to effectively account for

- parameter uncertainty, *e.g.*, imprecision on the mass properties or loads, inaccuracies on the torque constants of the actuators, friction, and so on.

- the presence of *unmodeled* dynamics, such as structural resonant modes, neglected time-delays (in the actuators, for instance), or finite sampling rate.

(ii) Quantify the resulting *modeling/performance trade-offs*, and in particular, the effect on tracking performance of discarding any particular term in the dynamic model.

The methodology is based on a notational simplification, which amounts to replacing an $n^{th}$ order tracking problem by a first order stabilization problem. Although "perfect" performance can in principle be achieved in the presence of arbitrary parameter inaccuracies, uncertainties in the model structure (*i.e.*, unmodeled dynamics) lead to a trade-off between tracking performance and parametric uncertainty, given the available "bandwidth." In practice, this corresponds to replacing a switching, chattering control law by its smooth approximation. In specific applications where control chattering is acceptable, the pure switching control laws can yield extremely high performance.

Sliding controller design provides a systematic approach to the problem of maintaining stability in the face of modeling imprecisions. Furthermore, it quantifies the modeling/performance trade-offs, and in that sense can illuminate the whole design and testing process. Finally, it offers the potential of simplifying higher-level programming, by accepting reduced information about both task and system.

# 7.6 Notes and References

The concept of a sliding surface originated in the Soviet literature [*e.g.*, Aizerman and Gantmacher, 1957; Emelyanov, 1957; Filippov, 1960] (see also [Tsypkin, 1955; Flugge-Lotz, *et al.*, 1958]), mostly in the context of "variable-structure" regulation of linear systems, see [Utkin, 1977] for a review (also [Young, 1978]).  Classical sliding mode control, however, had important drawbacks limiting its practical applicability, such as chattering and large control authority. The development of sections 7.1-7.3 is based on [Slotine, 1984].

The combination of sliding controllers with state observers is discussed in [Bondarev, *et al.*, 1985] in the linear case, and [Hedrick and Ragavan, 1990] in the nonlinear case. Observers based on sliding surfaces are discussed in [Drakunov, 1983; Slotine, *et al.*, 1986, 1987; Walcott and Zak, 1987].  The development of section 7.4 is adapted from [Slotine, 1985; Hedrick and Gopalswamy, 1989]. The reader is referred to, *e.g.*, [Luenberger, 1979] for a simple proof of the Frobenius-Perron theorem.  Some details on boundary layer interpolations in the multi-input case can be found in [Slotine, 1985].

Practical implementations of sliding control are described, *e.g.*, in [Yoerger, *et al.*, 1986] (underwater vehicles), [Hedrick, *et al.*, 1988] (automotive applications), [Harashima, *et al.*, 1988] (robot manipulators). The literature in the field has been extremely active in the past few years.

Related approaches to robust control include, *e.g.*, [Corless and Leitmann, 1981; Gutman and Palmor, 1982; Ha and Gilbert, 1985].

# 7.7 Exercises

**7.1**   Consider the underwater vehicle of Example 7.4, and assume there is an unmodeled "pendulum mode" of the vehicle at 2 Hz.  Choose $\lambda$ accordingly, and simulate the vehicle's performance on the same trajectory as in Example 7.4.  What is the minimum sampling rate required to implement your design?

Discuss the performance of the system on various trajectories, which you may want to generate using a reference model, as in equation (II.5), with $k_1 = 2\lambda$, $k_2 = \lambda^2$.

Simulate the unmodeled 2 Hz mode by first passing the control law through a second-order lowpass filter of unit d.c. gain before actually inputting it in the system. Tune $\lambda$ "experimentally" around the value given by (7.41), for different values of the filter's damping.

**7.2**   For the system

$$\dot{x}_1 = \sin x_2 + \sqrt{t+1}\ x_2$$

$$\dot{x}_2 = \alpha_1(t)\, x_1{}^4 \cos x_2 + \alpha_2(t)\, u$$

design a controller to track an arbitrary desired trajectory $x_{d1}(t)$. Assume that the state $[x_1 \ x_2]^T$ is measured, that $x_{d1}(t), \dot{x}_{d1}(t), \ddot{x}_{d1}(t)$ are all known, and that $\alpha_1(t)$ and $\alpha_2(t)$ are unknown time-varying functions verifying the known bounds

$$\forall \, t \geq 0, \, |\alpha_1(t)| \leq 10 \qquad 1 \leq \alpha_2(t) \leq 2$$

Write the full expression of the controller, as a function of the measured state $[x_1 \ x_2]^T$. Check your design in simple simulations. (*Hint*: First feedback linearize the system by differentiating the first equation.)

**7.3**   Consider again the system of Exercise 7.1, but define $s$ so as to use integral control (as discussed in section 7.1.3). Simulate the controller's performance and compare with the results of Exercise 7.1, on various trajectories. Show that, besides allowing the system to always start at $s = 0$, the integral control term keeps adjusting the control law as long as the effects of bounded disturbances are not compensated.

**7.4**   Consider a system of the form (7.1), but where $b$ is now *constant* and of known constant positive bounds. Divide both sides of the equation by $b$, and write the sliding condition as

$$\frac{1}{2} \frac{d}{dt} h \, s^2 \, \leq \, -\eta \, h \, |s|$$

where $h = 1/b$. By representing uncertainty on $h$ additively, design a simple switching controller to satisfy the above condition.

Smooth out the switching controller in boundary layers, and derive the corresponding balance conditions.

Show that the accuracy of the approximate "bandwidth" analysis in the boundary layer increases with $\lambda$.

**7.5**   Consider a system of the form (7.1), and assume that not only **x** but also $\dot{s}$ can be measured (*e.g.*, that $x^{(n)}$ can be measured). An additional term of the form $-\alpha \dot{s}$ (with $\alpha > 0$) can then be included in the control law.

Write the corresponding expression of $\dot{s}$. Assuming for simplicity that the gain margin $\beta$ is constant, show that, given the bandwidth $\lambda$ and the parametric uncertainty, the effect of the additional term in $u$ is to reduce the maximum value of $s$ by a factor $(1 + \alpha/\beta)$. Show that this implies that the tracking error $\tilde{x}$ can in principle be made arbitrarily small simply by increasing $\alpha$.

What are the limitations of this approach? In particular, assume that there is $1\%$ uncertainty on the value of $x^{(n)}$. How small must $1$ be to make the approach worthwhile? (Adapted from [Asada and Slotine, 1986].)

**7.6**   Show that a condition similar to (7.68) can also be obtained by requiring that the system take at least two sampling periods to cross the boundary layer.

Assume that sampling is the active limit on bandwidth, and that chattering is acceptable. Based on the above result, how do the tracking performances of a switching controller and a smooth sliding controller compare?

**7.7**   Design a switching controller for the system

$$\ddot{x} + \alpha_1(t)\,|x|\,\dot{x}^2 + \alpha_2(t)\,x^3 \cos 2x \; = \; 5\,\dot{u} + u$$

where $\alpha_1(t)$ and $\alpha_2(t)$ are unknown time-varying functions verifying the known bounds

$$\forall\, t \geq 0, \quad |\alpha_1(t)| \leq 1 \qquad -1 \leq \alpha_2(t) \leq 5$$

(Hint: let $v = 5\,\dot{u} + u$ . Discuss the effect of chattering in $v$ .)

**7.8**   In the context of section 7.2, define

$$s_\Delta \; = \; s \; - \; \Phi\,\mathrm{sat}\,(s/\Phi)$$

as a measure of distance to the boundary layer. Show that the time derivative of $s_\Delta^2$ is well defined and continuous in time, and, specifically, that one can write

$$\frac{1}{2}\frac{\mathrm{d}}{\mathrm{d}t}\,s_\Delta^2 \; = \; s_\Delta\,\dot{s} \; - \; |s_\Delta|\,\dot{\Phi}$$

Show that equation (7.26) can be written

$$\frac{1}{2}\frac{\mathrm{d}}{\mathrm{d}t}\,s_\Delta^2 \; \leq \; -\,\eta\,|s_\Delta|$$

(Adapted from [Slotine and Coetsee, 1986].)

**7.9**   In the context of tracking control, discuss alternate definitions of $s$ . For instance, consider choices based on Butterworth filters rather than (7.3), and how they would modify bounds (7.4).

**7.10**   Design a sliding controller for the system

$$x^{(3)} + \alpha_1(t)\,\ddot{x}^2 + \alpha_2(t)\,\dot{x}^5 \sin 4x \; = \; b(t)\,u$$

where $\alpha_1(t)$, $\alpha_2(t)$, and $b(t)$ are unknown time-varying functions verifying the known bounds

$$\forall\, t \geq 0, \quad |\alpha_1(t)| \leq 1 \qquad |\alpha_2(t)| \leq 2 \qquad 1 \leq b(t) \leq 4$$

Assume that the state is measured, and that the slowest unmodeled dynamics is the actuator dynamics, with a time-constant of about 1/50. Simulate the performance of the system on various trajectories (which you may want to generate using a reference model).

# Chapter 8
# Adaptive Control

Many dynamic systems to be controlled have *constant or slowly-varying uncertain parameters*. For instance, robot manipulators may carry large objects with unknown inertial parameters. Power systems may be subjected to large variations in loading conditions. Fire-fighting aircraft may experience considerable mass changes as they load and unload large quantities of water. Adaptive control is a approach to the control of such systems. The basic idea in adaptive control is to estimate the uncertain plant parameters (or, equivalently, the corresponding controller parameters) on-line based on the measured system signals, and use the estimated parameters in the control input computation. An adaptive control system can thus be regarded as a control system with on-line parameter estimation. Since adaptive control systems, whether developed for linear plants or for nonlinear plants, are inherently nonlinear, their analysis and design is intimately connected with the materials presented in this book, and in particular with Lyapunov theory.

Research in adaptive control started in the early 1950's in connection with the design of autopilots for high-performance aircraft, which operate at a wide range of speeds and altitudes and thus experience large parameter variations. Adaptive control was proposed as a way of automatically adjusting the controller parameters in the face of changing aircraft dynamics. But interest in the subject soon diminished due to the lack of insights and the crash of a test flight. It is only in the last decade that a coherent theory of adaptive control has been developed, using various tools from nonlinear control theory. These theoretical advances, together with the availability of cheap computation, have lead to many practical applications, in areas such as robotic manipulation, aircraft and rocket control, chemical processes, power systems, ship steering, and bioengineering.

The objective of this chapter is to describe the main techniques and results in adaptive control. We shall start with intuitive concepts, and then study more systematically adaptive controller design and analysis for linear and nonlinear systems. The study in this chapter is mainly concerned with single-input systems. Adaptive control designs for some complex multi-input physical systems are discussed in chapter 9.

# 8.1   Basic Concepts in Adaptive Control

In this section, we address a few basic questions, namely, why we need adaptive control, what the basic structures of adaptive control systems are, and how to go about designing adaptive control systems.

## 8.1.1  Why Adaptive Control ?

In some control tasks, such as those in robot manipulation, the systems to be controlled have parameter uncertainty at the beginning of the control operation. Unless such parameter uncertainty is gradually reduced on-line by an adaptation or estimation mechanism, it may cause inaccuracy or instability for the control systems. In many other tasks, such as those in power systems, the system dynamics may have well known dynamics at the beginning, but experience unpredictable parameter variations as the control operation goes on. Without continuous "redesign" of the controller, the initially appropriate controller design may not be able to control the changing plant well. Generally, the basic objective of adaptive control is to maintain consistent performance of a system in the presence of uncertainty or unknown variation in plant parameters. Since such parameter uncertainty or variation occurs in many practical problems, adaptive control is useful in many industrial contexts. These include:

• **Robot manipulation:** Robots have to manipulate loads of various sizes, weights, and mass distributions (Figure 8.1). It is very restrictive to assume that the inertial parameters of the loads are well known before a robot picks them up and moves them away. If controllers with constant gains are used and the load parameters are not accurately known, robot motion can be either inaccurate or unstable. Adaptive control, on the other hand, allows robots to move loads of unknown parameters with high speed and high accuracy.

• **Ship steering**: On long courses, ships are usually put under automatic steering. However, the dynamic characteristics of a ship strongly depend on many uncertain parameters, such as water depth, ship loading, and wind and wave conditions (Figure 8.2). Adaptive control can be used to achieve good control performance under varying operating conditions, as well as to avoid energy loss due to excessive rudder motion.

• **Aircraft control**: The dynamic behavior of an aircraft depends on its altitude, speed, and configuration. The ratio of variations of some parameters can lie between 10 to 50 in a given flight. As mentioned earlier, adaptive control was originally developed to achieve consistent aircraft performance over a large flight envelope.

• **Process control**: Models for metallurgical and chemical processes are usually complex and also hard to obtain. The parameters characterizing the processes vary from batch to batch. Furthermore, the working conditions are usually time-varying (*e.g.*, reactor characteristics vary during the reactor's life, the raw materials entering the process are never exactly the same, atmospheric and climatic conditions also tend to change). In fact, process control is one of the most important and active application areas of adaptive control.

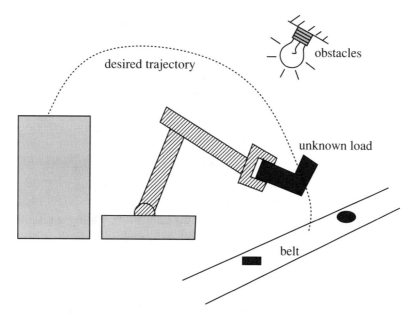

**Figure 8.1 :** A robot carrying a load of uncertain mass properties

Adaptive control has also been applied to other areas, such as power systems and biomedical engineering. Most adaptive control applications are aimed at handling inevitable parameter variation or parameter uncertainty. However, in some applications, particularly in process control, where hundreds of control loops may be present in a given system, adaptive control is also used to reduce the number of design parameters to be manually tuned, thus yielding an increase in engineering efficiency and practicality.

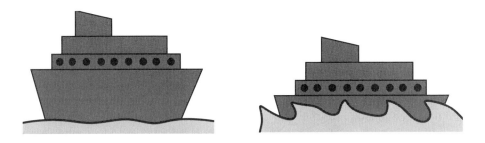

**Figure 8.2 :** A freight ship under various loadings and sea conditions

To gain insights about the behavior of the adaptive control systems and also to avoid mathematical difficulties, we shall assume the unknown plant parameters are *constant* in analyzing the adaptive control designs. In practice, the adaptive control systems are often used to handle time-varying unknown parameters. In order for the analysis results to be applicable to these practical cases, the time-varying plant parameters must vary considerably slower than the parameter adaptation. Fortunately, this is often satisfied in practice. Note that fast parameter variations may also indicate that the modeling is inadequate and that the dynamics causing the parameter changes should be additionally modeled.

Finally, let us note that robust control can also be used to deal with parameter uncertainty, as seen in chapter 7. Thus, one may naturally wonder about the differences and relations between the robust approach and the adaptive approach. In principle, adaptive control is superior to robust control in dealing with uncertainties in *constant or slowly-varying parameters*. The basic reason lies in the learning behavior of adaptive control systems: an adaptive controller improves its performance as adaptation goes on, while a robust controller simply attempts to keep consistent performance. Another reason is that an adaptive controller requires little or no *a priori* information about the unknown parameters, while a robust controller usually requires reasonable *a priori* estimates of the parameter bounds. Conversely, robust control has some desirable features which adaptive control does not have, such as its ability to deal with disturbances, quickly varying parameters, and unmodeled dynamics. Such features actually may be *combined* with adaptive control, leading to robust adaptive controllers in which uncertainties on constant or slowly-varying parameters is reduced by parameter adaptation and other sources of uncertainty are handled by robustification techniques. It is also important to point out that existing adaptive techniques for nonlinear systems generally require a *linear parametrization* of the plant dynamics, *i.e.*, that parametric uncertainty be expressed linearly in terms of a set of unknown parameters. In some cases, full linear parametrization and thus adaptive control cannot be achieved, but robust control (or adaptive control with robustifying terms) may be possible.

## 8.1.2 What Is Adaptive Control ?

An adaptive controller differs from an ordinary controller in that the controller parameters are variable, and there is a mechanism for adjusting these parameters on-line based on signals in the system. There are two main approaches for constructing adaptive controllers. One is the so-called model-reference adaptive control method, and the other is the so-called self-tuning method.

**MODEL-REFERENCE ADAPTIVE CONTROL (MRAC)**

Generally, a *model-reference adaptive control* system can be schematically represented by Figure 8.3. It is composed of four parts: a *plant* containing unknown parameters, a *reference model* for compactly specifying the desired output of the control system, a feedback *control law* containing adjustable parameters, and an *adaptation mechanism* for updating the adjustable parameters.

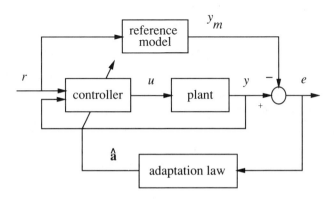

$\hat{\mathbf{a}}$ --- estimated parameters

**Figure 8.3 :** A model-reference adaptive control system

The *plant* is assumed to have a known structure, although the parameters are unknown. For linear plants, this means that the number of poles and the number of zeros are assumed to be known, but that the locations of these poles and zeros are not. For nonlinear plants, this implies that the structure of the dynamic equations is known, but that some parameters are not.

A *reference model* is used to specify the ideal response of the adaptive control system to the external command. Intuitively, it provides the ideal plant response which the adaptation mechanism should seek in adjusting the parameters. The choice of the reference model is part of the adaptive control system design. This choice has to satisfy two requirements. On the one hand, it should reflect the performance specification in the control tasks, such as rise time, settling time, overshoot or frequency domain characteristics. On the other hand, this ideal behavior should be achievable for the adaptive control system, *i.e.*, there are some inherent constraints on the structure of the reference model (*e.g.*, its order and relative degree) given the assumed structure of the plant model.

The *controller* is usually parameterized by a number of adjustable parameters (implying that one may obtain a family of controllers by assigning various values to the adjustable parameters). The controller should have *perfect tracking* capacity in order to allow the possibility of tracking convergence. That is, when the plant parameters are exactly known, the corresponding controller parameters should make the plant output identical to that of the reference model. When the plant parameters are not known, the adaptation mechanism will adjust the controller parameters so that perfect tracking is asymptotically achieved. If the control law is linear in terms of the adjustable parameters, it is said to be *linearly parameterized*. Existing adaptive control designs normally require linear parametrization of the controller in order to obtain adaptation mechanisms with guaranteed stability and tracking convergence.

The *adaptation* mechanism is used to adjust the parameters in the control law. In MRAC systems, the adaptation law searches for parameters such that the response of the plant under adaptive control becomes the same as that of the reference model, *i.e.*, the objective of the adaptation is to make the tracking error converge to zero. Clearly, the main difference from conventional control lies in the existence of this mechanism. The main issue in adaptation design is to synthesize an adaptation mechanism which will guarantee that the control system remains stable and the tracking error converges to zero as the parameters are varied. Many formalisms in nonlinear control can be used to this end, such as Lyapunov theory, hyperstability theory, and passivity theory. Although the application of one formalism may be more convenient than that of another, the results are often equivalent. In this chapter, we shall mostly use Lyapunov theory.

As an illustration of MRAC control, let us describe a simple adaptive control system for an unknown mass.

### Example 8.1: MRAC control of an unknown mass

Consider the control of a mass on a frictionless surface by a motor force $u$, with the plant dynamics being

$$m\ddot{x} = u \tag{8.1}$$

Assume that a human operator provides the positioning command $r(t)$ to the control system (possibly through a joystick). A reasonable way of specifying the ideal response of the controlled mass to the external command $r(t)$ is to use the following reference model

$$\ddot{x}_m + \lambda_1 \dot{x}_m + \lambda_2 x_m = \lambda_2 r(t) \tag{8.2}$$

with the positive constants $\lambda_1$ and $\lambda_2$ chosen to reflect the performance specifications, and the reference model output $x_m$ being the ideal output of the control system (*i.e.*, ideally, the mass should go to the specified position $r(t)$ like a well-damped mass-spring-damper system).

If the mass $m$ is known exactly, we can use the following control law to achieve perfect tracking

$$u = m(\ddot{x}_m - 2\lambda \dot{\tilde{x}} - \lambda^2 \tilde{x})$$

with $\tilde{x} = x(t) - x_m(t)$ representing the tracking error and $\lambda$ a strictly positive number. This control law leads to the exponentially convergent tracking error dynamics

$$\ddot{\tilde{x}} + 2\lambda \dot{\tilde{x}} + \lambda^2 \tilde{x} = 0$$

Now let us assume that the mass is *not known exactly*. We may use the following control law

$$u = \hat{m}(\ddot{x}_m - 2\lambda \dot{\tilde{x}} - \lambda^2 \tilde{x}) \tag{8.3}$$

which contains the adjustable parameter $\hat{m}$. Substitution of this control law into the plant dynamics leads to the closed-loop error dynamics

$$m\dot{s} + \lambda m s = \tilde{m} v \tag{8.4}$$

where $s$, a combined tracking error measure, is defined by

$$s = \dot{\tilde{x}} + \lambda \tilde{x} \tag{8.5}$$

the signal quantity $v$ by

$$v = \ddot{x}_m - 2\lambda \dot{\tilde{x}} - \lambda^2 \tilde{x}$$

and the parameter estimation error $\tilde{m}$ by

$$\tilde{m} = \hat{m} - m$$

Equation (8.4) indicates that the combined tracking error $s$ is related to the parameter error through a stable filter relation. One way of adjusting parameter $\hat{m}$ (for reasons to be seen later) is to use the following update law

$$\dot{\hat{m}} = -\gamma v s \qquad (8.6)$$

where $\gamma$ is a positive constant called the adaptation gain. One easily sees the nonlinear nature of the adaptive control system, by noting that the parameter $\hat{m}$ is adjusted based on system signals, and thus the controller (8.3) is nonlinear.

The stability and convergence of this adaptive control system can be analyzed using Lyapunov theory. For the closed-loop dynamics (8.4) and (8.6), with $s$ and $\tilde{m}$ as states, we can consider the following Lyapunov function candidate,

$$V = \frac{1}{2}[\, m s^2 + \frac{1}{\gamma}\tilde{m}^2\,] \qquad (8.7)$$

Its derivative can be easily shown to be

$$\dot{V} = -\lambda m s^2 \qquad (8.8)$$

Using Barbalat's lemma in chapter 4, one can easily show that $s$ converges to zero. Due to the relation (8.5), the convergence of $s$ to zero implies that of the position tracking error $\tilde{x}$ and the velocity tracking error $\dot{\tilde{x}}$.

For illustration, simulations of this simple adaptive control system are provided in Figures 8.4 and 8.5. The true mass is assumed to be $m=2$. The initial value of $\hat{m}$ is chosen to be zero, indicating no *a priori* parameter knowledge. The adaptation gain is chosen to be $\gamma = 0.5$, and the other design parameters are taken to be $\lambda_1 = 10, \lambda_2 = 25, \lambda = 6$. Figure 8.4 shows the results when the commanded position is $r(t) = 0$, with initial conditions being $\dot{x}(0) = \dot{x}_m(0) = 0$ and $x(0) = x_m(0) = 0.5$. Figure 8.5 shows the results when the desired position is a sinusoidal signal, $r(t) = \sin(4\,t)$. It is clear that the position tracking errors in both cases converge to zero, while the parameter error converge to zero only for the latter case. The reason for the non-convergence of parameter error in the first case can be explained by the simplicity of the tracking task: the asymptotic tracking of $x_m(t)$ can be achieved by many possible values of the estimated parameter $\hat{m}$, besides the true parameter. Therefore, the parameter adaptation law does not bother to find out the true parameter. On the other hand, the convergence of the parameter error in Figure 8.5 is because of the complexity of the tracking task, *i.e.*, tracking error convergence can be achieved only when the true mass is used in the control law. One may examine Equation (8.4) to see the mathematical evidence for these statements (also see Exercise 8.1). It is also helpful for the readers to sketch the specific structure of this adaptive mass control system. A more detailed discussion of parameter convergence is provided in section 8.2.                  □

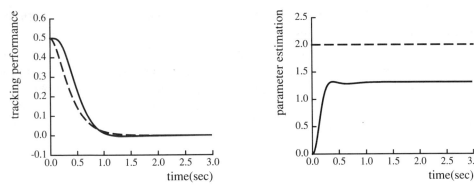

**Figure 8.4 :** Tracking Performance and Parameter Estimation for an Unknown Mass,
$r(t) = 0$

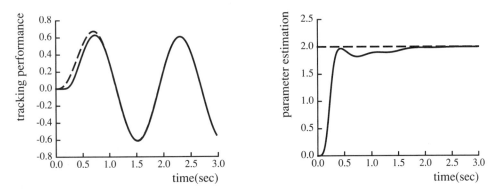

**Figure 8.5 :** Tracking Performance and Parameter Estimation for an Unknown Mass,
$r(t) = \sin 4t$

## SELF-TUNING CONTROLLERS (STC)

In non-adaptive control design (*e.g.*, pole placement), one computes the parameters of the controllers from those of the plant. If the plant parameters are not known, it is intuitively reasonable to replace them by their estimated values, as provided by a parameter estimator. A controller thus obtained by coupling a controller with an on-line (recursive) parameter estimator is called a self-tuning controller. Figure 8.6 illustrates the schematic structure of such an adaptive controller. Thus, a self-tuning controller is a controller which performs simultaneous identification of the unknown plant.

The operation of a self-tuning controller is as follows: at each time instant, the estimator sends to the controller a set of estimated plant parameters ($\hat{\mathbf{a}}$ in Figure 8.6), which is computed based on the past plant input $u$ and output $y$; the computer finds the corresponding controller parameters, and then computes a control input $u$ based on the controller parameters and measured signals; this control input $u$ causes a new plant output to be generated, and the whole cycle of parameter and input updates is repeated. Note that the controller parameters are computed from the estimates of the plant parameters as if they were the true plant parameters. This idea is often called the *certainty equivalence* principle.

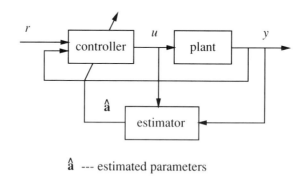

$\hat{\mathbf{a}}$ --- estimated parameters

**Figure 8.6 :** A self-tuning controller

Parameter estimation can be understood simply as the process of finding a set of parameters that *fits* the available input-output data from a plant. This is different from parameter adaptation in MRAC systems, where the parameters are adjusted so that the tracking errors converge to zero. For linear plants, many techniques are available to estimate the unknown parameters of the plant. The most popular one is the least-squares method and its extensions. There are also many control techniques for linear plants, such as pole-placement, PID, LQR (linear quadratic control), minimum variance control, or $H^\infty$ designs. By coupling different control and estimation schemes, one can obtain a variety of self-tuning regulators. The self-tuning method can also be applied to some nonlinear systems without any conceptual difference.

In the basic approach to self-tuning control, one estimates the plant parameters and then computes the controller parameters. Such a scheme is often called *indirect adaptive control*, because of the need to translate the estimated parameters into controller parameters. It is possible to eliminate this part of the computation. To do this, one notes that the control law parameters and plant parameters are related to each other for a specific control method. This implies that we may reparameterize the plant model using controller parameters (which are also unknown, of course), and then use standard estimation techniques on such a model. Since no translation is needed in this scheme, it is called a *direct adaptive control* scheme. In MRAC systems, one can similarly consider direct and indirect ways of updating the controller parameters.

**Example 8.2 : Self-tuning control of the unknown mass**

Consider the self-tuning control of the mass of Example 8.1. Let us still use the pole-placement (placing the poles of the tracking error dynamics) control law (8.3) for generating the control input, but let us now generate the estimated mass parameter using an estimation law.

Assume, for simplicity, that the acceleration can be measured by an accelerometer. Since the only unknown variable in Equation (8.1) is $m$, the simplest way of estimating it is to simply divide the control input $u(t)$ by the acceleration $\ddot{x}$ , *i.e.*,

$$\hat{m}(t) = \frac{u(t)}{\ddot{x}} \tag{8.9}$$

However, this is not a good method because there may be considerable noise in the measurement $\ddot{x}$, and, furthermore, the acceleration may be close to zero. A better approach is to estimate the parameter using a least-squares approach, *i.e.*, choosing the estimate in such a way that the total prediction error

$$J = \int_0^t e^2(r)\,dr \tag{8.10}$$

is minimal, with the prediction error $e$ defined as

$$e(t) = \hat{m}(t)\,\ddot{x}(t) - u(t)$$

The prediction error is simply the error in fitting the known input $u$ using the estimated parameter $\hat{m}$. This total error minimization can potentially average out the effects of measurement noise. The resulting estimate is

$$\hat{m} = \frac{\int_0^t w\,u\,dr}{\int_0^t w^2\,dr} \tag{8.11}$$

with $w = \ddot{x}$. If, actually, the unknown parameter $m$ is slowly time-varying, the above estimate has to be recalculated at every new time instant. To increase computational efficiency, it is desirable to adopt a recursive formulation instead of repeatedly using (8.11). To do this, we define

$$P(t) = \frac{1}{\int_0^t w^2\,dr} \tag{8.12}$$

The function $P(t)$ is called the estimation gain, and its update can be directly obtained by using

$$\frac{d}{dt}[P^{-1}] = w^2 \tag{8.13}$$

Then, differentiation of Equation (8.11) (which can be written $P^{-1}\hat{m} = \int_o^t w\,u\,dr$) leads to

$$\dot{\hat{m}} = -P(t)w\,e \tag{8.14}$$

In implementation, the parameter estimate $\hat{m}$ is obtained by numerically integrating equations (8.13) and (8.14), instead of using (8.11). Note that a number of other estimation methods can be used to provide the estimate of the mass. Such methods and their properties are discussed in more detail in section 8.7, together with a technique for avoiding the use of acceleration measurement $\ddot{x}$, using instead velocity or position measurements. □

It is seen from this example that, in self-tuning control, estimator design and controller design are separated. The estimation law (using $y$ and $u$) is independent of the choice of the control law, unlike in MRAC design where the parameter adaptation law is affected by the choice of the control law (it is also interesting to note that, in self-tuning control, saturation of the control input has no direct consequence on the convergence of parameter estimation). While this implies flexibility in design and simplicity in concept, the analysis of the convergence and stability of the self-tuning control system is usually more complicated.

## RELATIONS BETWEEN MRAC AND ST METHODS

As described above, MRAC control and ST control arise from different perspectives, with the parameters in MRAC systems being updated so as to minimize the tracking errors between the plant output and reference model output, and the parameters in ST systems being updated so as to minimize the data-fitting error in input-output measurements. However, there are strong relations between the two design methodologies. Comparing Figures 8.3 and 8.6, we note that the two kinds of systems both have an inner loop for control and an outer loop for parameter estimation. From a theoretical point of view, it can actually be shown that MRAC and ST controllers can be put under a unified framework.

The two methods can be quite different in terms of analysis and implementation. Compared with MRAC controllers, ST controllers are more flexible because of the possibility of coupling various controllers with various estimators (*i.e.*, the separation of control and estimation). However, the stability and convergence of self-tuning controllers are generally quite difficult to guarantee, often requiring the signals in the system to be sufficiently rich so that the estimated parameters converge to the true parameters. If the signals are not very rich (for example, if the reference signal is zero or a constant), the estimated parameters may not be close to the true pamareters, and the stability and convergence of the resulting control system may not be guaranteed. In this situation, one must either introduce perturbation signals in the input, or somehow modify the control law. In MRAC systems, however, the stability and tracking error convergence are usually guaranteed regardless of the richness of the signals.

Historically, the MRAC method was developed from optimal control of deterministic servomechanisms, while the ST method evolved from the study of stochastic regulation problems. MRAC systems have usually been considered in continuous-time form, and ST regulators in discrete time-form. In recent years, discrete-time version of MRAC controllers and continuous versions of ST controllers have also been developed. In this chapter, we shall mostly focus on MRAC systems in continuous form. Methods for generating estimated parameters for self-tuning control are discussed in section 8.7.

## 8.1.3  How To Design Adaptive Controllers ?

In conventional (non-adaptive) control design, a controller structure (*e.g.*, pole placement) is chosen first, and the parameters of the controller are then computed based on the known parameters of the plant. In adaptive control, the major difference is that the plant parameters are unknown, so that the controller parameters have to be provided by an adaptation law. As a result, the adaptive control design is more involved, with the additional needs of choosing an adaptation law and proving the stability of the system with adaptation.

The design of an adaptive controller usually involves the following three steps:

  • choose a control law containing variable parameters

  • choose an adaptation law for adjusting those parameters

  • analyze the convergence properties of the resulting control system

These steps are clearly seen in Example 8.1.

When one uses the self-tuning approach for linear systems, the first two steps are quite straightforward, with inventories of control and adaptation (estimation) laws available. The difficulty lies in the analysis. When one uses MRAC design, the adaptive controller is usually found by trial and error. Sometimes, the three steps are coordinated by the use of an appropriate Lyapunov function, or using some symbolic construction tools such as the passivity formalism. For instance, in designing the adaptive control system of Example 8.1, we actually start from guessing the Lyapunov function $V$ (as a representation of total error) in (8.7) and choose the control and adaptation laws so that $V$ decreases. Generally, the choices of control and adaptation laws in MRAC can be quite complicated, while the analysis of the convergence properties are relatively simple.

Before moving on to the application of the above procedure to adaptive control design for specific systems, let us derive a basic lemma which will be very useful in guiding our choice of adaptation laws for MRAC systems.

**Lemma 8.1**: *Consider two signals e and* $\phi$ *related by the following dynamic equation*

$$e(t) = H(p)[k\,\phi^T(t)\,\mathbf{v}(t)] \tag{8.15}$$

*where e(t) is a scalar output signal, H(p) is a strictly positive real transfer function, k is an unknown constant with known sign,* $\phi(t)$ *is a m×1 vector function of time, and* $\mathbf{v}(t)$ *is a measurable m×1 vector. If the vector* $\phi$ *varies according to*

$$\dot{\phi}(t) = -\operatorname{sgn}(k)\,\gamma\,e\,\mathbf{v}(t) \tag{8.16}$$

*with* $\gamma$ *being a positive constant, then e(t) and* $\phi(t)$ *are globally bounded. Furthermore, if* $\mathbf{v}$ *is bounded, then*

$$e(t) \to 0 \quad \text{as} \quad t \to \infty$$

Note that while (8.15) involves a mixture of time-domain and frequency-domain notations (with $p$ being the Laplace variable), its meaning is clear: $e(t)$ is the response of the linear system of SPR transfer function $H(p)$ to the input $[k\,\phi^T(t)\,\mathbf{v}(t)]$ (with arbitrary initial conditions). Such hybrid notation is common in the adaptive control literature, and later on it will save us the definition of intermediate variables.

In words, the above lemma means that if the input signal depends on the output in the form (8.16), then the whole system is globally stable (*i.e.*, all its states are bounded). Note that this is a feedback system, shown in Figure 8.7, where the plant dynamics, being SPR, have the unique properties discussed in section 4.6.1.

**Proof**: Let the state-space representation of (8.15) be

$$\dot{\mathbf{x}} = \mathbf{A}\,\mathbf{x} + \mathbf{b}\,[k\,\phi^T\,\mathbf{v}] \tag{8.17a}$$

$$e = \mathbf{c}^T\,\mathbf{x} \tag{8.17b}$$

Since $H(p)$ is SPR, it follows from the Kalman-Yakubovich lemma in chapter 4 that given a symmetric positive definite matrix $\mathbf{Q}$, there exists another symmetric positive definite matrix $\mathbf{P}$ such that

$$\mathbf{A}^T\mathbf{P} + \mathbf{P}\mathbf{A} = -\mathbf{Q}$$

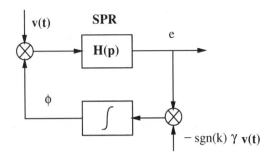

**Figure 8.7 :** A system containing a SPR transfer function

$$\mathbf{P}\,\mathbf{b} = \mathbf{c}$$

Let $V$ be a positive definite function of the form

$$V[\mathbf{x}, \boldsymbol{\phi}] = \mathbf{x}^T \mathbf{P} \mathbf{x} + \frac{|k|}{\gamma} \boldsymbol{\phi}^T \boldsymbol{\phi} \tag{8.18}$$

Its time derivative along the trajectories of the system defined by (8.17) and (8.16) is

$$\dot{V} = \mathbf{x}^T (\mathbf{P}\mathbf{A} + \mathbf{A}^T \mathbf{P})\mathbf{x} + 2\mathbf{x}^T \mathbf{P}\mathbf{b}\,(k\boldsymbol{\phi}^T \mathbf{v}) - 2\boldsymbol{\phi}^T (k e \mathbf{v})$$

$$= -\mathbf{x}^T \mathbf{Q}\mathbf{x} \le 0 \tag{8.19}$$

Therefore, the system defined by (8.15) and (8.16) is globally stable. The equations (8.18) and (8.19) also imply that $\mathbf{e}$ and $\boldsymbol{\phi}$ are globally bounded.

If the signal $\mathbf{v}(t)$ is bounded, $\dot{\mathbf{x}}$ is also bounded, as seen from (8.17a). This implies the uniform continuity of $\dot{V}$, since its derivative

$$\ddot{V} = -2\mathbf{x}\mathbf{Q}\dot{\mathbf{x}}$$

is then bounded. Application of Barbalat's lemma of chapter 4 then indicates the asymptotic convergence of $e(t)$ to zero. ☐

It is useful to point out that the system defined by (8.15) and (8.16) not only guarantees the boundedness of $e$ and $\boldsymbol{\phi}$, but also that of the whole state $\mathbf{x}$, as seen from (8.18). Note that the state-space realization in (8.17) can be non-minimal (implying the possibility of unobservable or uncontrollable modes) provided that the unobservable and uncontrollable modes be stable, according to the Meyer-Kalman-Yakubovich lemma. Intuitively, this is reasonable because stable hidden modes are not affected by the choice of $\boldsymbol{\phi}$.

In our later MRAC designs, the tracking error between the plant output and reference model output will often be related to the parameter estimation errors by an equation of the form (8.15). Equation (8.16) thus provides a technique for adjusting the controller parameters while guaranteeing system stability. Clearly, the tracking-error dynamics in (8.4) satisfy the conditions of Lemma 8.1 and the adaptation law is in the form of (8.16).

## 8.2  Adaptive Control of First-Order Systems

Let us now discuss the adaptive control of first-order plants using the MRAC method, as an illustration of how to design and analyze an adaptive control system.  The development can also have practical value in itself, because a number of simple systems of engineering interest may be represented by a first-order model. For example, the braking of an automobile, the discharge of an electronic flash, or the flow of fluid from a tank may be approximately represented by a first-order differential equation

$$\dot{y} = -a_p y + b_p u \tag{8.20}$$

where $y$ is the plant output, $u$ is its input, and $a_p$ and $b_p$ are constant plant parameters.

**PROBLEM SPECIFICATION**

In the adaptive control problem, the plant parameters $a_p$ and $b_p$ are assumed to be unknown.  Let the desired performance of the adaptive control system be specified by a first-order reference model

$$\dot{y}_m = -a_m y_m + b_m r(t) \tag{8.21}$$

where $a_m$ and $b_m$ are constant parameters, and $r(t)$ is a bounded external reference signal.  The parameter $a_m$ is required to be strictly positive so that the reference model is stable, and $b_m$ is chosen strictly positive without loss of generality.  The reference model can be represented by its transfer function $M$

$$y_m = M r$$

where

$$M = \frac{b_m}{p + a_m}$$

with $p$ being the Laplace variable. Note that $M$ is a SPR function.

The objective of the adaptive control design is to formulate a control law, and an adaptation law, such that the resulting model following error $y(t) - y_m$ asymptotically converges to zero. In order to accomplish this, we have to assume the sign of the parameter $b$ to be known. This is a quite mild condition, which is often satisfied in practice. For example, for the braking of a car, this assumption amounts to the simple physical knowledge that braking slows down the car.

## CHOICE OF CONTROL LAW

As the first step in the adaptive controller design, let us choose the control law to be

$$u = \hat{a}_r(t) r + \hat{a}_y(t) y \qquad (8.22)$$

where $\hat{a}_r$ and $\hat{a}_y$ are variable feedback gains. With this control law, the closed-loop dynamics are

$$\dot{y} = -(a_p - \hat{a}_y b_p) y + \hat{a}_r b_p \, r(t) \qquad (8.23)$$

The reason for the choice of control law in (8.22) is clear: it allows the possibility of perfect model matching. Indeed, if the plant parameters were known, the following values of control parameters

$$a_r^* = \frac{b_m}{b_p} \qquad\qquad a_y^* = \frac{a_p - a_m}{b_p} \qquad (8.24)$$

would lead to the closed-loop dynamics

$$\dot{y} = -a_m y + b_m r$$

which is identical to the reference model dynamics, and yields zero tracking error. In this case, the first term in (8.22) would result in the right d.c. gain, while the second term in the control law (8.22) would achieve the dual objectives of canceling the term $(-a_p y)$ in (8.20) and imposing the desired pole $-a_m y$.

In our adaptive control problem, since $a_p$ and $b_p$ are unknown, the control input will achieve these objectives adaptively, *i.e.*, the adaptation law will continuously search for the right gains, based on the tracking error $y - y_m$, so as to make $y$ tend to $y_m$ asymptotically. The structure of the adaptive controller is illustrated in Figure 8.8.

## CHOICE OF ADAPTATION LAW

Let us now choose the adaptation law for the parameters $\hat{a}_r$ and $\hat{a}_y$. Let

$$e = y - y_m$$

be the tracking error. The parameter errors are defined as the difference between the

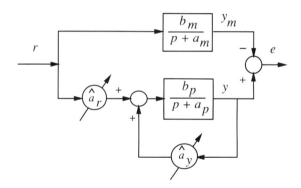

**Figure 8.8 :** A MRAC system for the first-order plant

controller parameter provided by the adaptation law and the ideal parameters, *i.e.*,

$$\tilde{\mathbf{a}}(t) = \begin{bmatrix} \tilde{a}_r \\ \tilde{a}_y \end{bmatrix} = \begin{bmatrix} \hat{a}_r - a_r^* \\ \hat{a}_y - a_y^* \end{bmatrix} \tag{8.25}$$

The dynamics of tracking error can be found by subtracting (8.23) from (8.21),

$$\dot{e} = - a_m(y - y_m) + (a_m - a_p + b_p \hat{a}_y)y + (b_p \hat{a}_r - b_m)r$$

$$= - a_m e + b_p(\tilde{a}_r r + \tilde{a}_y y) \tag{8.26}$$

This can be conveniently represented as

$$e = \frac{b_p}{p + a_m}(\tilde{a}_r r + \tilde{a}_y y) = \frac{1}{a_r^*} M(\tilde{a}_r r + \tilde{a}_y y) \tag{8.27}$$

with *p* denoting the Laplace operator.

  Relation (8.27) between the parameter errors and tracking error is in the familiar form given by (8.15). Thus, Lemma 8.1 suggests the following adaptation law

$$\dot{\hat{a}}_r = - \operatorname{sgn}(b_p)\, \gamma\, e\, r \tag{8.28a}$$

$$\dot{\hat{a}}_y = - \operatorname{sgn}(b_p)\, \gamma\, e\, y \tag{8.28b}$$

with $\gamma$ being a positive constant representing the adaptation gain. From (8.28), it is seen that $\operatorname{sgn}(b_p)$ determines the direction of the search for the proper controller parameters.

**TRACKING CONVERGENCE ANALYSIS**

With the control law and adaptation law chosen above, we can now analyze the system's stability and convergence behavior using Lyapunov theory, or equivalently Lemma 8.1. Specifically, the Lyapunov function candidate

$$V(e, \phi) = \frac{1}{2} e^2 + \frac{1}{2\gamma} |b_p| (\tilde{a}_r{}^2 + \tilde{a}_y{}^2) \tag{8.29}$$

can be easily shown to have the following derivative along system trajectories

$$\dot{V} = - a_m e^2$$

Thus, the adaptive control system is globally stable, *i.e.*, the signals $e$, $\tilde{a}_r$ and $\tilde{a}_y$ are bounded. Furthermore, the global asymptotic convergence of the tracking error $e(t)$ is guaranteed by Barbalat's lemma, because the boundedness of $e$, $\tilde{a}_r$ and $\tilde{a}_y$ implies the boundedness of $\dot{e}$ (according to (8.26)) and therefore the uniform continuity of $\dot{V}$.

It is interesting to wonder why the adaptation law (8.28) leads to tracking error convergence. To understand this, let us see intuitively how the control parameters *should* be changed. Consider, without loss of generality, the case of a positive $\text{sgn}(b_p)$. Assume that at a particular instant $t$ the tracking error $e$ is negative, indicating that the plant output is too small. From (8.20), the control input $u$ should be increased in order to increase the plant output. From (8.22), an increase of the control input $u$ can be achieved by increasing $\hat{a}_r$ (assuming that $r(t)$ is positive). Thus, the adaptation law, with the variation rate of $\hat{a}_r$ depending on the product of $\text{sgn}(b)$, $r$ and $e$, is intuitively reasonable. A similar reasoning can be made about $\hat{a}_y$ .

The behavior of the adaptive controller is demonstrated in the following simulation example.

**Example 8.3: A first-order plant**

Consider the control of the unstable plant

$$\dot{y} = y + 3 u$$

using the previously designed adaptive controller. The plant parameters $a_p = -1, b_p = 3$ are assumed to be unknown to the adaptive controller. The reference model is chosen to be

$$\dot{x}_m = -4 x_m + 4 r$$

*i.e.*, $a_m = 4$, $b_m = 4$. The adaptation gain $\gamma$ is chosen to be equal to 2. The initial values of both parameters of the controller are chosen to be 0, indicating no *a priori* knowledge. The initial conditions of the plant and the model are both zero.

Two different reference signals are used in the simulation:

- $r(t) = 4$. It is seen from Figure 8.9 that the tracking error converges to zero but the parameter error does not.

- $r(t) = 4 \sin(3t)$. It is seen from Figure 8.10 that both the tracking error and parameter error converge to zero. □

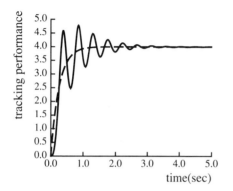 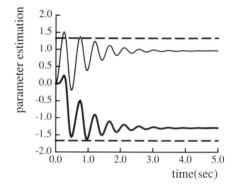

**Figure 8.9 :** Tracking performance and parameter estimation, $r(t) = 4$

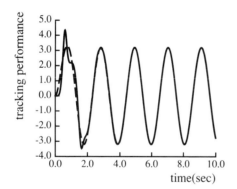 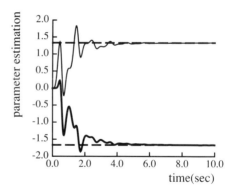

**Figure 8.10 :** Tracking performance and parameter estimation, $r(t) = 4 \sin(3t)$

Note that, in the above adaptive control design, although the stability and convergence of the adaptive controller is guaranteed for any positive $\gamma$, $a_m$ and $b_m$, the performance of the adaptive controller will depend critically on $\gamma$. If a small gain is chosen, the adaptation will be slow and the transient tracking error will be large. Conversely, the magnitude of the gain and, accordingly, the performance of the adaptive control system, are limited by the excitation of unmodeled dynamics, because too large an adaptation gain will lead to very oscillatory parameters.

## PARAMETER CONVERGENCE ANALYSIS

In order to gain insights about the behavior of adaptive control system, let us understand the convergence of estimated parameters. From the simulation results of Example 8.3, one notes that the estimated parameters converge to the exact parameter values for one reference signal but not for the other. This prompts us to speculate a relation between the features of the reference signals and parameter convergence, *i.e.*, the estimated parameters will not converge to the ideal controller parameters unless the reference signal $r(t)$ satisfies certain conditions.

Indeed, such a relation between the features of reference signal and convergence of estimated parameters can be intuitively understood. In MRAC systems, the objective of the adaptation mechanism is to find out the parameters which drive the tracking error $y - y_m$ to zero. If the reference signal $r(t)$ is very simple, such as a zero or a constant, it is possible for many vectors of controller parameters, besides the ideal parameter vector, to lead to tracking error convergence. Then, the adaptation law will not bother to find out the ideal parameters. Let $\Omega$ denote the set composed of all the parameter vectors which can guarantee tracking error convergence for a particular reference signal history $r(t)$. Then, depending on the initial conditions, the vector of estimated parameters may converge to any point in the set or wonder around in the set instead of converging to the true parameters. However, if the reference signal $r(t)$ is so complex that only the true parameter vector $\mathbf{a}^* = [a_r^* \ \ a_y^*]^T$ can lead to tracking error convergence, then we shall have parameter convergence.

Let us now find out the exact conditions for parameter convergence. We shall use simplified arguments to avoid tedious details. Note that the output of the stable filter in (8.27) converges to zero and that its input is easily shown to be uniformly continuous. Thus, $\tilde{a}_r r + \tilde{a}_y y$ must converge to zero. From the adaptation law (8.28) and the tracking error convergence, the rate of the parameter estimates converges to zero. Thus, when time $t$ is large, $\tilde{\mathbf{a}}$ is almost constant, and

$$r(t)\tilde{a}_r + y(t)\tilde{a}_y = 0$$

*i.e.*,

$$\mathbf{v}^T(t)\,\tilde{\mathbf{a}} = 0 \tag{8.30}$$

with

$$\mathbf{v} = [r \ \ y]^T \qquad\qquad \mathbf{a} = [a_r \ \ a_y]^T$$

Here we have one equation (with time-varying coefficients) and two variables. The issue of parameter convergence is reduced to the question of what conditions the vector $[r(t)\ y(t)]^T$ should satisfy in order for the equation to have a unique zero solution.

If $r(t)$ is a constant $r_o$, then for large $t$,

$$y(t) = y_m = \alpha\, r_o$$

with $\alpha$ being the d.c. gain of the reference model. Thus,

$$[r\ \ y] = [1\ \ \alpha] r_o$$

Equation (8.30) becomes

$$\tilde{a}_r + \alpha \tilde{a}_y = 0$$

Clearly, this implies that the estimated parameters, instead of converging to zero, converge to a straight line in parameter space. For Example 8.3, with $\alpha = 1$, the above equation implies that the steady state errors of the two parameters should be of equal magnitudes but opposite signs. This is obviously confirmed in Figure 8.9.

However, when $r(t)$ is such that the corresponding signal vector $\mathbf{v}(t)$ satisfies the so called "persistent excitation" condition, we can show that (8.28) will guarantee parameter convergence. By persistent excitation of $\mathbf{v}$, we mean that there exist strictly positive constants $\alpha_1$ and $T$ such that for any $t > 0$,

$$\int_t^{t+T} \mathbf{v}\mathbf{v}^T dr \geq \alpha_1 \mathbf{I} \tag{8.31}$$

To show parameter convergence, we note that multiplying (8.30) by $\mathbf{v}(t)$ and integrating the equation for a period of time $T$, leads to

$$\int_t^{t+T} \mathbf{v}\mathbf{v}^T dr\, \tilde{\mathbf{a}} = 0$$

Condition (8.31) implies that the only solution of this equation is $\tilde{\mathbf{a}} = 0$, *i.e.*, parameter error being zero. Intuitively, the persistent excitation of $\mathbf{v}(t)$ implies that the vectors $\mathbf{v}(t)$ corresponding to different times $t$ cannot always be linearly dependent.

The only remaining question is the relation between $r(t)$ and the persistent excitation of $\mathbf{v}(t)$. One can easily show that, in the case of the first order plant, the persistent excitation of $\mathbf{v}$ can be guaranteed, if $r(t)$ contains at least one sinusoidal component.

## EXTENSION TO NONLINEAR PLANTS

The same method of adaptive control design can be used for the nonlinear first-order plant described by the differential equation

$$\dot{y} = - a_p y - c_p f(y) + b_p u \tag{8.32}$$

where $f$ is any known nonlinear function. The nonlinearity in these dynamics is characterized by its linear parametrization in terms of the unknown constant $c$. Instead of using the control law (8.22), we now use the control law

$$u = \hat{a}_y y + \hat{a}_f f(y) + \hat{a}_r r \tag{8.33}$$

where the second term is introduced with the intention of adaptively canceling the nonlinear term.

Substituting this control law into the dynamics (8.32) and subtracting the resulting equation by (8.21), we obtain the error dynamics

$$e = \frac{1}{k_r^*} M (\tilde{a}_y y + \tilde{a}_f f(y) + \tilde{a}_r r)$$

where the parameter error $\tilde{a}_f$ is defined as

$$\tilde{a}_f = \hat{a}_f - \frac{c_p}{b_p}$$

By choosing the adaptation law

$$\dot{\hat{a}}_y = - \text{sgn}(b_p)\, \gamma\, e\, y \tag{8.34a}$$

$$\dot{\hat{a}}_f = - \text{sgn}(b_p)\, \gamma\, e\, f \tag{8.34b}$$

$$\dot{\hat{a}}_r = - \text{sgn}(b_p)\, \gamma\, e\, r \tag{8.34c}$$

one can similarly show that the tracking error $e$ converges to zero, and the parameter error remains bounded.

As for parameter convergence, similar arguments as before can reveal the convergence behavior of the estimated parameters. For constant reference input $r = r_o$, the estimated parameters converge to the line (with $\alpha$ still being the d.c. gain of the reference model)

$$r_o \tilde{a}_r + \tilde{a}_y (\alpha r_o) + \tilde{a}_f f(\alpha r_o) = 0$$

which is a straight line in the three-dimensional parameter space. In order for the parameters to converge to the ideal values, the signal vector $\mathbf{v} = [r(t)\ y(t)\ f(y)]^T$ should be persistently exciting, *i.e.*, there exists positive constants $\alpha_1$ and $T$ such that for any time $t \geq 0$,

$$\int_t^{t+T} \mathbf{v}\mathbf{v}^T dr \geq \alpha_1 \mathbf{I}$$

Generally speaking, for linear systems, the convergent estimation of $m$ parameters require at least $m/2$ sinusoids in the reference signal $r(t)$, as will be discussed in more detail in sections 8.3 and 8.4. However, for this nonlinear case, such simple relation may not be valid. Usually, the qualitative relation between $r(t)$ and $\mathbf{v}(t)$ is dependent on the particular nonlinear functions $f(y)$. It is unclear how many sinusoids in $r(t)$ are necessary to guarantee the persistent excitation of $\mathbf{v}(t)$.

The following example illustrates the behavior of the adaptive system for a nonlinear plant.

**Example 8.4: simulation of a first-order nonlinear plant**

Assume that a nonlinear plant is described by the equation

$$\dot{y} = y + y^2 + bu \tag{8.35}$$

This differs from the unstable plant in Example 8.3 in that a quadratic term is introduced in the plant dynamics.

Let us use the same reference model, initial parameters, and design parameters as in Example 8.3. For the reference signal $r(t) = 4$, the results are shown in Figure 8.11. It is seen that the tracking error converges to zero, but the parameter errors are only bounded. For the reference signal $r(t) = 4\sin(3t)$, the results are shown in Figure 8.12. It is noted that the tracking error and the parameter errors for the three parameters all converge to zero. ◻

In this example, it is interesting to note two points. The first point is that a single sinusoidal component in $r(t)$ allows three parameters to be estimated. The second point is that the various signals (including $\hat{a}$ and $y$) in this system are much more oscillatory than those in Example 8.3. Let us understand why. The basic reason is provided by the observation that nonlinearity usually generates more frequencies, and thus $\mathbf{v}(t)$ may contain more sinusoids than $r(t)$. Specifically, in the above example, with $r(t) = 4\sin(3t)$, the signal vector $\mathbf{v}$ converges to

$$\mathbf{v}(t) = [r(t)\ \ y_{ss}(t)\ \ f_{ss}(t)]$$

where $y_{ss}(t)$ is the steady-state response and $f_{ss}(t)$ the corresponding function value,

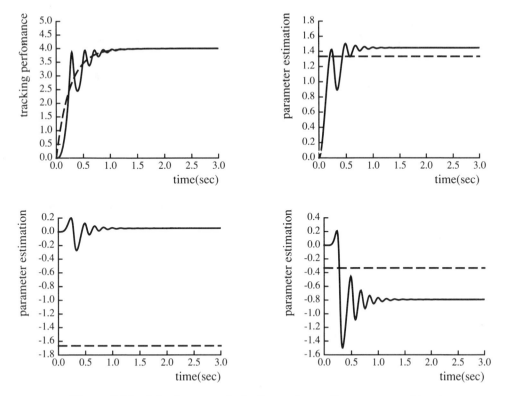

**Figure 8.11 :** Adaptive control of a first-order nonlinear system, $r(t) = 4$
upper left: tracking performance
upper right: parameter $\hat{a}_r$ ; lower left: parameter $\hat{a}_y$ ; lower right: parameter $\hat{a}_f$

$$y_{ss}(t) = y_m(t) = 4A \sin(3t + \phi)$$

$$f_{ss}(t) = y_{ss}^2 = 16A^2 \sin^2(3t + \phi) = 8 A^2 (1 - \cos(6t + 2\phi))$$

where $A$ and $\phi$ are the magnitude and phase shift of the reference model at $\omega = 3$. Thus, the signal vector $\mathbf{v}(t)$ contains *two* sinusoids, with $f(y)$ containing a sinusoid at twice the original frequency. Intuitively, this component at double frequency is the reason for the convergent estimation of the *three* parameters and the more oscillatory behavior of the estimated parameters.

## 8.3   Adaptive Control of Linear Systems With Full State Feedback

Let us now move on to adaptive control design for more general systems. In this section, we shall study the adaptive control of linear systems when the full state is measurable. We consider the adaptive control of $n^{\text{th}}$-order linear systems in

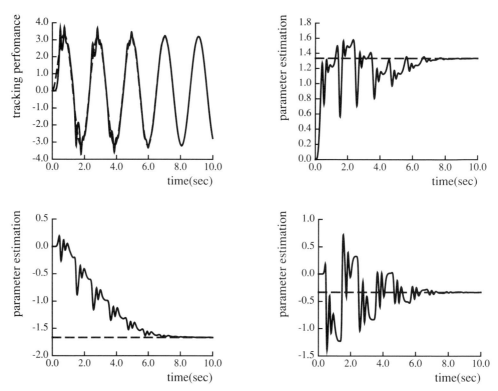

**Figure 8.12 :** Adaptive control of a first-order nonlinear system, $r(t) = 4\sin(3t)$
upper left: tracking performance
upper right: parameter $\hat{a}_r$ ; lower left: parameter $\hat{a}_y$ ;  lower right: parameter $\hat{a}_f$

companion form:

$$a_n y^{(n)} + a_{n-1} y^{(n-1)} + \ldots + a_o y = u \qquad (8.36)$$

where the state components $y, \dot{y}, \ldots, y^{(n-1)}$ are measurable.  We assume that the coefficient vector $\mathbf{a} = [\, a_n \ldots \quad a_1 \ a_o \,]^T$ is unknown, but that the sign of $a_n$ is assumed to be known.  An example of such systems is the dynamics of a mass-spring-damper system

$$m\ddot{y} + c\dot{y} + ky = u$$

where we measure position and velocity (possibly with an optical encoder for position measurement, and a tachometer for velocity measurement, or simply numerically differentiate the position signals).

The objective of the control system is to make $y$ closely track the response of a stable reference model

$$\alpha_n y_m^{(n)} + \alpha_{n-1} y_m^{(n-1)} + \ldots + \alpha_o y_m = r(t) \tag{8.37}$$

with $r(t)$ being a bounded reference signal.

## CHOICE OF CONTROL LAW

Let us define a signal $z(t)$ as follows

$$z(t) = y_m^{(n)} - \beta_{n-1} e^{(n-1)} - \ldots - \beta_o e \tag{8.38}$$

with $\beta_1, \ldots, \beta_n$ being positive constants chosen such that $p^n + \beta_{n-1} p^{n-1} + \ldots + \beta_o$ is a stable (Hurwitz) polynomial. Adding $(-a_n z(t))$ to both sides of (8.36) and rearranging, we can rewrite the plant dynamics as

$$a_n [y^{(n)} - z] = u - a_n z - a_{n-1} y^{(n-1)} - \ldots - a_o y$$

Let us choose the control law to be

$$u = \hat{a}_n z + \hat{a}_{n-1} y^{(n-1)} + \ldots + \hat{a}_o y = \mathbf{v}^T(t) \hat{\mathbf{a}}(t) \tag{8.39}$$

where $\mathbf{v}(t) = [z(t) \; y^{(n-1)} \; \ldots \; \dot{y} \; y]^T$ and

$$\hat{\mathbf{a}}(t) = [\hat{a}_n \; \hat{a}_{n-1} \; \ldots \; \hat{a}_1 \; \hat{a}_o]^T$$

denotes the estimated parameter vector. This represents a pole-placement controller which places the poles at positions specified by the coefficients $\beta_i$. The tracking error $e = y - y_m$ then satisfies the closed-loop dynamics

$$a_n [e^{(n)} + \beta_{n-1} e^{(n-1)} + \ldots + \beta_o e] = \mathbf{v}^T(t) \tilde{\mathbf{a}}(t) \tag{8.40}$$

where

$$\tilde{\mathbf{a}} = \hat{\mathbf{a}} - \mathbf{a}$$

## CHOICE OF ADAPTATION LAW

Let us now choose the parameter adaptation law. To do this, let us rewrite the closed-loop error dynamics (8.40) in state space form,

$$\dot{\mathbf{x}} = \mathbf{A}\mathbf{x} + \mathbf{b}[(1/a_n) \, \mathbf{v}^T \tilde{\mathbf{a}}] \tag{8.41a}$$

$$e = \mathbf{c}\mathbf{x} \tag{8.41b}$$

where

$$A = \begin{bmatrix} 0 & 1 & 0 & . & . & 0 \\ 0 & 0 & 1 & . & . & . \\ . & . & . & . & . & . \\ . & . & . & . & . & . \\ 0 & 0 & 0 & . & . & 1 \\ -\beta_o & -\beta_1 & -\beta_2 & . & . & -\beta_{n-1} \end{bmatrix} \qquad b = \begin{bmatrix} 0. \\ 0. \\ . \\ . \\ 0. \\ 1. \end{bmatrix}$$

$$c = [1 \ 0 \ . \ . \ 0 \ 0]$$

Consider the Lyapunov function candidate

$$V(\mathbf{x}, \tilde{\mathbf{a}}) = \mathbf{x}^T \mathbf{P} \mathbf{x} + \tilde{\mathbf{a}}^T \mathbf{\Gamma}^{-1} \tilde{\mathbf{a}}$$

where both $\mathbf{\Gamma}$ and $\mathbf{P}$ are symmetric positive definite constant matrices, and $\mathbf{P}$ satisfies

$$\mathbf{P}\mathbf{A} + \mathbf{A}^T \mathbf{P} = -\mathbf{Q} \qquad \mathbf{Q} = \mathbf{Q}^T > 0$$

for a chosen $\mathbf{Q}$. The derivative $\dot{V}$ can be computed easily as

$$\dot{V} = -\mathbf{x}^T \mathbf{Q} \mathbf{x} + 2 \tilde{\mathbf{a}}^T \mathbf{v} \mathbf{b}^T \mathbf{P} \mathbf{x} + 2 \tilde{\mathbf{a}}^T \mathbf{\Gamma}^{-1} \dot{\tilde{\mathbf{a}}}$$

Therefore, the adaptation law

$$\dot{\hat{\mathbf{a}}} = -\mathbf{\Gamma} \mathbf{v} \mathbf{b}^T \mathbf{P} \mathbf{x} \tag{8.42}$$

leads to

$$\dot{V} = -\mathbf{x}^T \mathbf{Q} \mathbf{x}$$

One can easily show the convergence of $\mathbf{x}$ using Barbalat's lemma. Therefore, with the adaptive controller defined by control law (8.39) and adaptation law (8.42), $e$ and its $(n-1)$ derivatives converge to zero. The parameter convergence condition can again be shown to be the persistent excitation of the vector $\mathbf{v}$. Note that a similar design can be made for nonlinear systems in the controllability canonical form, as discussed in section 8.5.

# 8.4  Adaptive Control of Linear Systems With Output Feedback

In this section, we consider the adaptive control of linear systems in the presence of only output measurement, rather than full state feedback. Design in this case is considerably more complicated than when the full state is available. This partly arises from the need to introduce dynamics in the controller structure, since the output only provides partial information about the system state. To appreciate this need, one can simply recall that in conventional design (no parameter uncertainty) a controller obtained by multiplying the state with constant gains (pole placement) can stabilize systems where all states are measured, while additional observer structures must be used for systems where only outputs are measured.

A linear time-invariant system can be represented by the transfer function

$$W(p) = k_p \frac{Z_p(p)}{R_p(p)} \tag{8.43}$$

where

$$R_p(p) = a_o + a_1 p + \ldots + a_{n-1} p^{n-1} + p^n$$

$$Z_p(p) = b_o + b_1 p + \ldots + b_{m-1} p^{m-1} + p^m$$

where $k_p$ is called the high-frequency gain. The reason for this term is that the plant frequency response at high frequency verifies

$$|W(j\omega)| \approx \frac{k_p}{\omega^{n-m}}$$

*i.e.*, the high frequency response is essentially determined by $k_p$. The relative degree $r$ of this system is $r = n - m$. In our adaptive control problem, the coefficients $a_i, b_j$ ($i = 0, 1, \ldots, n-1; j = 0, 1, \ldots, m-1$) and the high frequency gain $k_p$ are all assumed to be unknown.

The desired performance is assumed to be described by a reference model with transfer function

$$W_m(p) = k_m \frac{Z_m}{R_m} \tag{8.44}$$

where $Z_m$ and $R_m$ are monic Hurwitz polynomials of degrees $n_m$ and $m_m$, and $k_m$ is positive. It is well known from linear system theory that the relative degree of the reference model has to be larger than or equal to that of the plant in order to allow the possibility of perfect tracking. Therefore, in our treatment, we will assume that $n_m - m_m \geq n - m$.

The objective of the design is to determine a control law, and an associated adaptation law, so that the plant output $y$ asymptotically approaches $y_m$. In determining the control input, the output $y$ is assumed to be measured, but no differentiation of the output is allowed, so as to avoid the noise amplification associated with numerical differentiation. In achieving this design, we assume the following *a priori* knowledge about the plant:

- the plant order $n$ is known

- the relative degree $n - m$ is known

- the sign of $k_p$ is known

- the plant is minimum-phase

Among the above assumptions, the first and the second imply that the model structure of the plant is known. The third is required to provide the direction of parameter adaptation, similarly to (8.28) in section 8.2. The fourth assumption is somewhat restrictive. It is required because we want to achieve convergent tracking in the adaptive control design. Adaptive control of non-minimum phase systems is still a topic of active research and will not be treated in this chapter.

In section 8.4.1, we discuss output-feedback adaptive control design for linear plants with relative degree one, *i.e.*, plants having one more pole than zeros. Design for these systems is relatively straightforward. In section 8.4.2, we discuss output-feedback design for plants with higher relative degree. The design and implementation of adaptive controllers in this case is more complicated because it is not possible to use SPR functions as reference models.

## 8.4.1  Linear Systems With Relative Degree One

When the relative degree is 1, *i.e.*, $m = n - 1$, the reference model can be chosen to be SPR. This choice proves critical in the development of globally convergent adaptive controllers.

**CHOICE OF CONTROL LAW**

To determine the appropriate control law for the adaptive controller, we must first know what control law can achieve perfect tracking *when the plant parameters are perfectly known.* Many controller structures can be used for this purpose. The following one, although somewhat peculiar, is particularly convenient for later adaptation design.

### Example 8.5: A controller for perfect tracking

Consider the plant described by

$$y = \frac{k_p(p + b_p)}{p^2 + a_{p1}p + a_{p2}} u \tag{8.45}$$

and the reference model

$$y_m = \frac{k_m(p + b_m)}{p^2 + a_{m1}p + a_{m2}} r \tag{8.46}$$

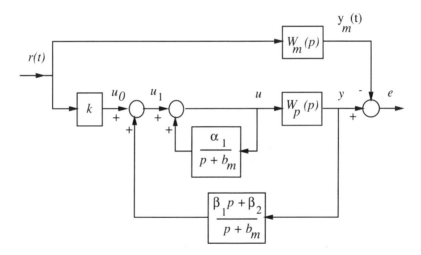

**Figure 8.13 :** A model-reference control system for relative degree 1

Let the controller be chosen as shown in Figure 8.13, with the control law being

$$u = \alpha_1 z + \frac{\beta_1 p + \beta_2}{p + b_m} y + k r \tag{8.47}$$

where $z = u/(p + b_m)$, *i.e.*, $z$ is the output of a first-order filter with input $u$, and $\alpha_1$, $\beta_1$, $\beta_2$, $k$ are controller parameters. If we take these parameters to be

$$\alpha_1 = b_p - b_m$$

$$\beta_1 = \frac{a_{m1} - a_{p1}}{k_p}$$

$$\beta_2 = \frac{a_{m2} - a_{p2}}{k_p}$$

$$k = \frac{k_m}{k_p}$$

one can straightforwardly show that the transfer function from the reference input $r$ to the plant output $y$ is

$$W_{ry} = \frac{k_m(p + b_m)}{p^2 + a_{m1}p + a_{m2}} = W_m(p)$$

Therefore, perfect tracking is achieved with this control law, *i.e.*, $y(t) = y_m(t)$, $\forall t \geq 0$.

It is interesting to see why the closed-loop transfer function can become exactly the same as that of the reference model. To do this, we note that the control input in (8.47) is composed of three parts. The first part in effect replaces the plant zero by the reference model zero, since the transfer function from $u_1$ to $y$ (see Figure 8.13) is

$$W_{u1,y} = \frac{p + b_m}{p + b_p} \frac{k_p(p + b_p)}{p^2 + a_{p1}p + a_{p2}} = \frac{k_p(p + b_m)}{p^2 + a_{p1}p + a_{p2}}$$

The second part places the closed-loop poles at the locations of those of the reference model. This is seen by noting that the transfer function from $u_o$ to $y$ is (Figure 8.13)

$$W_{u0,y} = \frac{W_{u1,y}}{1 + W_f W_{u1,y}} = \frac{k_p(p + b_m)}{p^2 + (a_{p1} + \beta_1 k_p)p + (a_{p2} + \beta_2 k_p)}$$

The third part of the control law $(k_m/k_p)r$ obviously replaces $k_p$, the high frequency gain of the plant, by $k_m$. As a result of the above three parts, the closed-loop system has the desired transfer function. $\qquad\qquad\square$

The controller structure shown in Figure 8.13 for second-order plants can be extended to any plant with relative degree one. The resulting structure of the control system is shown in Figure 8.14, where $k^*$, $\theta_1{}^*$, $\theta_2{}^*$ and $\theta_o{}^*$ represents controller parameters which lead to perfect tracking when the plant parameters are known.

The structure of this control system can be described as follows. The block for generating the filter signal $\omega_1$ represents an $(n-1)^{\text{th}}$ order dynamics, which can be described by

$$\dot{\omega}_1 = \Lambda \omega_1 + \mathbf{h} u$$

where $\omega_1$ is an $(n-1) \times 1$ state vector, $\Lambda$ is an $(n-1) \times (n-1)$ matrix, and $\mathbf{h}$ is a constant vector such that $(\Lambda, \mathbf{h})$ is controllable. The poles of the matrix $\Lambda$ are chosen to be the

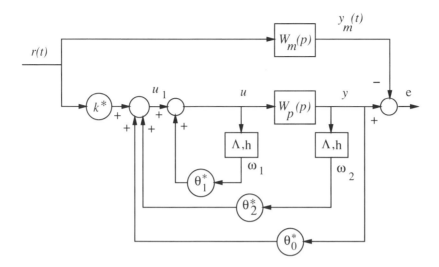

**Figure 8.14 :** A control system with perfect tracking

same as the roots of the polynomial $Z_m(p)$, *i.e.*,

$$\det[p\mathbf{I} - \mathbf{\Lambda}] = Z_m(p) \tag{8.48}$$

The block for generating the $(n-1)\times1$ vector $\boldsymbol{\omega}_2$ has the same dynamics but with y as input, *i.e.*,

$$\dot{\boldsymbol{\omega}}_2 = \mathbf{\Lambda}\boldsymbol{\omega}_2 + \mathbf{h}\,\mathbf{y}$$

It is straightforward to discuss the controller parameters in Figure 8.14. The scalar gain $k^*$ is defined to be

$$k^* = \frac{k_m}{k_p}$$

and is intended to modulate the high-frequency gain of the control system. The vector $\boldsymbol{\theta}_1^*$ contains $(n-1)$ parameters which intend to cancel the zeros of the plant. The vector $\boldsymbol{\theta}_2^*$ contains $(n-1)$ parameters which, together with the scalar gain $\boldsymbol{\theta}_o^*$ can move the poles of the closed-loop control system to the locations of the reference model poles. Comparing Figure 8.13 and Figure 8.14 will help the reader become familiar with this structure and the corresponding notations.

As before, the control input in this system is a linear combination of t' reference signal $r(t)$, the vector signal $\boldsymbol{\omega}_1$ obtained by filtering the control input $u$, signals $\boldsymbol{\omega}_2$ obtained by filtering the plant output $y$, and the output itself. The cor' input $u$ can thus be written, in terms of the adjustable parameters and the va'

signals, as

$$u^*(t) = k^* r + \theta_1^* \omega_1 + \theta_2^* \omega_2 + \theta_o^* y \tag{8.49}$$

Corresponding to this control law and *any* reference input $r(t)$, the output of the plant is

$$y(t) = \frac{B(p)}{A(p)} u^*(t) = W_m r(t) \tag{8.50}$$

since these parameters result in perfect tracking. At this point, one easily sees the reason for assuming the plant to be minimum-phase: this allows the plant zeros to be canceled by the controller poles.

In the adaptive control problem, the plant parameters are unknown, and the ideal control parameters described above are also unknown. Instead of (8.49), the control law is chosen to be

$$u = k(t)r + \theta_1(t)\omega_1 + \theta_2(t)\omega_2 + \theta_o(t)y \tag{8.51}$$

where $k(t)$, $\theta_1(t)$, $\theta_2(t)$ and $\theta_o(t)$ are controller parameters to be provided by the adaptation law.

## CHOICE OF ADAPTATION LAW

For notational simplicity, let $\theta$ be the $2n \times 1$ vector containing all the controller parameters, and $\omega$ be the $2n \times 1$ vector containing the corresponding signals, *i.e.*,

$$\theta(t) = [\, k(t) \quad \theta_1(t) \quad \theta_2(t) \quad \theta_o(t)\,]^T$$

$$\omega(t) = [\, r(t) \quad \omega_1(t) \quad \omega_2(t) \quad y(t)\,]^T$$

Then the control law (8.51) can be compactly written as

$$u = \theta^T(t)\,\omega(t) \tag{8.52}$$

Denoting the ideal value of $\theta$ by $\theta^*$ and the error between $\theta(t)$ and $\theta^*$ by $\phi(t) = \theta(t) - \theta^*$, the estimated parameters $\theta(t)$ can be represented as

$$\theta(t) = \theta^* + \phi(t)$$

Therefore, the control law (8.52) can also be written as

$$u = \theta^{*T}\omega + \phi^T(t)\omega$$

In order to choose an adaptation law so that the tracking error $e$ converges to zero, we

have to first find out how the tracking error is related to the parameter error. Let us use a simple technique for this purpose.

With the control law given in (8.52), the control system with variable gains can be equivalently represented as shown in Figure 8.15, with $\phi^T(t)\omega/k^*$ regarded as an external signal. Since the ideal parameter vector $\theta^*$ is such that the plant output in Figure 8.15 is given by (8.50), the output here must be

$$y(t) = W_m(p)r + W_m(p)[\phi^T(t)\omega/k^*] \tag{8.53}$$

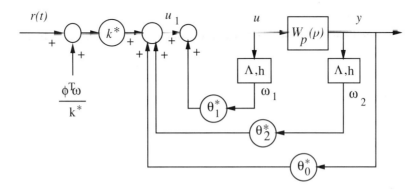

**Figure 8.15 :** An equivalent control system for time-varying gains

Since $y_m(t) = W_m(p)r$, the tracking error is seen to be related to the parameter error by the simple equation

$$e(t) = W_m(p)[\phi^T(t)\omega(t)/k^*] \tag{8.54}$$

Since this is the familiar equation seen in Lemma 8.1, the following adaptation law is chosen

$$\dot{\theta} = -\operatorname{sgn}(k_p)\,\gamma\,e(t)\,\omega(t) \tag{8.55}$$

where $\gamma$ is a positive number representing the adaptation gain and we have used the fact that the sign of $k^*$ is the same as that of $k_p$, due to the assumed positiveness of $k_m$.

Based on Lemma 8.1 and through a straightforward procedure for establishing signal boundedness, one can show that the tracking error in the above adaptive control system converges to zero asymptotically.

## 8.4.2 Linear Systems With Higher Relative Degree

The design of adaptive controller for plants with relative degree larger than 1 is both similar to, and different from, that for plants with relative degree 1. Specifically, the choice of control law is quite similar but the choice of adaptation law is very different. This difference comes from the fact that the reference model now cannot be SPR.

### CHOICE OF CONTROL LAW

We can show that the controller part of the system in Figure 8.15 is also applicable to plants with relative degree larger than 1, leading to exact tracking when the plant parameters are exactly known. Let us again start from a simple example.

**Example 8.6:** Consider the second order plant described by the transfer function

$$y = \frac{k_p u}{p^2 + a_{p1}p + a_{p2}}$$

and the reference model

$$y_m = \frac{k_m r}{p^2 + a_{m1}p + a_{m2}}$$

which are similar to those in Example 8.5, but now contain no zeros.

Let us consider the control structure shown in Figure 8.16 which is a slight modification of the controller structure in Figure 8.13. Note that $b_m$ in the filters in Figure 8.13 has been replaced by a positive number $\lambda_o$. Of course, the transfer functions $W_p$ and $W_m$ in Figure 8.16 now have relative degree 2.

The closed-loop transfer function from the reference signal $r$ to the plant output $y$ is

$$W_{ry} = k \frac{\dfrac{p + \lambda_o}{p + \lambda_o + \alpha_1} \dfrac{k_p}{p^2 + a_{p1}p + a_{p2}}}{1 + \dfrac{p + \lambda_o}{p + \lambda_o + \alpha_1} \dfrac{\beta_1 p + \beta_2}{p + \lambda_o} \dfrac{k_p}{p^2 + a_{p1}p + a_{p2}}}$$

$$= \frac{k k_p (p + \lambda_o)}{(p + \lambda_o + \alpha_1)(p^2 + a_{p1}p + a_{p2}) + k_p(\beta_1 p + \beta_2)}$$

Therefore, if the controller parameters $\alpha_1$, $\beta_1$, $\beta_2$, and $k$ are chosen such that

$$(p + \lambda_o + \alpha_1)(p^2 + a_{p1}p + a_{p2}) + k_p(\beta_1 p + \beta_2) = (p + \lambda_o)(p^2 + a_{m1}p + a_{m2})$$

and

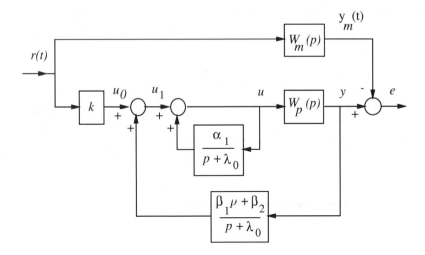

**Figure 8.16 :** A model-reference control system for relative degree 2

$$k = \frac{k_m}{k_p}$$

then the closed loop transfer function $W_{ry}$ becomes identically the same as that of the reference model. Clearly, such choice of parameters exists and is *unique*.  □

For general plants of relative degree larger than 1, the same control structure as given in Figure 8.14 is chosen. Note that the order of the filters in the control law is still $(n-1)$. However, since the model numerator polynomial $Z_m(p)$ is of degree smaller than $(n-1)$, it is no longer possible to choose the poles of the filters in the controller so that $\det[p\mathbf{I} - \mathbf{\Lambda}] = Z_m(p)$ as in (8.48). Instead, we now choose

$$\lambda(p) = Z_m(p)\,\lambda_1(p) \tag{8.57}$$

where $\lambda(p) = \det[p\mathbf{I} - \mathbf{\Lambda}]$ and $\lambda_1(p)$ is a Hurwitz polynomial of degree $(n-1-m)$. With this choice, the desired zeros of the reference model can be imposed.

Let us denote the transfer function of the feedforward part $(u/u_1)$ of the controller by $\lambda(p)/(\lambda(p) + C(p))$, and that of the feedback part by $D(p)/\lambda(p)$, where the polynomial $C(p)$ contains the parameters in the vector $\mathbf{\theta}_1$, and the polynomial $D(p)$ contains $\theta_o$ and the parameters in the vector $\mathbf{\theta}_2$. Then, the closed-loop transfer function is easily found to be

$$W_{ry} = \frac{k\,k_p Z_p \lambda_1(p) Z_m(p)}{R_p(p)[\lambda(p) + C(p)] + k_p Z_p D(p)} \tag{8.58}$$

The question now is whether in this general case there exist choice of values for $k$, $\theta_o$, $\theta_1$ and $\theta_2$ such that the above transfer function becomes exactly the same as $W_m(p)$, or equivalently

$$R_p(\lambda(p) + C(p)) + k_p Z_p D(p) = \lambda_1 Z_p R_m(p) \tag{8.59}$$

The answer to this question can be obtained from the following lemma:

**Lemma 8.2**: *Let $A(p)$ and $B(p)$ be polynomials of degree $n_1$ and $n_2$, respectively. If $A(p)$ and $B(p)$ are relatively prime, then there exist polynomials $M(p)$ and $N(p)$ such that*

$$A(p)M(p) + B(p)N(p) = A^*(p) \tag{8.60}$$

*where $A^*(p)$ is an arbitrary polynomial.*

This lemma can be used straightforwardly to answer our question regarding (8.59). By regarding $R_p$ as $A(p)$ in the lemma, $k_p Z_p$ as $B(P)$ and $\lambda_1(p) Z_p R_m$ as $A^*(p)$, we conclude that there exist polynomials $(\lambda(p) + C(p))$ and $D(p)$ such that (8.59) is satisfied. This implies that a proper choice of the controller parameters

$$k = k^* \qquad \theta_o = \theta_o{}^* \qquad \theta_1 = \theta_1{}^* \qquad \theta_2 = \theta_2{}^*$$

exists so that exact model-following is achieved.

## CHOICE OF ADAPTATION LAW

When the plant parameters are unknown, we again use a control law of the form (8.52), *i.e.,*

$$u = \theta^T(t)\,\omega(t) \tag{8.61}$$

with the $2n$ controller parameters in $\theta(t)$ provided by the adaptation law. Using a similar reasoning as before, we can again obtain the output $y$ in the form of (8.53) and the tracking error in the form of (8.54), *i.e.,*

$$e(t) = W_m(p)[\phi^T \omega / k^*] \tag{8.62}$$

However, the choice of adaptation law given by (8.55) cannot be used, because now the reference model transfer function $W_m(p)$ is no longer SPR. A famous technique called *error augmentation* can be used to avoid the difficulty in finding an adaptation law for (8.62). The basic idea of the technique is to consider a so-called augmented error $\varepsilon(t)$ which correlates to the parameter error $\phi$ in a more desirable way than the tracking error $e(t)$.

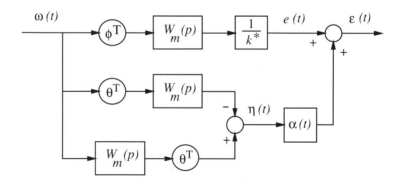

**Figure 8.17** : The augmented error

Specifically, let us define an *auxiliary error* $\eta(t)$ by

$$\eta(t) = \boldsymbol{\theta}^T(t) W_m(p)[\boldsymbol{\omega}] - W_m(p)[\boldsymbol{\theta}^T(t)\boldsymbol{\omega}(t)] \tag{8.63}$$

as shown in Figure 8.17. It is useful to note two features about this error. First, $\eta(t)$ can be computed on-line, since the estimated parameter vector $\boldsymbol{\theta}(t)$ and the signal vector $\boldsymbol{\omega}(t)$ are both available. Secondly, this error is caused by the time-varying nature of the estimated parameters $\boldsymbol{\theta}(t)$, in the sense that when $\boldsymbol{\theta}(t)$ is replaced by the true (constant) parameter vector $\boldsymbol{\theta}^*$, we have

$$\boldsymbol{\theta}^{*T} W_m(p)[\boldsymbol{\omega}] - W_m(p)[\boldsymbol{\theta}^{*T}\boldsymbol{\omega}(t)] = 0$$

This also implies that $\eta$ can be written

$$\eta(t) = \boldsymbol{\phi}^T W_m(\boldsymbol{\omega}) - W_m(\boldsymbol{\phi}^T\boldsymbol{\omega})$$

Now let us define an *augmented error* $\varepsilon(t)$, by combining the tracking error $e(t)$ with the auxiliary error $\eta(t)$ as

$$\varepsilon(t) = e(t) + \alpha(t)\,\eta(t) \tag{8.64}$$

where $\alpha(t)$ is a time-varying parameter to be determined by adaptation. Note that $\alpha(t)$ is not a controller parameter, but only a parameter used in forming the new error $\varepsilon(t)$. For convenience, let us write $\alpha(t)$ in the form

$$\alpha(t) = \frac{1}{k^*} + \phi_\alpha(t)$$

where $\phi_\alpha = \alpha(t) - \frac{1}{k^*}$. Substituting (8.62) and (8.63) into (8.64), we obtain

$$\varepsilon(t) = \frac{1}{k^*} \phi^T(t) \, \underline{\omega} + \phi_\alpha \eta(t) \tag{8.65}$$

where

$$\underline{\omega}(t) = W_m(p)[\,\omega\,] \tag{8.66}$$

This implies that the augmented error can be *linearly* parameterized by the parameter errors $\phi(t)$ and $\phi_\alpha$. Equation (8.65) thus represents a form commonly seen in system identification. A number of standard techniques to be discussed in section 8.7, such as the gradient method or the least-squares method, can be used to update the parameters for equations of this form. Using the gradient method with normalization, the controller parameters $\theta(t)$ and the parameter $\alpha(t)$ for forming the augmented error are updated by

$$\dot{\theta} = -\frac{\text{sgn}(k_p) \gamma \varepsilon \, \underline{\omega}}{1 + \underline{\omega}^T \underline{\omega}} \tag{8.67a}$$

$$\dot{\alpha} = -\frac{\gamma \varepsilon \eta}{1 + \underline{\omega}^T \underline{\omega}} \tag{8.67b}$$

With the control law (8.61) and adaptation law (8.67), global convergence of the tracking error can be shown. The proof is mathematically involved and will be omitted here.

Finally, note that there exist other techniques to get around the difficulty associated with equation (8.62). In particular, it can be shown that an alternative technique is to generate a different augmented error, which is related to the parameter error $\theta$ through a properly selected SPR transfer function.

# 8.5  Adaptive Control of Nonlinear Systems

There exists relatively little general theory for the adaptive control of nonlinear systems. However, adaptive control has been successfully developed for some important classes of nonlinear control problems. Such problems usually satisfy the following conditions:

1. the nonlinear plant dynamics can be linearly parameterized

2. the full state is measurable

3. nonlinearities can be canceled stably (*i.e.*, without unstable hidden modes or dynamics) by the control input if the parameters are known

In this section, we describe one such class of SISO systems to suggest how to design adaptive controllers for nonlinear systems (as an extension of the technique in section 8.3). In chapter 9, we shall study in detail the adaptive control of special classes of MIMO nonlinear physical systems.

## PROBLEM STATEMENT

We consider $n^{th}$-order nonlinear systems in companion form

$$y^{(n)} + \sum_{i=1}^{n} \alpha_i f_i(\mathbf{x}, t) = bu \tag{8.68}$$

where $\mathbf{x} = [y \;\; \dot{y} \;\; .... \;\; y^{(n-1)}]^T$ is the state vector, the $f_i$ are known nonlinear functions of the state and time, and the parameters $\alpha_i$ and $b$ are unknown constants. We assume that the state is measured, and that the sign of $b$ is known. One example of such dynamics is

$$m\ddot{x} + cf_1(\dot{x}) + kf_2(x) = u \tag{8.69}$$

which represents a mass-spring-damper system with nonlinear friction and nonlinear damping.

The objective of the adaptive control design to make the output asymptotically tracks a desired output $y_d(t)$ despite the parameter uncertainty. To facilitate the adaptive controller derivation, let us rewrite equation (8.68) as

$$hy^{(n)} + \sum_{i=1}^{n} a_i f_i(\mathbf{x}, t) = u \tag{8.70}$$

by dividing both sides by the unknown constant $b$, where $h = 1/b$ and $a_i = \alpha_i/b$.

## CHOICE OF CONTROL LAW

Similarly to the sliding control approach of chapter 7, let us define a combined error

$$s = e^{(n-1)} + \lambda_{n-2} e^{(n-2)} + ... + \lambda_o e = \Delta(p)e$$

where $e$ is the output tracking error and $\Delta(p) = p^{n-1} + \lambda_{n-2} p^{(n-2)} + ..... + \lambda_o$ is a stable (Hurwitz) polynomial in the Laplace variable $p$. Note that $s$ can be rewritten as

$$s = y^{(n-1)} - y_r^{(n-1)}$$

where $y_r^{(n-1)}$ is defined as

$$y_r^{(n-1)} = y_d^{(n-1)} - \lambda_{n-2} e^{(n-2)} - ... - \lambda_o e$$

Consider the control law

$$u = h y_r^{(n)} - ks + \sum_{i=1}^{n} a_i f_i(\mathbf{x}, t) \tag{8.71}$$

where $k$ is a constant of the same sign as $h$, and $y_r^{(n)}$ is the derivative of $y_r^{(n-1)}$, i.e.,

$$y_r^{(n)} = y_d^{(n)} - \lambda_{n-2} e^{(n-1)} - \ldots - \lambda_o \dot{e}$$

Note that $y_r^{(n)}$, the so-called "reference" value of $y^{(n)}$, is obtained by modifying $y_d^{(n)}$ according to the tracking errors.

If the parameters are all known, this choice leads to the tracking error dynamics

$$h\dot{s} + k s = 0$$

and therefore gives exponential convergence of $s$, which, in turn, guarantees the convergence of $e$.

## CHOICE OF ADAPTATION LAW

For our adaptive control, the control law (8.71) is replaced by

$$u = \hat{h} y_r^{(n)} - ks + \sum_{i=1}^{n} \hat{a}_i f_i(\mathbf{x}, t) \tag{8.72}$$

where $h$ and the $a_i$ have been replaced by their estimated values. The tracking error from this control law can be easily shown to be

$$h\dot{s} + k s = \tilde{h} y_r^{(n)} + \sum_{i=1}^{n} \tilde{a}_i f_i(\mathbf{x}, t) \tag{8.73}$$

This can be rewritten as

$$s = \frac{1/h}{p + (k/h)} [\tilde{h} y_r^{(n)} + \sum_{i=1}^{n} \tilde{a}_i f_i(\mathbf{x}, t)] \tag{8.74}$$

Since this represents an equation in the form of (8.15) with the transfer function obviously being SPR, Lemma 8.1 suggests us to choose the following adaptation law

$$\dot{\hat{h}} = - \gamma \operatorname{sgn}(h) \, s \, y_r^{(n)}$$

$$\dot{\hat{a}}_i = - \gamma \operatorname{sgn}(h) \, s f_i$$

Specifically, using the Lyapunov function candidate

$$V = |h|s^2 + \gamma^{-1}[\tilde{h}^2 + \sum_{i=1}^{n} \tilde{a}_i^2]$$

it is straightforward to verify that

$$\dot{V} = -2|k|s^2$$

and therefore the global tracking convergence of the adaptive control system can be easily shown.

Note that the formulation used here is very similar to that in section 8.3. However, due to the use of the compact tracking error measure $s$, the derivation and notation here is much simpler. Also, one can easily show that global tracking convergence is preserved if a different adaptation gain $\gamma_i$ is used for each unknown parameter.

The sliding control ideas of chapter 7 can be used further to create controllers that can adapt to constant unknown parameters while being robust to unknown but bounded fast-varying coefficients or disturbances, as in systems of the form

$$h\, y^{(n)} + \sum_{i=1}^{n} [a_i + a_{iv}(t)]\, f_i(\mathbf{x}, t) = u$$

where the $f_i$ are known nonlinear functions of the state and time, $h$ and the $a_i$ are unknown constants, and the time-varying quantities $a_{iv}(t)$ are unknown but of known (possibly state-dependent or time-varying) bounds (Exercise 8.8).

# 8.6  Robustness of Adaptive Control Systems

The above tracking and parameter convergence analysis has provided us with considerable insight into the behavior of the adaptive control system. The analysis has been carried out assuming that no other uncertainties exist in the control system besides parametric uncertainties. However, in practice, many types of non-parametric uncertainties can be present. These include

- high-frequency unmodeled dynamics, such as actuator dynamics or structural vibrations

- low-frequency unmodeled dynamics, such as Coulomb friction and stiction

- measurement noise

- computation roundoff error and sampling delay

Since adaptive controllers are designed to control real physical systems and such non-parametric uncertainties are unavoidable, it is important to ask the following questions concerning the non-parametric uncertainties:

  • what effects can they have on adaptive control systems ?

  • when are adaptive control systems sensitive to them ?

  • how can adaptive control systems be made insensitive to them ?

While precise answers to such questions are difficult to obtain, because adaptive control systems are nonlinear systems, some qualitative answers can improve our understanding of adaptive control system behavior in practical applications.  Let us now briefly discuss these topics.

Non-parametric uncertainties usually lead to performance degradation, *i.e.*, the increase of model following error.  Generally, small non-parametric uncertainties cause small tracking error, while larger ones cause larger tracking error.  Such relations are universal in control systems and intuitively understandable. We can naturally expect the adaptive control system to become unstable when the non-parametric uncertainties become too large.

## PARAMETER DRIFT

When the signal **v** is persistently exciting, both simulations and analysis indicate that the adaptive control systems have some robustness with respect to non-parametric uncertainties.  However, when the signals are not persistently exciting, even small uncertainties may lead to severe problems for adaptive controllers. The following example illustrates this situation.

### Example 8.7: Rohrs's Example

The sometimes destructive consequence of non-parametric uncertainties is clearly shown in the well-known example by Rohrs, which consists of an adaptive first-order control system containing unmodeled dynamics and measurement noise. In the adaptive control design, the plant is assumed to have a the following nominal model

$$H_o(p) = \frac{k_p}{p + a_p}$$

The reference model has the following SPR function

$$M(p) = \frac{k_m}{p + a_m} = \frac{3}{p + 3}$$

The real plant, however, is assumed to have the transfer function relation

$$y = \frac{2}{p+1} \; \frac{229}{p^2 + 30p + 229} \, u$$

This means that the real plant is of third order while the nominal plant is of only first order. The unmodeled dynamics are thus seen to be $229/(p^2 + 30p + 229)$, which are high-frequency but *lightly-damped* poles at $(-15 + j)$ and $(-15 - j)$.

Besides the unmodeled dynamics, it is assumed that there is some measurement noise $n(t)$ in the adaptive system. The whole adaptive control system is shown in Figure 8.18. The measurement noise is assumed to be $n(t) = 0.5 \sin(16.1\,t)$.

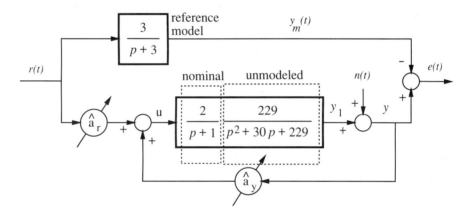

**Figure 8.18 :** Adaptive control with unmodeled dynamics and measurement noise

Corresponding to the reference input $r(t) = 2$, the results of the adaptive control system are shown in Figure 8.19. It is seen that the output $y(t)$ initially converges to the vicinity of $y = 2$, then operates with a small oscillatory error related to the measurement noise, and finally diverges to infinity. $\qquad\qquad\qquad\qquad\qquad\qquad\qquad\qquad\qquad\qquad\qquad\qquad\qquad\quad$ $\square$

In view of the *global* tracking convergence proven in the absence of non-parametric uncertainties and the *small* amount of non-parametric uncertainties present in the above example, the observed instability can seem quite surprising. However, one can gain some insight into what is going on in the adaptive control system by examining the parameter estimates in Figure 8.19. It is seen that the parameters drift slowly as time goes on, and suddenly diverge sharply. The simplest explanation of the parameter drift problem is that the constant reference input contains insufficient parameter information and the parameter adaptation mechanism has difficulty distinguishing the parameter information from noise. As a result, the parameters drift in a direction along which the tracking error remains small. Note that even though the

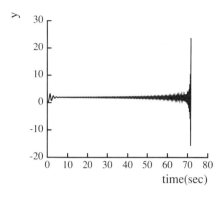
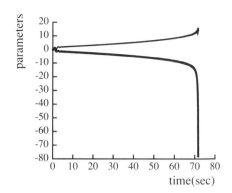

**Figure 8.19 :** Instability and parameter drift

tracking error stays at the same level when the parameters drift, the poles of the closed-loop system continuously shift (since the parameters vary very slowly, the adaptive control system may be regarded as a linear time-invariant system with three poles). When the estimated parameters drift to the point where the closed-loop poles enter the right-half complex plane, the whole system becomes unstable. The above reasoning can be confirmed mathematically.

In general, the following points can be made about parameter drift. Parameter drift occurs when the signals are not persistently exciting; it is mainly caused by measurement noise; it does not affect tracking accuracy until instability occurs; it leads to sudden failure of the adaptive control system (by exciting unmodeled dynamics).

Parameter drift is a major problem associated with non-parametric uncertainties (noise and disturbance). But there are possibly other problems. For example, when the adaptation gain or the reference signal are very large, adaptation becomes fast and the estimated parameters may be quite oscillatory. If the oscillations get into the frequency range of unmodeled dynamics, the unmodeled dynamics may be excited and the parameter adaptation may be based on meaningless signals, possibly leading to instability of the control system. For parameter oscillation problems, techniques such as normalization of signals (divide $\mathbf{v}$ by $1 + \mathbf{v}^T \mathbf{v}$ ) or the composite adaptation in section 8.8 can be quite useful.

## DEAD-ZONE

Even though the possibility of small disturbances leading to instability is quite undesirable, it does not mean that adaptive control is impractical. A number of techniques for modifying the adaptation law are available to avoid the parameter drift problem. The simplest is called the "dead-zone" technique. Because of its simplicity and effectiveness, it is most frequently used.

The dead-zone technique is based on the observation that small tracking errors contain mostly noise and disturbance, therefore, one should shut the adaptation mechanism off for small tracking errors. Specifically, we should replace an adaptation law

$$\dot{\hat{\mathbf{a}}} = -\gamma \mathbf{v} e \tag{8.75}$$

by

$$\dot{\hat{\mathbf{a}}} = \begin{cases} -\gamma \mathbf{v} e & |e| > \Delta \\ 0 & |e| < \Delta \end{cases} \tag{8.76}$$

where $\Delta$ is the size the dead-zone. As the following example shows, such a simple modification can greatly reduce the effects of the disturbances.

### Example 8.8: Use of Dead-Zone

Consider again the adaptive control system of Example 8.7, but modify the adaptation law by incorporating a dead-zone of $\Delta = 0.7$. The results are shown in Figure 8.20. It is seen that tracking error stays around the ideal response of $y = 2$, with an oscillation due to the measurement noise. The parameters now do not have any indication of drifting. It is interesting to point out that the oscillation appears very fast because the time-scale in the figure is large and the noise itself is of quite high frequency. $\quad\quad\quad\quad\Box$

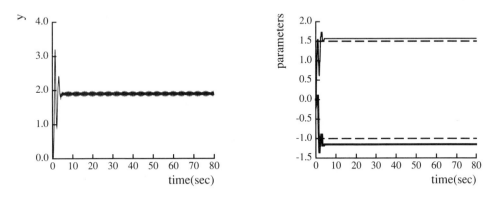

**Figure 8.20 :** Adaptive control with dead-zone

A number of other techniques also exist to relieve the problem of parameter drift. One involves the so-called σ-modification, which approximates the original integrator in the adaptation law by a lowpass filter. Another is the "regressor replacement" technique. By "regressor", we mean the vector $\mathbf{v}(t)$ in (8.75). Note that $\mathbf{v}(t)$ is usually computed based on the plant measurement $y$ and thus affected by measurement noise $n(t)$. Since the adaptation law (8.75) involves the multiplication of

$\mathbf{v}(t)$ by $e(t)$, the update rate is related to the square of the measurement noise and causes parameter drift. For example, in the presence of measurement noise $n(t)$, (8.28b) can be written

$$\dot{\hat{a}}_y = -\,\text{sgn}(b)\,\gamma\,(y_1 + n - y_m)\,(y_1 + n)$$

$$= -\,\text{sgn}(b)\,\gamma\,[(y_1 - y_m)\,y_1 + n\,(\,2y_1 - y_m) + n^2]$$

where $y_1$ is the true plant output. It is noted that the first term truly contains parameter information, the second term tends to average out, and the third term $-\,\text{sgn}(b)\,\gamma\,n^2$ is the reason for the drifting of $\hat{a}_y$ in Figure 8.19 ($\hat{a}_r$ drifts accordingly so that the tracking error remains small). As a result of this observation, one can relieve parameter drift by replacing $y$ in (8.28b) by $y_m$ which is independent of $n$. It is desirable to start this replacement after the tracking error has converged well.

# 8.7   * On-Line Parameter Estimation

When there is parameter uncertainty in a dynamic system (linear or nonlinear), one way to reduce it is to use parameter estimation, *i.e.*, inferring the values of the parameters from the measurements of input and output signals of the system. Parameter estimation can be done either on-line or off-line. Off-line estimation may be preferable if the parameters are constant and there is sufficient time for estimation before control. However, for parameters which vary (even though slowly) during operation, on-line parameter estimation is necessary to keep track of the parameter values. Since problems in the adaptive control context usually involve slowly time-varying parameters, on-line estimation methods are thus more relevant.

In this section, we study a few basic methods of on-line estimation. Unlike most discussions of parameter estimation, we use a continuous-time formulation rather than a discrete-time formulation. This is motivated by the fact that nonlinear physical systems are continuous in nature and are hard to meaningfully discretize. Furthermore, digital control systems may be treated as continuous-time systems in analysis and design if high sampling rates are used. The availability of cheap computation generally allows high sampling rates and thus continuous-time models to be used.

Note that, although the main purpose of the on-line estimators may be to provide parameter estimates for self-tuning control, they can also be used for other purposes, such as load monitoring or failure detection.

# 8.7.1 Linear Parametrization Model

The essence of parameter estimation is to extract parameter information from available data concerning the system. Therefore, we need an estimation model to relate the available data to the unknown parameters, similarly to the familiar experimental data fitting scenario, where we need to hypothesize the form of a curve before finding specific coefficients describing it, based on the data. This estimation model may or may not be the same as the model used for the control purpose. A quite general model for parameter estimation applications is in the linear parametrization form

$$\mathbf{y}(t) = \mathbf{W}(t)\,\mathbf{a} \tag{8.77}$$

where the $n$-dimensional vector $\mathbf{y}$ contains the "outputs" of the system, the $m$-dimensional vector $\mathbf{a}$ contains unknown parameters to be estimated, and the $n{\times}m$ matrix $\mathbf{W}(t)$ is a signal matrix. Note that both $\mathbf{y}$ and $\mathbf{W}$ are required to be known from the measurements of the system signals, and thus the only unknown quantities in (8.77) are the parameters in $\mathbf{a}$. This means that (8.77) is simply a linear equation in terms of the unknown $\mathbf{a}$. For every time instant $t$, there is such an equation. So if we are given the continuous measurements of $\mathbf{y}(t)$ and $\mathbf{W}(t)$ throughout a time interval, we have an infinite number of equations in the form of (8.77). If we are given the values of $\mathbf{y}(t)$ and $\mathbf{W}(t)$ at $k$ sampling instants, we have $k$ sets of such equations instead. The objective of parameter estimation is to simply solve these redundant equations for the $m$ unknown parameters. Clearly, in order to be able to estimate $m$ parameters, we need at least a total of $m$ equations. However, in order to estimate the parameters $\mathbf{a}$ well in the presence of inevitable noise and modeling error, more data points are preferable.

In off-line estimation, one collects the data of $\mathbf{y}$ and $\mathbf{W}$ for a period of time, and solves the equations once and for all. In on-line estimation, one solves the equation recursively, implying that the estimated value of $\hat{\mathbf{a}}$ is updated once a new set of data $\mathbf{y}$ and $\mathbf{W}$ is available.

How well and how fast the parameters $\mathbf{a}$ are estimated depends on two aspects, namely, the estimation method used and the information content (persistent excitation) of the data $\mathbf{y}$ and $\mathbf{W}$. Our primary objective in this section is to examine the properties of some standard estimation methods. The generation of informative data is a complex issue discussed extensively in the system identification literature. While we shall not study this issue in detail, the relation between the signal properties and estimation results will be discussed.

Model (8.77), although simple, is actually quite general. Any linear system can be rewritten in this form after filtering both sides of the system dynamics equation through an exponentially stable filter of proper order, as seen in the following example.

### Example 8.9: Filtering Linear Dynamics

Let us first consider the first-order dynamics

$$\dot{y} = -a_1 y + b_1 u \tag{8.78}$$

Assume that $a_1$ and $b_1$ in the model are unknown, and that only the output $y$ and the input $u$ are available. The above model cannot be directly used for estimation, because the derivative of $y$ appears in the above equation (note that numerically differentiating $y$ is usually undesirable because of noise considerations). To eliminate $\dot{y}$ in the above equation, let us filter (multiply) both sides of the equation by $1/(p + \lambda_f)$ (where $p$ is the Laplace operator and $\lambda_f$ is a known positive constant). Rearranging, this leads to the form

$$y(t) = y_f (\lambda_f - a_1) + u_f b_1 \tag{8.79}$$

where

$$y_f = \frac{y}{p + \lambda_f} \qquad u_f = \frac{u}{p + \lambda_f}$$

with the subscript $f$ denoting filtered quantities. Note that, as a result of the filtering operation, the only unknown quantities in (8.79) are the parameters $(\lambda_f - a_1)$ and $b_1$.

Note that the above filtering introduces a d.c. gain of $1/\lambda_f$, *i.e.*, the magnitudes of $y_f$ and $u_f$ are smaller than those of $y$ and $u$ by a factor of $\lambda_f$ at low frequencies. Since smaller signals may lead to slower estimation, one may multiply both sides of (8.79) by a constant number, *e.g.*, $\lambda_f$.

Generally, for a linear single-input single-output system, its dynamics can be described by

$$A(p) y = B(p) u \tag{8.80}$$

with

$$A(p) = a_o + a_1 p + \ldots + a_{n-1} p^{n-1} + p^n$$

$$B(p) = b_o + b_1 p + \ldots + b_{n-1} p^{n-1}$$

Let us divide both sides of (8.80) by a *known* monic polynomial of order $n$, leading to

$$y = \frac{A_o(p) - A(p)}{A_o(p)} y + \frac{B(p)}{A_o(p)} u \tag{8.81}$$

where

$$A_o = \alpha_o + \alpha_1 p + .... + \alpha_{n-1} p^{n-1} + p^n$$

has known coefficients. In view of the fact that

$$A_o(p) - A(p) = (\alpha_o - a_o) + (\alpha_1 - a_1)p + .... + (\alpha_{n-1} - a_{n-1})p^{n-1}$$

we can write (8.81) in the basic form

$$y = \boldsymbol{\theta}^T \mathbf{w}(t) \tag{8.82}$$

with $\boldsymbol{\theta}$ containing $2n$ unknown parameters, and $\mathbf{w}$ containing the filtered versions of the input and output, defined by

$$\boldsymbol{\theta} = [(\alpha_o - a_o) \quad (\alpha_1 - a_1) \quad ... \quad (\alpha_{n-1} - a_{n-1}) \quad b_o \quad ... \quad b_{n-1}]^T$$

$$\mathbf{w} = [\frac{y}{A_o} \quad \frac{py}{A_o} \quad ... \quad \frac{p^{n-1}y}{A_o} \quad \frac{u}{A_o} \quad ... \quad \frac{p^{n-1}u}{A_o}]^T$$

Note that $\mathbf{w}$ can be computed on-line based on the available values of $y$ and $u$. $\qquad\square$

The dynamics of many nonlinear systems can also be put into the form (8.77). A simple example is the nonlinear mass-spring-damper system (8.69), for which the input force is obviously linear in terms of mass, friction coefficient, and spring coefficient. For some more complicated nonlinear dynamics, proper filtering and parameter transformation may be needed to put the dynamics into the form of (8.77), as we now show.

### Example 8.10: Linear parametrization of robot dynamics

Consider the nonlinear dynamics (6.9) of the two-link robot of Example 6.2. Clearly, the joint torque vector $\mathbf{q}$ is nonlinear in terms of joint positions and velocities. It is also nonlinear in terms of the physical parameters $l_{c1}$, $l_{c2}$, and so on. However, it can be put into the form of (8.77) by a proper reparametrization and a filtering operation.

Consider the reparametrization first. Let us define

$$a_1 = m_2$$

$$a_2 = m_2 l_{c_2}$$

$$a_3 = I_1 + m_1 l_{c_1}^2$$

$$a_4 = I_2 + m_2 l_{c_2}^2$$

Then, one can show that each term on the left-hand side of (6.9) is linear in terms of the equivalent inertia parameters $\mathbf{a} = [a_1 \quad a_2 \quad a_3 \quad a_4]^T$. Specifically

$$H_{11} = a_3 + a_4 + a_1 \, l_1^{\,2} + 2 \, a_2 \, l_1 \cos q_2$$

$$H_{22} = a_4$$

$$H_{12} = H_{21} = a_2 \, l_1 \cos q_2 + a_4$$

Note that $l_1$ and $l_2$ are kinematic parameters, which are assumed to be known (they can be similarly treated if not known). The above expressions indicate that the inertia torque terms are linear in terms of **a**. It is easy to show that the other terms are also linear in **a**. Thus, we can write

$$\tau = \mathbf{Y}_1(\mathbf{q}, \dot{\mathbf{q}}, \ddot{\mathbf{q}}) \, \mathbf{a} \tag{8.83}$$

with the matrix $\mathbf{Y}_1$ expressed (nonlinearly) as a function of only $\mathbf{q}$, $\dot{\mathbf{q}}$ and $\ddot{\mathbf{q}}$; and **a** being a $m \times 1$ vector of equivalent parameters. This linear parametrization property actually applies to any mechanical system, including multiple-link robots.

Relation (8.83) cannot be directly used for parameter estimation, because of the presence of the unmeasurable joint acceleration $\ddot{\mathbf{q}}$. To avoid the joint acceleration in this relation, we can use the above filtering technique. Specifically, let $w(t)$ be the impulse response of a stable, proper filter (for example, for the first-order filter $\lambda/(p + \lambda)$, the impulse response is $e^{-\lambda t}$). Then, convolving both sides of (6.9) by $w$ yields

$$\int_o^t w(t-r)\,\tau(r)\,dr = \int_o^t w(t-r) \,[\, \mathbf{H}\ddot{\mathbf{q}} + \mathbf{C}\dot{\mathbf{q}} + \mathbf{G} \,]\,dr \tag{8.84}$$

Using partial integration, the first term on the right-hand side of (8.84) can be rewritten as

$$\int_o^t w(t-r)\,[\,\mathbf{H}\,\ddot{\mathbf{q}}\,]\,dr = w(t-r)\,\mathbf{H}\,\dot{\mathbf{q}}\big|_0^t - \int_o^t \frac{d}{dr}[w\,\mathbf{H}]\,\dot{\mathbf{q}}\,dr$$

$$= w(0)\,\mathbf{H}(\mathbf{q})\dot{\mathbf{q}} - w(0)\,\mathbf{H}[\mathbf{q}(0)]\,\dot{\mathbf{q}}(0) - \int_o^t [w(t-r)\,\dot{\mathbf{H}}\,\dot{\mathbf{q}} - \dot{w}(t-r)\,\mathbf{H}\,\dot{\mathbf{q}}]\,dr$$

This means that equation (8.84) can be rewritten as

$$\mathbf{y}(t) = \mathbf{W}(\mathbf{q}, \dot{\mathbf{q}})\,\mathbf{a} \tag{8.85}$$

where **y** is the filtered torque and **W** is the filtered version of $\mathbf{Y}_1$. Thus, the matrix **W** can be computed from available measurements of **q** and $\dot{\mathbf{q}}$. The filtered torque **y** can also be computed (assuming no actuator dynamics) because the torque signals issued by the computer are known. ☐

It is obvious at this point that the "output" **y** in model (8.77) does not have to be the same as the output in a control problem. In the above robotic example, the "output" **y** is actually the filtered version of the physical input to the robot. From a parameter estimation point of view, what we want is just a linear relation between known data and the unknown parameters. The following example further demonstrates this point.

### Example 8.11: An alternative estimation model for robots

Consider again the above 2-link robot. Using the principle of energy conversation, one sees that the rate of change of mechanical energy is equal to the power input from the joint motors, *i.e.*,

$$\tau^T \dot{\mathbf{q}} = \frac{dE}{dt} \tag{8.86}$$

where $E(\mathbf{q}, \dot{q})$ is the total mechanical energy of the robot. Since it is easy to show that the mechanical energy can be linearly parameterized, *i.e.*,

$$E = \mathbf{v}(\mathbf{q}, \dot{\mathbf{q}}) \mathbf{a}$$

one can write the energy relation (8.86) as

$$\tau^T \dot{\mathbf{q}} = \frac{d\mathbf{v}(\mathbf{q}, \dot{\mathbf{q}})}{dt} \mathbf{a}$$

with $\mathbf{v}$ computable from the measurement of $\mathbf{q}$ and $\dot{\mathbf{q}}$. To eliminate the joint acceleration $\ddot{\mathbf{q}}$ in this relation, we can again use filtering by a first-order filter $1/(p + \lambda_f)$ . This leads to

$$\frac{1}{p + \lambda_f} [\tau \dot{\mathbf{q}}] = [\mathbf{v} - \frac{\lambda_f \mathbf{v}}{p + \lambda_f}] \mathbf{a}$$

This is in the form of (8.77), with

$$y = \frac{1}{p + \lambda_f} [\tau^T \dot{\mathbf{q}}]$$

$$\mathbf{w} = \mathbf{v} - \frac{\lambda_f \mathbf{v}}{p + \lambda_f} \qquad \qquad \square$$

In the above example, note that the control model and the estimation model are drastically different. Actually, while the control model (6.9) has two equations and two outputs, the estimation model (8.86) has only one equation and one output (actually, the scalar model (8.86) is applicable to robots with any number of links because the energy relation is always scalar). With the energy relation (8.86), the computation of $\mathbf{w}$ is greatly simplified compared with that of $\mathbf{W}$ in (8.85), because there is no need to compute the complex centripetal and Coriolis forces (which may contain hundreds of terms for multiple-DOF robots).

## 8.7.2  Prediction-Error-Based Estimation Methods

Before studying the various methods for parameter estimation, let us first discuss the concept of prediction error. Assume that the parameter vector in (8.77) is unknown, and is estimated to be $\hat{\mathbf{a}}(t)$ at time $t$. One can predict the value of the output $\mathbf{y}(t)$ based on the parameter estimate and the model (8.77),

$$\hat{\mathbf{y}}(t) = \mathbf{W}(t)\hat{\mathbf{a}}(t) \tag{8.87}$$

where $\hat{\mathbf{y}}$ is called the predicted output at time $t$. The difference between the predicted output and the measured output $y$ is called the prediction error, denoted by $\mathbf{e}_1$, *i.e.*,

$$\mathbf{e}_1(t) = \hat{\mathbf{y}}(t) - \mathbf{y}(t) \tag{8.88}$$

The on-line estimation methods to be discussed in this section are all based on this error, *i.e.*, the parameter estimation law is driven by $\mathbf{e}_1$. The resulting estimators belong to the so-called prediction-error based estimators, a major class of on-line parameter estimators. The prediction error is related to the parameter estimation error, as can be seen from

$$\mathbf{e}_1 = \mathbf{W}\hat{\mathbf{a}} - \mathbf{W}\mathbf{a} = \mathbf{W}\tilde{\mathbf{a}} \tag{8.89}$$

where $\tilde{\mathbf{a}} = \hat{\mathbf{a}} - \mathbf{a}$ is the parameter estimation error.

In the following, we shall discuss the motivation, formulation and properties of the following methods:

- Gradient estimation

- Standard least-squares estimation

- Least-squares with exponential forgetting

- A particular method of variable exponential forgetting

Note that in the convergence analysis of these estimators we shall assume that the true parameters are constant, so that insights concerning the estimator's behavior can be obtained. However, in the back of our mind, we will always be aware of the task of handling time-varying parameters.

## 8.7.3  The Gradient Estimator

The simplest on-line estimator is the gradient estimator. Let us discuss its formulation, convergence properties, and robustness properties.

## FORMULATION AND CONVERGENCE

The basic idea in gradient estimation is that the parameters should be updated so that the prediction error is reduced. This idea is implemented by updating the parameters in the converse direction of the gradient of the squared prediction error with respect to the parameters, *i.e.*,

$$\dot{\hat{\mathbf{a}}} = -p_o \frac{\partial [\mathbf{e}_1^T \mathbf{e}_1]}{\partial \hat{\mathbf{a}}}$$

where $p_o$ is a positive number called the estimator gain. In view of (8.88) and (8.87), this can be written as

$$\dot{\hat{\mathbf{a}}} = -p_o \mathbf{W} \mathbf{e}_1 \qquad\qquad (8.90)$$

To see the properties of this estimator, we use (8.90) and (8.89) to obtain

$$\dot{\tilde{\mathbf{a}}} = -p_o \mathbf{W}^T \mathbf{W} \tilde{\mathbf{a}}$$

Using the Lyapunov function candidate

$$V = \tilde{\mathbf{a}}^T \tilde{\mathbf{a}}$$

its derivative is easily found to be

$$\dot{V} = -2p_o \tilde{\mathbf{a}}^T \mathbf{W}^T \mathbf{W} \tilde{\mathbf{a}} \le 0$$

This implies that the gradient estimator is always stable. By noting that $V$ is actually the squared parameter error, we see that the magnitude of the parameter error is always decreasing.

However, the convergence of the estimated parameters to the true parameters depends on the excitation of the signals. To gain some insights on that point, let us consider the estimation of a single parameter.

**Example 8.12: Gradient estimation of a single parameter**

Consider the estimation of one parameter from the model

$$y = w a$$

(with the mass estimation in Example 8.2 being such a case). The gradient estimation law is

$$\dot{\hat{a}} = -p_o w e_1$$

This implies that

$$\dot{\tilde{a}} = -p_o w^2 \tilde{a}$$

which can be solved as

$$\tilde{a}(t) = \tilde{a}(0) \exp[-\int_o^t p_o w^2(r) dr]$$

This implies that the parameter error will converge to zero if the signal $w$ is such that

$$\lim_{t \to \infty} \int_o^t w^2(r) dr = \infty$$

Note that $\tilde{a}$ will exponentially converge to zero if $w$ is persistently exciting, *i.e.*, if there exist positive constants $T$ and $\alpha_1$ such that for all $t \geq 0$ ,

$$\int_t^{t+T} w^2 dr \geq \alpha_1$$

In fact, the convergence rate is easily found to be $p_o \alpha_1/T$.

Clearly, in this case, a constant non-zero $w$ can guarantee the exponential convergence of $\tilde{a}$. However, if the signal $w$ decays too fast (*e.g.*, $w = e^{-t}$ ), one easily shows that the parameter error does not converge to zero.                                                      □

The above convergence result concerning the one-parameter case can by extended to the estimation of multiple parameters. Specifically, if the matrix $\mathbf{W}$ is persistently exciting, *i.e.*, there exist positive constants $\alpha_1$ and $T$ such that $\forall T \geq 0$

$$\int_t^{t+T} \mathbf{W}^T \mathbf{W} dr \geq \alpha_1 \mathbf{I} \tag{8.91}$$

then the parameter error $\tilde{\mathbf{a}}$ will converge exponentially, as shown in [Anderson, 1977; Morgan and Narendra, 1977] through a fairly involved procedure. Such a condition for parameter convergence is easily understandable, analogously to the case of model-reference adaptive control. In the case of linear systems, as described by (8.80), it is easy to verify that $m$ sinusoids in the input signal $u$ can guarantee the estimation of up to $2m$ parameters. In the case of nonlinear systems, the relation between the number of sinusoids and the number of parameters which can be estimated is not so clear. It is possible to estimate more than $2m$ parameters with an input $u$ containing $m$ sinusoids, as explained in section 8.2.

A slightly more general version of the gradient estimator is obtained by replacing the scalar gain by a positive definite matrix gain $\mathbf{P}_o$,

$$\dot{\hat{\mathbf{a}}} = -\mathbf{P}_O \frac{\partial [\mathbf{e}_1{}^T \mathbf{e}_1]}{\partial \hat{\mathbf{a}}}$$

The global stability of the algorithm can be shown using the Lyapunov function

$$V = \tilde{\mathbf{a}}^T \mathbf{P}_O^{-1} \tilde{\mathbf{a}}$$

The convergence properties are very similar.

## EFFECTS OF ESTIMATION GAIN

The choice of estimation gain $p_O$ has a fundamental influence on the convergence behavior of the estimator. For the single-parameter estimation case, one easily sees that a larger $p_O$ implies faster parameter convergence. In fact, the convergence rate is linearly related to the estimation gain $p_O$. For the multiple-parameter case, however, the relation between the magnitude of $p_O$ and the convergence rate of the estimated parameters is not as simple. Generally speaking, in a small range, increasing estimation gain leads to faster parameter convergence. But beyond some point, further increasing the estimation gain leads to more oscillatory and slower convergence, as can be demonstrated with the estimation of the four load parameters in the robot example 8.10. This phenomenon is caused by the gradient nature of the estimation, similarly to what happens in the gradient search method in optimization: within a small range, increase in step size in the gradient direction leads to faster convergence; but beyond some point, larger size leads to more oscillatory and possibly slower convergence.

Besides the effect on convergence speed, the choice of $p_O$ also has implications on the ability of the estimator to track time-varying parameters and withstand disturbances, as will be discussed soon.

## ROBUSTNESS PROPERTIES

The above analysis and simulations have been based on the assumed absence of parameter variation and non-parametric uncertainties. In order for an estimator to have practical value, however, it must have some robustness, *i.e.*, maintain reasonably good parameter estimation in the presence of parameter variation, measurement noise, disturbances, *etc*.

The quality of the parameter estimates in a gradient estimator depends on a number of factors, mainly,

- the level of persistent excitation of the signal $\mathbf{W}$

- the rate of parameter variation and the level of non-parametric uncertainties

- the magnitude of the estimator gain $p_o$

The level of persistent excitation in **W** is decided by the control task or experiment design. Persistent excitation is essential for the robustness of the estimator. If the signals in the original design are not persistently exciting, parameters will not converge even in the absence of non-parametric uncertainties. In the presence of non-parametric uncertainties, the estimator may possibly become unstable, *i.e.*, the parameters may diverge. One may have to add some perturbation signals to the control input to obtain good parameter estimation. The specific details in the data generation may be complicated, but the bottom line is that one should produce as much persist excitation as allowed by the involved constraints.

How fast the true parameters vary and how large the non-parametric uncertainties also affect the quality of the parameter estimates. Obviously, if the true parameters vary faster, it is harder for the parameter estimator to estimate accurately. If a lot of noise and unmodeled disturbances and dynamics are present, the estimates also become poor. To see this, let us consider the mass estimation problem, but now in the presence of disturbance (unmodeled Coulomb friction or measurement noise) and time-varying mass. The mass dynamics becomes

$$\tau(t) = m(t) w(t) + d(t)$$

where $d(t)$ is the disturbance and $w$ is the acceleration. The prediction error in this case is

$$e_1 = \hat{\tau} - \tau = \tilde{m} w - d$$

By substituting this into (8.90), the parameter error is easily shown to satisfy

$$\frac{d\tilde{a}}{dt} = -p_o w^2 \tilde{a} - \dot{a} + p_o w d \tag{8.92}$$

This can be interpreted as a time-varying filter with "output" $\tilde{a}$ and "input" $(-\dot{a} + p_o w d)$. Clearly, larger parameter variation rate $\dot{a}$ and larger disturbance $d(t)$ leads to larger parameter error $\tilde{a}$. If parameter $a$ varies too fast or the non-parametric uncertainties have too large values, one should consider using more accurate models, *i.e.*, modeling the dynamics of the parameter variation and also the disturbance.

The magnitude of the estimator gain also has a considerable influence on the robustness of the estimator. If $p_o$ is chosen to be large, the "bandwidth" of the filter $p_o w^2$ becomes large (so that higher-frequency noise can pass) and the input component $p_o w d$ to the filter becomes larger. This means that parameter estimation error due to disturbance will become larger. The increase of estimation gain has the opposite effect on the estimator ability to estimate time-varying parameters. Consider the case of w being a constant for example. The steady-state error of the estimated parameter in (8.92) is $\dot{m}/(p_o w^2)$, which shows that the parameter error is decreased by the increase of the estimation gain.

The following simple simulation illustrates the behavior of the gradient estimator in the presence of measurement noise.

### Example 8.13: Gradient estimation of a constant mass

Consider the problem of mass estimation for the dynamics

$$u = mw(t) + d(t)$$

with $w(t) = \sin(t)$, and $d(t)$ is interpreted as either disturbance or measurement noise. The true mass is assumed to be $m = 2$. When the disturbance $d(t) = 0$, the estimation results are shown in the left plot in Figure 8.21. It is seen that larger gain corresponds to faster convergence, as expected. When the disturbance is $d(t) = 0.5 \sin(20t)$, the estimation results are shown on the right plot in Figure 8.21. It is seen that larger estimation gain leads to larger estimation error.

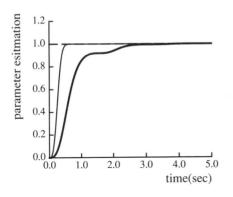 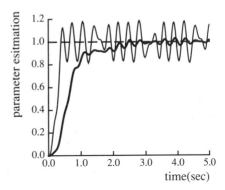

**Figure 8.21 :** gradient method, left: without noise, right: with noise

The following simple simulation illustrates the behavior of the gradient estimator in the presence of both parameter variation and measurement noise.

### Example 8.14: Gradient estimation of a time-varying mass

Now suppose that the true parameter is slowly time-varying, with $m(t) = 1 + 0.5\sin(0.5t)$. Let us take $p_o = 1$. The estimation results in the absence of disturbance is shown in the left plot in Figure 8.22. In the presence of the disturbance of the previous example, the parameter estimation result is shown in the right plot. It is seen that the gradient method works quite well in the presence of parameter variation and disturbance.                                    ☐

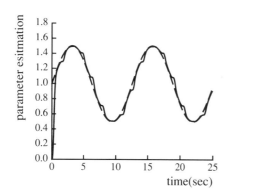 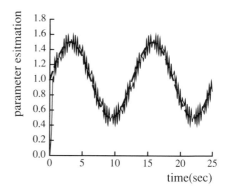

**Figure 8.22** : Time-varying gradient

## 8.7.4  The Standard Least-Squares Estimator

We all have some experience with least squares estimation (data fitting).  In this subsection, we shall formalize the technique and carefully study its properties.

**FORMULATION**

In the standard least-squares method, the estimate of the parameters is generated by minimizing the total prediction error

$$J = \int_o^t \| \mathbf{y}(r) - \mathbf{W}(r)\hat{\mathbf{a}}(t) \|^2 dr \tag{8.93}$$

with respect to $\hat{\mathbf{a}}(t)$. Since this implies the fitting of all past data, this estimate potentially has the advantage of averaging out the effects of measurement noise.  The estimated parameter $\hat{\mathbf{a}}$ satisfies

$$[\int_o^t \mathbf{W}^T \mathbf{W}\, dr\,] \, \hat{\mathbf{a}}(t) = \int_o^t \mathbf{W}^T \mathbf{y}\, dr \tag{8.94}$$

Define

$$\mathbf{P}(t) = [\,\int_o^t \mathbf{W}^T(r)\,\mathbf{W}(r)\, dr\,]^{-1} \tag{8.95}$$

To achieve computational efficiency, it is desirable to compute $\mathbf{P}$ recursively, instead of evaluating the integral at every time instant.  This amounts to replacing the above equation by the differential equation

$$\frac{d}{dt}[\mathbf{P}^{-1}(t)\,] = \mathbf{W}^T(t)\,\mathbf{W}(t) \tag{8.96}$$

Differentiating (8.94) and using (8.95) and (8.96), we find that the parameter update satisfies

$$\dot{\hat{\mathbf{a}}} = -\mathbf{P}(t)\mathbf{W}^T\mathbf{e}_1 \tag{8.97}$$

with $\mathbf{P}(t)$ being called the estimator gain matrix, similarly to the case of gradient estimation. In the implementation of the estimator, it is desirable to update the gain $\mathbf{P}$ directly, rather than using (8.96) and then inverting the matrix $\mathbf{P}^{-1}$. By using the identity

$$\frac{d}{dt}[\mathbf{P}\mathbf{P}^{-1}] = \dot{\mathbf{P}}\mathbf{P}^{-1} + \mathbf{P}\frac{d}{dt}[\mathbf{P}^{-1}] = 0$$

we obtain

$$\dot{\mathbf{P}} = -\mathbf{P}\mathbf{W}^T\mathbf{W}\mathbf{P} \tag{8.98}$$

In using (8.97) and (8.98) for on-line estimation, we have to provide an initial parameter value and an initial gain value. But there is a difficulty with this initialization, because (8.97) and (8.98) imply that $\mathbf{P}$ should be infinity, while $\hat{\mathbf{a}}$ is initially undefined. To avoid this problem, we shall provide finite values to initialize $\mathbf{P}$ and $\hat{\mathbf{a}}$. Clearly, one should use the best guess to initialize the $\hat{\mathbf{a}}$. The choice of the initial gain $\mathbf{P}(0)$ should be chosen as high as allowed by the noise sensitivity. $\mathbf{P}(0)$ can be chosen to be diagonal, for simplicity.

It is useful to remark that the above least-squares estimator can be interpreted in a Kalman filter framework, with $\mathbf{a}$ being the state to be estimated and $\mathbf{P}$ being the estimation covariance matrix. Based on this perspective, the initial gain $\mathbf{P}(0)$ should be chosen to represent the covariance of the initial parameter estimates $\hat{\mathbf{a}}(0)$.

## PARAMETER CONVERGENCE

The convergence property of the estimator can be best understood by solving the differential equations (8.96) and (8.97), assuming the absence of noise and parameter variation. From (8.96), (8.97) and (8.98), one easily shows that

$$\mathbf{P}^{-1}(t) = \mathbf{P}^{-1}(0) + \int_o^t \mathbf{W}^T(r)\mathbf{W}(r)\,dr \tag{8.99}$$

$$\frac{d}{dt}[\mathbf{P}^{-1}(t)\tilde{\mathbf{a}}(t)] = \mathbf{0}$$

Thus,

$$\tilde{\mathbf{a}}(t) = \mathbf{P}(t)\,\mathbf{P}^{-1}(0)\,\tilde{\mathbf{a}}(0) \tag{8.100}$$

If **W** is such that

$$\lambda_{min}\{\int_o^t \mathbf{W}^T \mathbf{W}\, dr\} \;\to\; \infty \qquad \text{as} \qquad t \to \infty \tag{8.101}$$

where $\lambda_{min}[\cdot]$ denotes the smallest eigenvalue of its argument, then the gain matrix converges to zero, and the estimated parameters asymptotically (but usually not exponentially) converge to the true parameters. Note that the "infinite-integral" condition (8.101) is a weaker condition than the persistent excitation (8.91). Indeed, for any positive integer $k$,

$$\int_o^{k\delta+\delta} \mathbf{W}^T \mathbf{W}\, dr = \sum_{i=0}^{k} \int_{i\delta}^{i\delta+\delta} \mathbf{W}^T \mathbf{W}\, dr \geq k\alpha_1 \mathbf{I} \tag{8.102}$$

Thus, if **W** is persistently exciting, (8.102) is satisfied, $\mathbf{P} \to \mathbf{0}$ and $\widetilde{\mathbf{a}} \to \mathbf{0}$.

The effects of the initial gain and initial parameter estimates can be easily seen from (8.100) and (8.99). Obviously, a small initial parameter error $\widetilde{\mathbf{a}}(0)$ results in a small parameter error all the time. A large initial gain $\mathbf{P}(0)$ also leads to a small parameter error. This is particularly clear if we choose $\mathbf{P}(0) = p_o\mathbf{I}$, which leads to

$$\widetilde{\mathbf{a}}(t) = [\mathbf{I} + p_o\int_o^t \mathbf{W}^T(r)\mathbf{W}(r)dr]^{-1}\widetilde{\mathbf{a}}(0)$$

## ROBUSTNESS

Roughly speaking, the least-squares method has good robustness with respect to noise and disturbance, but poor ability in tracking time-varying parameters. The reason for the good noise-rejection property is easy to understand: noise, particularly high-frequency noise, is averaged out. The estimator's inability in tracking time-varying parameters can also be understood intuitively, from two points of views. From a mathematical point of view, $\mathbf{P}(t)$ converges to zero when **W** is persistently exciting, *i.e.*, the parameter update is essentially shut off after some time, and the changing parameters cannot be estimated any more. From an information point of view, the least-square estimate attempts to fit *all* the data up to the current time, while, in reality, the old data is generated by old parameters.

The following example illustrates the behavior of the least-squares estimator in the presence of measurement noise.

**Example 8.15: Least-squares estimation of one parameter**

Consider again the estimation of the mass system, now with a least-square method. The parameter estimation results in the absence of disturbance, are shown in the left plot of Figure 8.23, with the two curves corresponding to the initial gain $p_o = 2$ and $p_o = 10$, respectively. It is seen that the parameter converges faster with larger initial gain. Another feature is that least square converge fast initially, but slowly afterward.

The parameter estimation results in the presence of disturbance are presented in the right plot of Figure 8.23. The same initial gain values are used. It is seen that the estimated parameters are much smoother than those from the gradient method. □

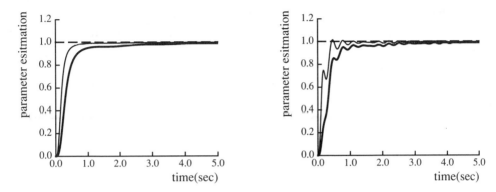

**Figure 8.23** : Least-squares method; left: without noise ; right: with noise

The inability of the least-squares method to estimate time-varying parameters is demonstrated in the following example.

**Example 8.16: Least-squares estimation of time-varying parameter**

Now the true parameter $m = 1$ is replaced by $m(t) = 1 + 0.5\sin(0.5\,t)$. The initial gain value is chosen to be $P(0) = 1$. The true and estimated parameters in the absence of disturbance are shown in the left plot of Figure 8.24. Those in the presence of disturbance are shown in the right plot. It is seen that the estimated parameters cannot follow the true parameter variation, regardless of the disturbance. □

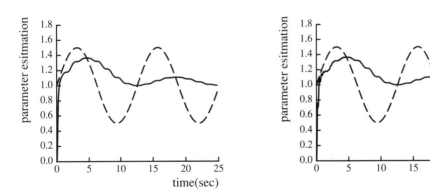

**Figure 8.24** : Least-squares time-varying gradient

## 8.7.5 Least-Squares With Exponential Forgetting

Exponential forgetting of data is a very useful technique in dealing with time-varying parameters. Its intuitive motivation is that past data are generated by past parameters and thus should be discounted when being used for the estimation of the current parameters. In this subsection, we describe the general formulation of least square method with time-varying forgetting factor. In the next subsection, a desirable form of forgetting factor variation will be discussed.

If exponential forgetting of data is incorporated into least-square estimation, one minimizes

$$J = \int_o^t \exp[-\int_s^t \lambda(r)\,dr] \| \mathbf{y}(s) - \mathbf{W}(s)\,\hat{\mathbf{a}}(t)\|^2 ds$$

instead of (8.93), where $\lambda(t) \geq 0$ is the time-varying forgetting factor. Note that the exponential term in the integral represents the weighting for the data. One easily shows that the parameter update law is still of the same form,

$$\dot{\hat{\mathbf{a}}} = -\,\mathbf{P}(t)\,\mathbf{W}^T \mathbf{e}_1 \qquad (8.103)$$

but that the gain update law is now

$$\frac{d}{dt}[\mathbf{P}^{-1}] = -\lambda(t)\mathbf{P}^{-1} + \mathbf{W}^T(t)\mathbf{W}(t) \qquad (8.104)$$

In implementation, it is more efficient to use the following form of the gain update

$$\frac{d}{dt}[\mathbf{P}] = \lambda(t)\mathbf{P} - \mathbf{P}\mathbf{W}^T(t)\mathbf{W}(t)\mathbf{P} \qquad (8.105)$$

To understand the convergence properties of the estimator, let us again solve for the gain and parameter errors explicitly. The estimator gain can be explicitly solved from (8.104) to be

$$\mathbf{P}^{-1}(t) = \mathbf{P}^{-1}(0)\exp[-\int_o^t \lambda(r)\,dr] + \int_o^t \exp[-\int_r^t \lambda(v)\,dv]\mathbf{W}^T(r)\mathbf{W}(r)\,dr \quad (8.106)$$

To solve the explicit form of the parameter error, let us note that one can easily obtain

$$\frac{d}{dt}[\mathbf{P}^{-1}\tilde{\mathbf{a}}] = -\lambda\mathbf{P}^{-1}\tilde{\mathbf{a}} \qquad (8.107)$$

Therefore,

$$\tilde{\mathbf{a}}(t) = \exp[-\int_o^t \lambda(r)\,dr]\,\mathbf{P}(t)\,\mathbf{P}^{-1}(0)\,\tilde{\mathbf{a}}(0) \qquad (8.108)$$

That is,

$$\tilde{\mathbf{a}}(t) = [\mathbf{P}^{-1}(0) + \int_o^t \exp(\int_o^r \lambda(v)\,dv)\,\mathbf{W}^T(r)\,\mathbf{W}(r)\,dr]^{-1}\mathbf{P}^{-1}(0)\,\tilde{\mathbf{a}}(0)$$

Comparing this and (8.100), and noting that

$$\exp[\int_o^r \lambda(v)\,dv] \geq 1$$

one sees that exponential forgetting always improves parameter convergence over standard least-squares. It also shows that the "infinite integral" condition (8.101) for standard least-squares still guarantees the asymptotic convergence of the estimated parameters. This also implies that persistent excitation of the signal $\mathbf{W}$ can guarantee the convergence of the estimated parameters.

It is easy to see that exponential forgetting leads to exponential convergence of the estimated parameters, provided that $\lambda(t)$ is chosen to be larger than or equal to a positive constant and that signals are *p.e.* Specifically, assume that $\lambda(t) \geq \lambda_o$, with $\lambda_o$ being a positive constant. Then, we have from (8.104)

$$\frac{d}{dt}[\mathbf{P}^{-1}] = -\lambda_o \mathbf{P}^{-1} + \mathbf{W}^T\mathbf{W} + (\lambda(t) - \lambda_o)\mathbf{P}^{-1}$$

$$\mathbf{P}^{-1}(t) = \mathbf{P}^{-1}(0)e^{-\lambda_o t} + \int_o^t e^{-\lambda_o(t-r)}\{\mathbf{W}^T\mathbf{W} + (\lambda(t) - \lambda_o)\mathbf{P}^{-1}\}\,dr$$

This guarantees from (8.92) that $\mathbf{P}^{-1}(t) \geq e^{-\lambda_o \delta}\alpha_1 \mathbf{I}$ and, accordingly, that

$$\mathbf{P}(t) \leq \frac{e^{\lambda_o \delta}}{\alpha_1}\mathbf{I}$$

for $t \geq \delta$. This and (8.108) show the exponential convergence of $\tilde{\mathbf{a}}(t)$ to zero with a rate of at least $\lambda_o$.

It is interesting to remark that, if the forgetting factor is constant, then the exponential convergence rate of the estimated parameters is the same as the forgetting factor. However, a constant forgetting factor may lead to diminishing magnitude in certain directions of $\mathbf{P}^{-1}$ (and accordingly, unbounded magnitude in certain directions of $\mathbf{P}$) in the absence of *p.e.*, due to the exponential decaying components in (8.106). Unboundedness (or even large magnitude) of the gain matrix is undesirable, since it implies that the disturbance and noise in the prediction error may, through the update law (8.103), lead to violent oscillations of the estimated parameters.

The following one-parameter example demonstrates the desirable and undesirable properties for exponential forgetting with a constant forgetting factor.

**Example 8.17 : Estimating one parameter with a constant forgetting factor**

Consider the estimation of parameter $a$ from the equation $y = w a$ using the least-squares method with a constant forgetting factor. The cost to be minimized is

$$J = \int_0^t e^{-\lambda_o (t-r)} \, [\, y(r) - w(r) \hat{a}(t) \,]^2 \, dr$$

The parameter update law is

$$\dot{\hat{a}}(t) = - P(t) w(t) e(t)$$

and the gain update law is

$$\frac{d}{dt} [P^{-1}] = - \lambda_o P^{-1} + w w$$

The gain update is easily solved to be

$$P^{-1}(t) = P^{-1}(0) e^{-\lambda_o t} + \int_0^t e^{-\lambda_o (t-r)} w^2 dr$$

and the parameter error is

$$\tilde{a}(t) = e^{-\lambda_o t} P(t) P^{-1}(0) \tilde{a}(0)$$

If the signal $w$ is persistently exciting, $P(t)$ will be upper bounded and $\tilde{a}$ will converge to zero exponentially with a rate $\lambda_o$. If $w$ is not persistently exciting (for example, for $w = e^{-t}$, $P^{-1}(t)$ may converge to zero and thus $P(t)$ to infinity).  ☐

Thus, we see that caution has to be exercised in choosing the forgetting factor. A zero forgetting factor leads to vanishing gain (standard least-squares and thus inability of tracking time-varying parameters) in the presence of persistent excitation, while a constant positive factor leads to exploding gain in the absence of persistent excitation. Since an estimator may encounter signals with different levels of persistent excitation, an automatic tuning method for the forgetting factor is necessary.

## 8.7.6 Bounded-Gain Forgetting

To keep the benefits of data forgetting (parameter tracking ability) while avoiding the possibility of gain unboundedness, it is desirable to tune the forgetting factor variation so that data forgetting is activated when **w** is persistently exciting, and suspended when **w** is not. We now discuss such a tuning technique and the resulting estimator convergence and robustness properties.

## A FORGETTING FACTOR TUNING TECHNIQUE

Since the magnitude of the gain matrix **P** is an indicator of the excitation level of **W**, it is reasonable to correlate the forgetting factor variation with $\|\mathbf{P}(t)\|$. A specific technique for achieving this purpose is to choose

$$\lambda(t) = \lambda_o \left( 1 - \frac{\|\mathbf{P}\|}{k_o} \right) \tag{8.109}$$

with $\lambda_o$ and $k_o$ being positive constants representing the maximum forgetting rate and prespecified bound for gain matrix magnitude, respectively. The forgetting factor in (8.109) implies forgetting the data with a factor $\lambda_o$ if the norm of **P** is small (indicating strong *p.e.*), reducing the forgetting speed if the norm of **P** becomes larger and suspends forgetting if the norm reaches the specified upper bound. Since a larger value of $\lambda_o$ means faster forgetting (which implies both stronger ability in following parameter variations, but also more oscillations in the estimated parameters due to shorter-time "averaging" of noisy data points), the choice of $\lambda_o$ represents a tradeoff between the speed of parameter tracking and the level of estimated parameter oscillation. The gain bound $k_o$ affects the speed of parameter update and also the effects of disturbance in the prediction error, thus involving a similar tradeoff. To be consistent with our gain-bounding intention, we choose $\|\mathbf{P}(0)\| \leq k_o$ (hence $\mathbf{P}(0) \leq k_o \mathbf{I}$). We shall refer to the least-squares estimator with the forgetting factor (8.109) as the bounded-gain-forgetting (BGF) estimator, because the norm of the gain matrix can be shown to be upper bounded by the pre-specified constant $k_o$ regardless of persistent excitation.

## CONVERGENCE PROPERTIES

We now show that the form (8.109) of forgetting factor variation guarantees that the resulting gain matrix **P**(t) is upper bounded regardless of the persistent excitation of **W**, unlike the case of constant forgetting factor. With the forgetting factor form (8.109), the gain update equation (8.104) can be expressed as

$$\frac{d}{dt}[\mathbf{P}^{-1}] = -\lambda_o \mathbf{P}^{-1} + (\lambda_o/k_o)\|\mathbf{P}\|\mathbf{P}^{-1} + \mathbf{W}^T\mathbf{W} \tag{8.110}$$

This leads to

$$\mathbf{P}^{-1}(t) = \mathbf{P}^{-1}(0)e^{-\lambda_o t} + \int_o^t e^{-\lambda_o(t-r)}[\frac{\lambda_o}{k_o}\|\mathbf{P}\|\mathbf{P}^{-1} + \mathbf{W}^T\mathbf{W}]\, dr$$

$$\geq (\mathbf{P}^{-1}(0) - k_o^{-1}\mathbf{I})e^{-\lambda_o t} + (1/k_o)\mathbf{I} + \int_o^t e^{-\lambda_o(t-r)}\mathbf{W}^T\mathbf{W}\, dr \qquad (8.111)$$

where we used the inequality $\|\mathbf{P}(t)\|\,\|\mathbf{P}^{-1}(t)\| \geq \mathbf{I}$, obtained from the fact that

$$\|\mathbf{P}\|\mathbf{P}^{-1} - \mathbf{I} = \mathbf{P}^{-1/2}[\|\mathbf{P}\|\mathbf{I} - \mathbf{P}]\mathbf{P}^{-1/2} \geq \mathbf{0}$$

Note that $\|\mathbf{P}(0)\| \leq k_o$ guarantees the positive-definiteness of $(\mathbf{P}^{-1}(0) - k_o^{-1}\mathbf{I})$, and therefore for all $t \geq 0$,

$$\mathbf{P}^{-1}(t) \geq \frac{1}{k_o}\mathbf{I}$$

so that $\mathbf{P}(t) \leq k_o\mathbf{I}$. Note that this implies, from (8.109), that

$$\lambda(t) \geq 0$$

Thus, we have shown the boundedness of $\mathbf{P}$ and the non-negative nature of $\lambda(t)$.

If $\mathbf{W}(t)$ is *p.e.* as defined by (8.91), we can further show that $\lambda(t) \geq \lambda_1 > 0$, and, thus, the estimated parameters are exponentially convergent. To show this, note that, from (8.111) and (8.91),

$$\mathbf{P}^{-1}(t) \geq [\frac{1}{k_o} + e^{-\lambda_o\delta}\alpha_1]\mathbf{I}$$

$$\mathbf{P}(t) \leq \frac{k_o}{1 + k_o\alpha_1 e^{-\lambda_o\delta}}\mathbf{I} \qquad (8.112)$$

This, in turn, leads to the uniform lower boundedness of the forgetting factor by a positive constant,

$$\lambda(t) = \frac{\lambda_o}{k_o}(k_o - \|\mathbf{P}\|) \geq \frac{\lambda_o k_o\alpha_1 e^{-\lambda_o\delta}}{1 + k_o\alpha_1 e^{-\lambda_o\delta}} = \lambda_1 \qquad (8.113)$$

which in turn implies the exponential convergence of the estimated parameters. Note that, if $\mathbf{W}$ is strongly *p.e.*, *i.e.*, $\alpha_1$ is very large, $\lambda(t) \approx \lambda_o$.

Under *p.e.*, one can also show that $\mathbf{P}(t)$ is uniformly lower bounded by a constant *p.d.* matrix, a property which is desirable for estimating time-varying parameters. Indeed, from (8.106) and (8.113),

$$\mathbf{P}^{-1}(t) \le \mathbf{P}^{-1}(0) + \int_{o}^{t} \exp[-\lambda_1(t - \tau)] \, \mathbf{W}^T \mathbf{W} \, dr$$

The second term on the right-hand side can be regarded as the output of the stable filter

$$\dot{\mathbf{M}} + \lambda_1 \mathbf{M} = \mathbf{W}^T \mathbf{W} \tag{8.114}$$

$\mathbf{M}$ is bounded if $\mathbf{W}$ is bounded. Thus, from (8.114) and (8.112), if $\mathbf{W}$ is *p.e.* and upper bounded, $\mathbf{P}$ will be upper bounded and lower bounded uniformly, *i.e.*,

$$k_2 \mathbf{I} \le \mathbf{P}(t) \le k_1 \mathbf{I}$$

where $0 < k_2 < k_1 < k_o$.

The properties of the BGF estimator are summarized below:

**Theorem 8.1** *In the bounded-gain-forgetting estimator, the parameter errors and the gain matrix are always upper bounded; If $\mathbf{W}$ is persistently exciting, then the estimated parameters converge exponentially and $\mathbf{P}(t)$ is upper and lower bounded uniformly by positive definite matrices.*

The advantage of the BGF method over the gradient method is that the estimated parameters are smooth. This implies, given the allowable level of estimated parameter oscillation, that much larger gain bound can be used in BGF method. As a result, faster parameter convergence can be achieved.

## AN ALTERNATIVE TECHNIQUE OF DATA FORGETTING

It is interesting to note that the bounded-gain forgetting is not the only way to keep data forgetting while avoiding gain explosion. In view of the fact that the problem with the non-zero forgetting factor is the possibility of $\mathbf{P}^{-1}$ diminishing, we may, as an alternative to variable-forgetting approach, use the following simple gain update modification

$$\frac{d}{dt} \mathbf{P}^{-1} = -\lambda(t)(\mathbf{P}^{-1} - \mathbf{K}_o^{-1}) + \mathbf{W}^T \mathbf{W} \tag{8.115}$$

where $\mathbf{K}_o$ is a symmetric *p.d.* matrix specifying the upper bound of the gain matrix $\mathbf{P}$, and $\lambda(t)$ is a constant or time-varying positive forgetting factor independent of $\mathbf{P}$. $\lambda(t)$ is chosen to satisfy $\lambda_o \le \lambda(t) \le \lambda_1$ with $\lambda_o$ and $\lambda_1$ denoting two positive constants. The parameter update law is still (8.103). One can easily show that this gain update leads to similar properties as the bounded-gain forgetting.

## ROBUSTNESS PROPERTIES

Let us now look at the behavior of the estimator in the presence of parameter variations and disturbances. Consider again the single-parameter estimation problem of Example 8.12. The gain update is

$$\frac{d}{dt} P^{-1}(t) = -\lambda(t)P^{-1} + w^2$$

One notes that the presence of noise and parameter variation does not affect the gain value.

To see how the parameter error behaves, let us note that

$$\frac{d}{dt}[P^{-1}\tilde{a}] + \lambda(t)P^{-1}\tilde{a} = -P^{-1}\dot{a} + wd \tag{8.116}$$

This describes a first-order filter of "bandwidth" $\lambda(t)$ with $(-P^{-1}\dot{a} + wd)$ as input and $P^{-1}\tilde{a}$ as output. We can easily draw some conclusions about the estimator behavior based on this relation, assuming that the signal $w$ is persistently exciting, the disturbance $d$ is bounded, and the rate of parameter variation is bounded. Since $P^{-1}$ has been shown to be upper bounded, the input to the filter is bounded. Furthermore, since the forgetting factor $\lambda(t)$ has been shown to be lower bounded, equation (8.116) represents an exponentially stable filter and its output $P^{-1}\tilde{a}$ is guaranteed to be bounded. The upper boundedness of $P$ further indicates the boundedness of $\tilde{a}$. Note that the bound of $\tilde{a}$ will depend on the magnitude of the disturbance and the parameter variation rate $\dot{a}$. It also depends on the persistent excitation level of the signal $w(t)$ (which affects $\lambda(t)$).

If the signal $w$ is not persistently exciting, then the estimator can no longer guarantee the boundedness of parameter errors in the presence of parameter variation and disturbances. We can only say that $P(t)$ (being independent to the additive disturbances and parameter variations) is bounded and the parameter errors cannot diverge too fast.

**Example 8.18: BGF estimation of the single parameter**

The BGF estimator is used to estimate the time-varying parameter. In the absence of measurement noise, the estimated and true parameters are shown in the left plot of Figure 8.25. In the presence of measurement noise, they are shown in the right-hand side plot of Figure 8.25.    ☐

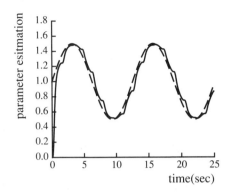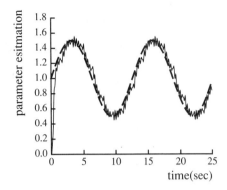

**Figure 8.25 :** BGF least-squares estimation; left: without noise; right: with noise

**USE OF DEAD-ZONE**

Similarly to the MRAC case, one should use a small dead-zone in estimating parameters by the previous methods. It is particularly important when the signal $W(t)$ is not persistently exciting, leading to the avoidance of parameter drift. The dead-zone should now be placed on the prediction error. The size of the dead-zone should be chosen to be larger than the range of noise and disturbance in the prediction error. The motivation is still that small error signals are dominated by noise and cannot be used for reliable estimation.

## 8.7.7  Concluding Remarks and Implementation Issues

We have discussed a number of estimators in this section. Persistent excitation is essential for good estimation. Most estimators do not have both fast convergence and good robustness. The gradient estimator is simple but has slow convergence. The standard least-squares estimator is robust to noise but cannot estimate time-varying parameters. Least-squares estimation with exponential forgetting factor has the ability to track time-varying parameters but there is a possibility of gain windup in the absence of persistent excitation.

A particular technique of gain tuning, called bounded-gain-forgetting, is developed to maintain the benefits of data forgetting while avoiding gain-windup. The resulting estimator has both fast convergence and good noise sensitivity. The computation and analysis of this estimator is reasonably simple.

In order to have good estimation performance, many implementation issues have to be considered carefully, including

- choice of the bandwidth of the filter for generating (8.77)

- choice of initial parameter and initial gain matrix

- choice of forgetting rate and gain bound

- choice of excitation signals

Tradeoffs and judgement have to be used in making the above choices. The filter bandwidth should be chosen to be larger than the plant bandwidth, so that the system signals are able to pass through. But it should be smaller than frequency range (usually high frequency) of the noise. Of course, the initial parameter estimates should be chosen to be as accurate as possible. The initial gain matrix should be chosen to obtain the proper convergence speed and noise robustness. The forgetting factor should be chosen to be large enough so that parameter variation can be tracked sufficiently accurately. However, it cannot be chosen too large lest the gain matrix is too large or too oscillatory. The gain bound is chosen based on the knowledge of the noise magnitude and the allowable level of estimated parameter oscillation, with larger noise leading to smaller gain bound. The excitation signals should contain sufficient spectrum lines to allow parameter convergence, and their frequencies should be within the bandwidth of the plant so as to be able to excite the plant. Note that, although unknown parameters can be time-varying, the speed of the variation should be much smaller than the plant bandwidth, otherwise the parameter dynamics should be modeled.

# 8.8  Composite Adaptation

In the MRAC controllers developed in sections 8.2-8.4, the adaptation laws extract information about the parameters from the tracking errors $e$. However, the tracking error is not the only source of parameter information. The prediction error $e_1$ also contains parameter information, as reflected in the parameter estimation schemes of section 8.7. This section examines whether the different sources of parameter information can be combined for parameter adaptation, and whether such a combined use of information sources can indeed improve the performance of the adaptive controller.

These questions are answered positively by a new adaptation method, called composite adaptation, which drives the parameter adaptation using both tracking error and prediction error. As we shall see, such an adaptation scheme not only maintains the global stability of the adaptive control system, but also leads to fast parameter convergence and smaller tracking errors. Indeed, the fundamental feature of composite adaptation is its ability to achieve quick adaptation.

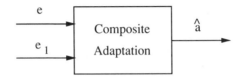

**Figure 8.26 :** Composite adaptation: $e$ - tracking error　$e_1$ - prediction error

In the following, we first describe the concept of composite adaptation using the simple mass control example, and then extend the results to more general linear and nonlinear systems.

## COMPOSITE ADAPTATION FOR A MASS

Although the benefits of composite adaptation are most obvious in adaptive control problems involving multiple parameters, its mechanism is most easily explained first on the simple example of controlling a mass. Consider the system $m\ddot{x} = u$ . A MRAC controller has been designed in Example 8.1 for this system, with the control law being

$$u = \hat{m} v \tag{8.117}$$

and the adaptation law being

$$\dot{\hat{m}} = -\gamma v s \tag{8.118}$$

If, instead, a prediction-error based estimator is used to estimate the parameter $m$, then the parameter update law is

$$\dot{\hat{m}} = -\gamma w e_1 \tag{8.119}$$

Note that while the adaptation law (8.118) is driven by the tracking error $s$, the estimation law (8.119) is driven by $e_1$.

In the so-called composite adaptive control, the control is the same as (8.117), but the adaptation law is now a "combination" of the earlier adaptation law and the prediction-error based estimation law

$$\dot{\hat{m}} = -P[vs + we_1]$$

Note that this adaptation law now is driven by both $s$ and $e_1$ . Also note that the gain $P(t)$ is provided by any of the gain update laws in section 8.7.

To show the stability of the resulting adaptive control system, let us still use the Lyapunov function candidate

$$V = \frac{1}{2}[\, ms^2 + P^{-1}\widetilde{m}^2 \,]$$

This is essentially the same as the $V$ in (8.7) except for the replacement of $\gamma$ by $P(t)$.

If $P(t)$ is chosen to be a constant, the derivative of $V$ is easily found to be

$$\dot{V} = -\lambda ms^2 - w^2\widetilde{m}^2$$

Note that the expression of $\dot{V}$ now contains a desirable additional quadratic (only semi-negative) term in $\widetilde{m}$. Using a similar reasoning as that in Example 8.1, one can now show that *both $s \to 0$ and $e_1 \to 0$ as $t \to \infty$*.

If $P$ is time-varying and generated as in BGF estimator, *i.e.*, using equation (8.104), we obtain

$$\dot{V} = -\lambda ms^2 - \frac{1}{2}w^2\widetilde{m}^2 - \frac{1}{2}\lambda(t)P^{-1}\widetilde{m}^2$$

This can also be shown to lead to $s \to 0$ and $e \to 0$ as $t \to \infty$. Furthermore, if $w$ is persistently exciting, one can easily show that both $s$ and $\widetilde{m}$ are exponentially convergent to zero, *i.e.*, the adaptive controller has exponential convergence.

## COMPOSITE ADAPTIVE CONTROL OF FIRST-ORDER SYSTEMS

The advantages of composite adaptation are most clearly seen in systems with more than one unknown parameters. In such cases, the composite adaptation law allows high adaptation gain to be used without getting the oscillatory behavior and slow convergence observed for the tracking-error-based adaptation laws (see section 8.2). Let us use the first-order system in section 8.2 to illustrate the performance features of composite adaptive controllers.

### Tracking-Error Based Adaptive Control

Using the MRAC controller in section 8.2, the control law is of the form

$$u = \mathbf{v}^T \mathbf{a} \qquad\qquad (8.120)$$

The parameter adaptation law is

$$\dot{\hat{\mathbf{a}}} = -\,\mathrm{sgn}(k_p)\gamma\,\mathbf{v}e \qquad\qquad (8.121)$$

with

$$\mathbf{a} = [a_r \ \ a_y]^T \qquad\qquad \mathbf{v} = [r \ \ y]^T$$

**Prediction-Error Based Estimation**

The estimation methods of section 8.7 can also be used to estimate $a_r$ and $a_y$. But before doing this we have to parameterize the plant in terms of these parameters. By adding $a_m y$ to both sides of (8.20), we obtain

$$\dot{y} + a_m y = -(a_p - a_m)y + b_p u$$

This leads to

$$u = \frac{1}{b_p}(p + a_m)y + \frac{a_p - a_m}{b_p}y$$

Since the ideal controller parameters and the plant parameters are related by (8.24), we obtain

$$u = a_r \frac{(p + a_m)y}{b_m} + a_y y$$

Multiplying both sides by $1/(p + a_m)$, we then obtain

$$\frac{u}{p + a_m} = a_r \frac{y}{b_m} + a_y \frac{y}{p + a_m}$$

This can be compactly expressed in the linear parametrization form

$$u_1 = \mathbf{w}\,\mathbf{a}$$

with

$$u_1 = \frac{u}{p + a_m} \qquad \mathbf{a} = [\, a_r \quad a_y \,]^T \qquad \mathbf{w} = [\, \frac{y}{b_m} \quad \frac{y}{p + a_m} \,]$$

Therefore, we can straightforwardly use the estimation algorithms of section 8.7 to obtain the following estimation laws

$$\dot{\hat{\mathbf{a}}} = -\mathbf{P}\mathbf{w}e_1 \tag{8.122}$$

where $e_1$ is the prediction error on $u_1$.

**Composite Adaptive Control**

The composite adaptation laws can be easily constructed, from (8.121) and (8.122), as

$$\dot{\hat{\mathbf{a}}} = -\mathbf{P}[\,\mathrm{sgn}(b_p)\mathbf{v}e + \alpha(t)\mathbf{w}e_1\,] \tag{8.123}$$

The controller is the same as in (8.120).

To analyze the stability and convergence of the composite adaptive controller, one can still use the Lyapunov function in (8.29). If **P** is chosen to be constant, one has

$$\dot{V} = -a_m e^2 - e_1^2$$

One can easily show that $e \to 0$ and $e_1 \to 0$ as $t \to \infty$.

If **P** is updated by the least-squares type algorithms, one can obtain similar results. For example, if **P** is updated by (8.104), we can use the Lyapunov function candidate

$$V = \frac{1}{2}[\,e^2 + \tilde{\mathbf{a}}^T \mathbf{P}^{-1} \tilde{\mathbf{a}}\,]$$

whose derivative is

$$\dot{V} = -a_m e^2 - \frac{1}{2} e_1^2 - \frac{1}{2}\lambda(t)\tilde{\mathbf{a}}^T \mathbf{P}^{-1}\tilde{\mathbf{a}}$$

From this, we can infer asymptotic and exponential convergence results, similarly to the previous mass-control case.

### Simulation Results

Simulations are used to illustrate the behavior of the composite adaptive controller.

**Example 8.19: Composite adaptive control of first-order systems**

The same first-order system as in section 8.2 is used for composite adaptive control. Everything is as in section 8.2, except that now the prediction error term is incorporated into the adaptation law. For the case of $r(t) = 4$, the tracking and estimation results are shown in Figure 8.27. The results for $r(t) = 4\sin 3t$ are shown in Figure 8.28. It is seen the parameter and tracking results are both smooth. One can further increase the adaptation gain to obtain smaller tracking errors without incurring much oscillation in the estimated parameters. $\square$

The advantage of the composite adaptive controller essentially comes from the smoothness of the results. This has significant implications on the adaptive control performance. Due to the possibility of using high adaptation gain, we can obtain smaller tracking error, faster parameter convergence without exciting high-frequency unmodeled dynamics. In fact, simulations show that composite adaptive controllers perform much better than standard adaptive controllers when unmodeled dynamics are present.

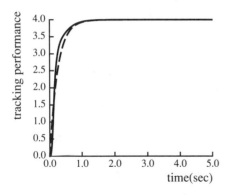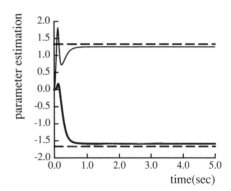

**Figure 8.27 :** Tracking performance and parameter estimation
$r(t) = 4$, composite adaptation

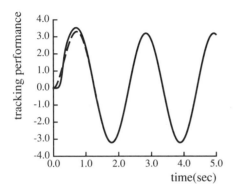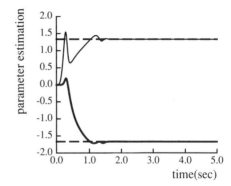

**Figure 8.28 :** Tracking performance and parameter estimation
$r(t) = 4 \sin 3t$, composite adaptation

### Interpretation of Composite Adaptation

To gain an intuitive interpretation of the composite adaptation law, let us for simplicity choose $\mathbf{P}(t) = \gamma \mathbf{I}$. Using (8.123) and (8.89), the composite adaptation law can be expressed as

$$\dot{\tilde{\mathbf{a}}} + \gamma \mathbf{w}^T \mathbf{w} \tilde{\mathbf{a}} = - \gamma \mathbf{v}^T e \tag{8.125}$$

Without the prediction term, we obtain the original MRAC adaptation law (8.28)

$$\dot{\tilde{\mathbf{a}}} = - \gamma \mathbf{v}^T e \tag{8.126}$$

Comparing (8.125) and (8.126), one notes that (8.126) represents an integrator while (8.125) a time-varying low-pass filter. Both tend to attenuate the high-frequency components in $e$, and thus parameters in both cases tend to be smoother than the tracking errors. However, an integrator also attenuates low frequency components, as can be seen from its Bode plot. The stable filter in (8.125), on the other hand, has much less attenuation on the low frequency components in $e$ (for example, if $w$ is a constant, (8.125) is a stable filter with a bandwidth of $\gamma w^2$). Therefore, the parameter search in the composite adaptation goes along an "average" or "filtered" direction as specified by the low frequency components in the tracking error $e$. This explains the smoothness of the parameter estimates in composite adaptive control.

As shown in this section, the principle of composite adaptation is to combine a tracking-error based adaptation law and a prediction-error based parameter estimation law. This idea can be straightforwardly extended to adaptive control in general. Note that one must parameterize both the plant model and the controller using a common set of parameters in order to be able to combine the two types of errors for adaptation. In chapter 9, we shall discuss the application of this idea to the adaptive control of robot manipulators, an important class of multi-input multi-output nonlinear systems.

## 8.9  Summary

Adaptive control is an appealing approach for controlling uncertain dynamic systems. In principle, the systems can be uncertain in terms of its dynamic structure or its parameters. So far, however, adaptive control can only deal with parameter-uncertain systems. Furthermore, existing adaptation methods generally require linear parametrization of the control law or the system dynamics.

Systematic theories exist on how to design adaptive controllers for general linear systems. If the full state is available, adaptive control design and implementation is quite simple. If only output feedback is available, however, adaptive control design is much more involved because of the need to introduce dynamics into the controller. Some classes of nonlinear systems can also be adaptively controlled.

In MRAC systems, the adaptation law extracts parameter information from the tracking errors. In self-tuning controllers, the parameter estimator extracts information from prediction errors. In a new technique called composite adaptive control, the adaptation law extracts parameter information from both sources. This new adaptation method leads to faster adaptation without incurring significant oscillation in the estimated parameters, thus yielding improved performance.

# 8.10   Notes and References

Analyses and discussions on the material of this chapter can be found, *e.g.*, in the recent books [Narendra and Annasswamy, 1989], which provides a systematic treatment of model-reference adaptive control theory for linear systems, and [Aström and Wittenmark, 1989], which includes extensive studies of self-tuning algorithms as well as discussions of adaptive control implementation issues. The basic formulation of a Lyapunov formalism for model-reference adaptive control is due to [Parks, 1966]. Detailed discussions of discrete versions and self-tuning control can be found in [Goodwin and Sin, 1984]. Passivity interpretations of model reference adaptive controllers are discussed in [Landau, 1979]. Detailed studies of robustness can be found, *e.g.*, in [Ioannou and Kokotovic, 1983; Anderson, *et al.*, 1986; Sastry and Bodson, 1989] and references therein.

The theoretical framework in section 8.4 is based on [Narendra and Annasswamy, 1989], from which Lemmas 8.1 and 8.2 and Figures 8.14 and 8.17 are adapted. Example 8.7 and Figure 8.18 are based on [Rohrs, *et al.*, 1985].

The development of section 8.5 is adapted from [Slotine and Coetsee, 1986], which also details "robust" combinations with sliding control terms. Extensions and applications to specific classes of nonlinear MIMO systems (robot manipulators) are developed in [Slotine and Li, 1986, 1987], and will be further detailed in chapter 9. [Taylor, *et al.*, 1988, 1989] study extensions to general MIMO systems under certain matching conditions, and most interestingly a singular perturbation analysis of the effects of unmodeled dynamics. [Kanellakopoulos, *et al.*, 1989] develop adaptive nonlinear controllers under some less stringent, "extended" matching conditions. [Sastry and Isidori, 1989] consider output tracking under the two assumptions of uniformly exponentially stable zero-dynamics and of global Lipschitzness of the regressor. [Bastin and Campion, 1989; Pomet and Praly, 1989] study indirect approaches to adaptive control of nonlinear systems.

Discrete-time parameter estimation is extensively studied in [Ljung, 1987]. The development of section 8.7 is adapted from [Li and Slotine, 1987]. Related discussions can be found in [Middleton, *et al.*, 1988]. [Nicolo and Katende, 1982; Atkeson, *et al.*, 1985; Khoshla and Kanade, 1985] detail linear parametrizations for robot dynamics.

Our discussion of section 8.8 is based on [Slotine and Li, 1987d, 1989] (see also chapter 9). A closely related approach is presented in [Narendra and Duarte, 1988, 1989] using separate controller and regressor parametrizations.

# 8.11   Exercises

**8.1**   Draw the block diagram for the adaptive mass control system in Example 8.1. Discuss the intuitive reasonableness of the adaptation law.

**8.2** Simulate the adaptive control system for the nonlinear first order plant in section 8.2, but with the nonlinearity being $f(y) = \sin y$.

**8.3** Simulate the adaptive control system for the second order plant

$$y = \frac{p + b_p}{p^2 + a_{p1}p + a_{p2}}$$

with $a_{p1} = 0.1$, $a_{p2} = -4, b_p = 2$

**8.4** Consider a second-order linear system with a constant unknown disturbance $d$. Compare the structure of a P.I.D. controller with that of a P.D. controller with adaptive compensation of $d$.

**8.5** Carry out the detailed proof of tracking error convergence for general linear systems with relative degree 1 (section 8.4.1).

**8.6** Simulate the adaptive control system for the second order plant

$$y = \frac{b_p}{p^2 + a_{p1}p + a_{p2}}$$

with $a_{p1} = 0.1$, $a_{p2} = -4, b_p = 2$ for two cases. In the first case, the output $y$ and its derivative $\dot{y}$ are assumed to be available. In the second case, only $y$ is assumed to be available.

**8.7** For the system

$$\dot{x}_1 = \sin x_2 + \sqrt{t + 1}\ x_2$$

$$\dot{x}_2 = \alpha_1 x_1^4 \cos x_2 + \alpha_2 x_1 x_2 \sin x_2 + u$$

design an adaptive controller to track an arbitrary desired trajectory $x_{d1}(t)$. Assume that the state $[x_1\ \ x_2]^T$ is measured, that $x_{d1}(t), \dot{x}_{d1}(t), \ddot{x}_{d1}(t)$ are all known, and that $\alpha_1$ and $\alpha_2$ are unknown constants.

Write the full expression of the controller, as a function of the measured state $[x_1\ \ x_2]^T$. Check your design in simple simulations. (*Hint*: First feedback linearize the system by differentiating the first equation.)

**8.8** Consider systems of the form

$$h\, y^{(n)} + \sum_{i=1}^{n} [a_i + a_{iv}(t)] f_i(\mathbf{x}, t) = u$$

where the state vector $\mathbf{x} = [y\ \ \dot{y}\ \ .... \ y^{(n-1)}]^T$ is measured, the $f_i$ are known nonlinear functions of the state and time, $h$ and the $a_i$ are unknown constants (or very slowly varying "average" coefficients), with $h > 0$, and the time-varying quantities $a_{iv}(t)$ are unknown but of known bounds

(possibly state-dependent or time-varying)

$$|a_{iv}(t)| \le A_i(t)$$

Show that the ideas of chapter 7 and section 8.5 can be combined to yield a robust adaptive controller. To do so, essentially replace $s$ by $s_\Delta = s - \Phi \operatorname{sat}(s/\Phi)$ in the Lyapunov function of section 8.5, *i.e.*, use

$$V = h s_\Delta^2 + [\gamma_o^{-1} \tilde{h}^2 + \sum_{i=1}^{n} \gamma_i^{-1} \tilde{a}_i^2]$$

For simplicity, you may first choose $\Phi$ to be constant.

Show that, assuming that the noise level in the system is small, the boundary layer concept leads to a natural choice of adaptation dead-zone (*Hint*: Choose the sliding control terms and the boundary layer thickness $\Phi$ as if the *constant* parameters $a_i$ were *known*, and note that $s_\Delta = 0$ in the boundary layer).

Illustrate your design in simple simulations. (Adapted from [Slotine and Coetsee, 1986]).

**8.9**    For the nonlinear adaptive controller of section 8.5, show that if *a priori* bounds are known on certain parameters, tracking convergence is preserved by temporarily stopping adaptation on a given parameter estimate if the adaptation law would otherwise drive the estimate outside the known bounds.

**8.10**    Discuss intuitively why parameter estimation can be easier on unstable systems than on stable systems.

**8.11**    Derive a composite adaptive controller for the nonlinear first-order plant in section 8.2. Simulate its behavior using different gains.

# Chapter 9
# Control of Multi-Input Physical Systems

As seen in the previous chapters, feedback control is well understood for large classes of nonlinear systems with single inputs or uncoupled multiple inputs. For general multi-input nonlinear systems, however, feedback control and especially robustness issues are still research topics, the urgency of which has been rendered more acute by the recent development of machines with challenging nonlinear dynamics, such as robot manipulators, high-performance aircraft, or advanced underwater and space vehicles.

In this chapter, we show that the new, richer control problems posed by nonlinear systems may benefit from taking some distance from purely mathematical approaches and having a closer look at the physics of the systems. The discussion may be viewed as a renewed emphasis on the original motivation of Lyapunov theory, which, as we saw in chapter 3, evolved from physical considerations.

In the following, we first motivate the approach by considering a specific example, the trajectory control of robot manipulators. We then discuss generalizations to other classes of controller design problems.

392

# 9.1   Robotics as a Prototype

Robot manipulators are familiar examples of trajectory-controllable mechanical systems. However, their nonlinear dynamics present a challenging control problem, since traditional linear control approaches do not easily apply. For a while, the difficulty was mitigated by the fact that manipulators were highly geared, thereby strongly reducing the interactive dynamic effects between links (it can be easily shown that, in geared manipulators, the nonlinear or time-varying dynamic effects are divided either by the transmission ratio $r$ or by $r^2$, with typical values of $r$ being about 100). However, in recent years, as the need for achieving greater accuracy in tracking and force control made designs evolve towards "cleaner" approaches, such as gear-free direct-drive arms (featuring reduced friction and avoiding backlash altogether) or low-reduction cable-driven mechanisms, explicit account of the nonlinear dynamic effects became critical in order to exploit the full dynamic potential of the new high-performance manipulator arms.

Consider, for instance, a planar, two-link, articulated manipulator (Figure 9.1), whose position can be described by a 2-vector $\mathbf{q}$ of joint angles, and whose actuator inputs consist of a 2-vector $\boldsymbol{\tau}$ of torques applied at the manipulator joints. The dynamics of this simple manipulator is strongly nonlinear, and can be written in the general form

$$\mathbf{H}(\mathbf{q})\ddot{\mathbf{q}} + \mathbf{C}(\mathbf{q},\dot{\mathbf{q}})\dot{\mathbf{q}} + \mathbf{g}(\mathbf{q}) = \boldsymbol{\tau} \tag{9.1}$$

where $\mathbf{H}(\mathbf{q})$ is the 2×2 manipulator inertia matrix (which is symmetric positive definite), $\mathbf{C}(\mathbf{q},\dot{\mathbf{q}})\dot{\mathbf{q}}$ is a 2-vector of centripetal and Coriolis torques (with $\mathbf{C}(\mathbf{q},\dot{\mathbf{q}})$ a 2×2 matrix), and $\mathbf{g}(\mathbf{q})$ is the 2-vector of gravitational torques. The feedback control problem for such a system is to compute the required actuator inputs to perform desired tasks (*e.g.*, follow a desired trajectory), given the measured system state, namely the vector $\mathbf{q}$ of joint angles, and the vector $\dot{\mathbf{q}}$ of joint velocities.

Note that the inertia matrix $\mathbf{H}$ depends on the joint position $\mathbf{q}$, consistently with physical insight (for instance, as viewed from its "shoulder", a stretched arm has a larger inertia than a folded arm). The centripetal torques vary with the square of individual joint velocities, while the Coriolis torques vary with the products of velocities at two different joints.

Also, note that the kinetic energy of the manipulator is (similarly to $1/2\, m\, v^2$ for a one degree-of-freedom system)

$$T = \frac{1}{2}\dot{\mathbf{q}}^T \mathbf{H}(\mathbf{q})\dot{\mathbf{q}}$$

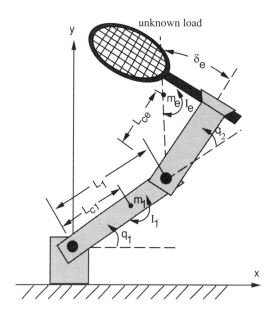

**Figure 9.1 :** An articulated two-link manipulator

This expression accounts for the positive definiteness of the inertia matrix $\mathbf{H}(\mathbf{q})$. Indeed, the kinetic energy must be strictly positive for any joint position $\mathbf{q}$ and *any* non-zero joint velocity $\dot{\mathbf{q}}$. More precisely, in later developments, we shall use the fact that $\mathbf{H}(\mathbf{q})$ is *uniformly* positive definite, *i.e.*, that there exists a constant $\alpha > 0$ such that, for all positions $\mathbf{q}$ in the robot's workspace,

$$\mathbf{H}(\mathbf{q}) \geq \alpha \mathbf{I}$$

where $\mathbf{I}$ is the identity matrix. This property can be seen by noticing that, if there did not exist such an $\alpha > 0$, then there would be a position $\mathbf{q}$ in the workspace where the inertia matrix has a zero eigenvalue, because the workspace is a closed set; letting $\mathbf{v}_o$ be the eigenvector associated with this zero eigenvalue, the robot arm could move with a unit velocity $(\mathbf{v}_o / \|\mathbf{v}_o\|)$ but with zero kinetic energy, an impossibility. Thus, $\mathbf{H}(\mathbf{q})$ is indeed uniformly positive definite.

## 9.1.1 Position Control

Let us assume that the manipulator of Figure 9.1 is in the horizontal plane $(\mathbf{g}(\mathbf{q}) \equiv \mathbf{0})$, and that the task is simply to move it to a given final position, as specified by a constant vector $\mathbf{q}_d$ of desired joint angles. It is physically clear (as already mentioned in chapter 3) that a joint proportional-derivative (P.D.) controller, namely a feedback control law that selects each actuator input independently, based on the local measurements of position errors $\tilde{q}_j = q_j - q_{dj}$ and joint velocities $\dot{q}_j$ ($j = 1,2$)

$$\tau_j = -k_{Pj}\,\tilde{q}_j - k_{Dj}\,\dot{q}_j \tag{9.2}$$

will achieve the desired position control task. Indeed, the control law (9.2) , where $k_{Pj}$ and $k_{Dj}$ are strictly positive constants, is simply mimicking the effect of equipping each of the manipulator's joints with a passive mechanical device composed of a coil-spring and a damper, and having the desired $q_{dj}$ as its rest position. The resulting passive physical system would simply exhibit damped oscillations towards the rest position $\mathbf{q}_d$ .

Yet, writing the system dynamics in the " $\mathbf{f} = m\,\mathbf{a}$ " (*i.e.*, Newtonian) form (9.1) does not easily capture this simple fact. In order to formalize the above discussion, a more appropriate approach is to rewrite the system's dynamics *in terms of energy transfers* (*i.e.*, in a Hamiltonian form). We can write conservation of energy in the form

$$\frac{1}{2}\frac{d}{dt}[\,\dot{\mathbf{q}}^T\,\mathbf{H}\,\dot{\mathbf{q}}\,] = \dot{\mathbf{q}}^T\cdot\tau \tag{9.3}$$

where the left-hand side is the derivative of the manipulator's kinetic energy, and the right-hand side represents the power input from the actuators. Expression (9.3) does not mean, of course, that the Coriolis and centripetal terms of (9.1) have disappeared, but simply that they are now accounted for implicitly, since they stem from the time-variation of the inertia matrix $\mathbf{H}$. The stability and convergence proof for the P.D. controller above can then be derived very simply. Let us actually take the control input in a form slightly more general than (9.2), namely

$$\tau = -\mathbf{K}_P\,\tilde{\mathbf{q}} - \mathbf{K}_D\,\dot{\mathbf{q}} \tag{9.4}$$

where $\mathbf{K}_P$ and $\mathbf{K}_D$ are constant symmetric positive definite matrices (the usual P.D. controller (9.2) corresponds to having $\mathbf{K}_P$ and $\mathbf{K}_D$ diagonal), and let us consider the total mechanical energy $V$ that would be associated with the system if control law (9.4) were implemented using physical springs and dampers, namely

$$V = \frac{1}{2}[\,\dot{\mathbf{q}}^T\mathbf{H}\dot{\mathbf{q}} + \tilde{\mathbf{q}}^T\mathbf{K}_P\tilde{\mathbf{q}}\,] \tag{9.5}$$

To analyze the closed-loop behavior of the controlled system, *we shall use this virtual mechanical energy V as our Lyapunov function*, very similarly to what we chose for the nonlinear mass-spring-damper systems of Chapter 3. The time-derivative of $V$ can be written, given (9.3), as

$$\dot{V} = \dot{\mathbf{q}}^T ( \boldsymbol{\tau} + \mathbf{K}_P \tilde{\mathbf{q}} )$$

which, using control law (9.4), simplifies as

$$\dot{V} = - \dot{\mathbf{q}}^T \mathbf{K}_D \dot{\mathbf{q}} \leq 0$$

Not surprisingly, we recognize $\dot{V}$ as the power dissipated by the virtual dampers. We now only need to check that the system cannot get "stuck" at a stage where $\dot{V}$ equal 0 while $\mathbf{q}$ does not equal $\mathbf{q}_d$, or, to put it more technically, invoke the invariant-set theorem. Since $\dot{V} = 0$ implies that $\dot{\mathbf{q}} = \mathbf{0}$, which in turn implies that $\ddot{\mathbf{q}} = \mathbf{H}^{-1} \mathbf{K}_P \tilde{\mathbf{q}}$, one has that $\dot{V}$ is identically 0 only if $\tilde{\mathbf{q}}$ equals $\mathbf{0}$. Therefore the system does converge to the desired state, as the physical reasoning suggested.

**Example 9.1**   Let us illustrate the above result in a simple simulation. Consider the two-link manipulator of Figure 9.1, whose dynamics can be written explicitly as

$$\begin{bmatrix} H_{11} & H_{12} \\ H_{21} & H_{22} \end{bmatrix} \begin{bmatrix} \ddot{q}_1 \\ \ddot{q}_2 \end{bmatrix} + \begin{bmatrix} -h\dot{q}_2 & -h(\dot{q}_1 + \dot{q}_2) \\ h\dot{q}_1 & 0 \end{bmatrix} \begin{bmatrix} \dot{q}_1 \\ \dot{q}_2 \end{bmatrix} = \begin{bmatrix} \tau_1 \\ \tau_2 \end{bmatrix}$$

where

$$H_{11} = a_1 + 2 a_3 \cos q_2 + 2 a_4 \sin q_2$$

$$H_{12} = H_{21} = a_2 + a_3 \cos q_2 + a_4 \sin q_2$$

$$H_{22} = a_2$$

$$h = a_3 \sin q_2 - a_4 \cos q_2$$

with

$$a_1 = I_1 + m_1 l_{c1}^2 + I_e + m_e l_{ce}^2 + m_e l_1^2$$

$$a_2 = I_e + m_e l_{ce}^2$$

$$a_3 = m_e l_1 l_{ce} \cos \delta_e$$

$$a_4 = m_e l_1 l_{ce} \sin \delta_e$$

In the simulation, we use

$$m_1 = 1 \quad l_1 = 1 \quad m_e = 2 \quad \delta_e = 30^o \quad I_1 = .12 \quad l_{c1} = 0.5 \quad I_e = 0.25 \quad l_{ce} = 0.6$$

The robot, initially at rest at $(q_1 = 0, q_2 = 0)$, is commanded a step to $(q_{d1} = 60^o, q_{d2} = 90^o)$. The corresponding transient position errors (translated in degrees, while all control computations and gains use radians) and control torques are plotted in Figure 9.2, with

$\mathbf{K}_D = 100\ \mathbf{I}, \mathbf{K}_P = 20\ \mathbf{K}_D$. We see that indeed the P.D. controller achieves position control after a stable transient. ◻

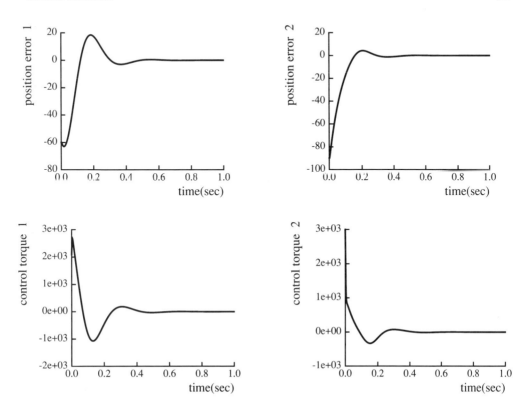

**Figure 9.2a :** Position errors (in degrees) and control torques under P.D. control

## 9.1.2  Trajectory Control

We now consider the case when the manipulator is actually required to follow a desired trajectory, rather than merely reach a desired position. Note that a trajectory control problem may arise even when the task is merely to move a load from its initial position to a final desired position. This may be due to requirements on the maximum time to complete the task, or may be necessary to avoid overshooting or otherwise bumping into obstacles during the task. These needs in turn specify the required "tracking" accuracy.

The simple P.D. controller above cannot be expected to handle the dynamic demands of trajectory tracking effectively. We consider first the use of feedback linearization. We then discuss the extension of the robustness results of Chapter 7 to this multi-input case. In section 9.2, we consider adaptive versions of the robust controller.  Note that the development is applicable to manipulators with an arbitrary number of degrees of freedom, and having revolute or translational joints (or both).

## FEEDBACK LINEARIZATION

Given the simple structure of Equation (9.1), the derivation of a feedback linearizing transformation is straightforward. Taking $\tau$ of the form

$$\tau = \mathbf{H}(\mathbf{q})\,\mathbf{v} + \mathbf{C}(\mathbf{q},\dot{\mathbf{q}})\,\dot{\mathbf{q}} + \mathbf{g}(\mathbf{q}) \tag{9.6}$$

where $\mathbf{v}$ is the new control input, leads to

$$\ddot{\mathbf{q}} = \mathbf{v}$$

Expression (9.6) is known as the "computed torque" in the robotics literature.  Let us define $\tilde{\mathbf{q}} = \mathbf{q} - \mathbf{q}_d$ as the tracking error. Letting

$$\mathbf{v} = \ddot{\mathbf{q}}_d - 2\,\lambda\,\dot{\tilde{\mathbf{q}}} - \lambda^2\,\tilde{\mathbf{q}} \qquad (\lambda > 0)$$

then leads to the exponentially stable closed-loop dynamics

$$\ddot{\tilde{\mathbf{q}}} + 2\,\lambda\,\dot{\tilde{\mathbf{q}}} + \lambda^2\,\tilde{\mathbf{q}} = \mathbf{0}$$

This assumes, of course, that the dynamic model, used in (9.6), is exact.

## ROBUST CONTROL

Consider now the derivation of a robust trajectory controller. In the multi-input case, many possibilities are available to specify the analog of sliding condition (7.5). One may require sliding conditions of the form

$$\frac{1}{2}\frac{d}{dt}s_i^2 \leq -\eta_i\,|s_i| \qquad (\eta_i > 0)$$

to be verified individually along each degree of freedom $i$, with $s_i = \dot{\tilde{q}}_i + \lambda_i\,\tilde{q}_i$ , as in section 7.4.  One may also interpret $s^2$ in (7.5) as the squared norm of the distance to the surface $s = 0$, and require in the multi-input case that

$$\frac{1}{2}\frac{d}{dt}\mathbf{s}^T\mathbf{s} \leq -\eta\,(\mathbf{s}^T\mathbf{s})^{1/2} \qquad (\eta > 0)$$

with, generalizing the single-input case (7.3), the vector $\mathbf{s}$ defined as,

$$\mathbf{s} = \dot{\tilde{\mathbf{q}}} + \mathbf{\Lambda}\tilde{\mathbf{q}} \tag{9.7}$$

where $\mathbf{\Lambda}$ is a symmetric positive definite matrix, or more generally a matrix such that $-\mathbf{\Lambda}$ is Hurwitz.

While both of these approaches are reasonable, it seems plausible, given the results of the previous section, that the design may be simpler using in the sliding condition the generalized squared norm $\mathbf{s}^T \mathbf{H} \mathbf{s}$ (obtained formally by replacing $\dot{\mathbf{q}}$ by $\mathbf{s}$ in the expression of the kinetic energy) rather than merely $\mathbf{s}^T \mathbf{s}$. Indeed, in this second-order system, we may interpret $\mathbf{s}$ of (9.7) as a "velocity error" term

$$\mathbf{s} = \dot{\tilde{\mathbf{q}}} + \mathbf{\Lambda}\tilde{\mathbf{q}} = \dot{\mathbf{q}} - \dot{\mathbf{q}}_r \tag{9.8}$$

where

$$\dot{\mathbf{q}}_r = \dot{\mathbf{q}}_d - \mathbf{\Lambda}\tilde{\mathbf{q}}$$

The "reference velocity" vector $\dot{\mathbf{q}}_r$, is formed by shifting the desired velocities $\dot{\mathbf{q}}_d$ according to the position error $\tilde{\mathbf{q}}$. It simply represents a notational manipulation which allows one to translate energy-related properties (expressed in terms of the actual joint velocity vector $\dot{\mathbf{q}}$) into trajectory control properties (expressed in terms of the virtual velocity error vector $\mathbf{s}$). Similarly to the single-input case, the vector $\mathbf{s}$ conveys information about boundedness and convergence of $\mathbf{q}$ and $\dot{\mathbf{q}}$, since the definition (9.8) of $\mathbf{s}$ can also be viewed as a stable first-order differential equation in $\tilde{\mathbf{q}}$, with $\mathbf{s}$ as an input. Thus, assuming bounded initial conditions, showing the boundedness of $\mathbf{s}$ also shows the boundedness of $\tilde{\mathbf{q}}$ and $\dot{\tilde{\mathbf{q}}}$, and therefore of $\mathbf{q}$ and $\dot{\mathbf{q}}$; similarly, if $\mathbf{s}$ tends to $\mathbf{0}$ as $t$ tends to infinity, so do the vectors $\tilde{\mathbf{q}}$ and $\dot{\tilde{\mathbf{q}}}$.

Our previous discussion of energy conservation also needs to be formalized a step further. In the presence of external gravity torques $\mathbf{g}(\mathbf{q})$, energy conservation can be written (generalizing (9.3) ) as

$$\dot{\mathbf{q}}^T (\mathbf{\tau} - \mathbf{g}) = \frac{1}{2}\frac{d}{dt}[\dot{\mathbf{q}}^T \mathbf{H} \dot{\mathbf{q}}] \tag{9.9}$$

Differentiating the right-hand side explicitly

$$\dot{\mathbf{q}}^T (\mathbf{\tau} - \mathbf{g}) = \dot{\mathbf{q}}^T \mathbf{H} \ddot{\mathbf{q}} + \frac{1}{2}\dot{\mathbf{q}}^T \dot{\mathbf{H}} \dot{\mathbf{q}}$$

and expanding the term $\mathbf{H}\ddot{\mathbf{q}}$ using the system dynamics (9.1), we conclude, after simplification of the input terms $(\mathbf{\tau} - \mathbf{g})$, that for all $\dot{\mathbf{q}}$

$$\dot{\mathbf{q}}^T (\dot{\mathbf{H}} - 2\mathbf{C}) \dot{\mathbf{q}} = 0 \tag{9.10}$$

This result *suggests* that the matrix $(\dot{\mathbf{H}} - 2\mathbf{C})$ is *skew-symmetric* − this is not quite a direct consequence of (9.10), since the matrix itself depends on $\dot{\mathbf{q}}$. Actually, while the Coriolis and centripetal torque vector $\mathbf{C}\dot{\mathbf{q}}$ and the joint velocity vector $\dot{\mathbf{q}}$ are uniquely defined physical quantities, this does not define uniquely the matrix $\mathbf{C}$. However, as detailed below, with **C** defined component-wise by

$$C_{ij} = \frac{1}{2} \sum_{k=1}^{n} \frac{\partial H_{ij}}{\partial q_k} \dot{q}_k + \frac{1}{2} \sum_{k=1}^{n} \left( \frac{\partial H_{ik}}{\partial q_j} - \frac{\partial H_{jk}}{\partial q_i} \right) \dot{q}_k \qquad (9.11)$$

the matrix $(\dot{\mathbf{H}} - 2\mathbf{C})$ is indeed *skew-symmetric*, a fact which thus can be viewed as a matrix expression of energy conservation.

**Proof:** Showing the above technical result (9.11) requires a more detailed look at the derivation of the manipulator's dynamic equations. The total kinetic energy of the manipulator is

$$T = \frac{1}{2} \dot{\mathbf{q}}^T \mathbf{H}(\mathbf{q}) \dot{\mathbf{q}}$$

while the potential (gravitational) energy $G$ associated with the robot links is a function only of the manipulator position $\mathbf{q}$,

$$G = G(\mathbf{q})$$

Thus, the system's Lagragian is

$$L = T - G = \frac{1}{2} \dot{\mathbf{q}}^T \mathbf{H}(\mathbf{q}) \dot{\mathbf{q}} - G(\mathbf{q})$$

Lagrange's equations,

$$\frac{d}{dt} \frac{\partial L}{\partial \dot{\mathbf{q}}} - \frac{\partial L}{\partial \mathbf{q}} = \tau$$

then yield the dynamic equations of the manipulator, in the form

$$\mathbf{H} \ddot{\mathbf{q}} + \mathbf{b}(\mathbf{q}, \dot{\mathbf{q}}) + \mathbf{g}(\mathbf{q}) = \tau \qquad (9.12)$$

where the gravitational torque vector $\mathbf{g}(\mathbf{q})$ is simply

$$\mathbf{g}(\mathbf{q}) = \frac{\partial G}{\partial \mathbf{q}}$$

and the $n \times 1$ vector $\mathbf{b}$

$$\mathbf{b}(\mathbf{q}, \dot{\mathbf{q}}) = \dot{\mathbf{H}} \dot{\mathbf{q}} - \frac{\partial T}{\partial \mathbf{q}}$$

is the vector of Coriolis and centripetal torques.

Let us examine the vector $\mathbf{b}(\mathbf{q}, \dot{\mathbf{q}})$ more closely. First, we note that each component of $\mathbf{b}$ is a quadratic function of the joint velocity components $\dot{q}_i$. This can be seen by noting that the $i^{\text{th}}$ component of $\mathbf{b}$ is

$$b_i(\mathbf{q}, \dot{\mathbf{q}}) = \sum_{j=1}^{j=n} \dot{H}_{ij} \dot{q}_j - \frac{1}{2} \dot{\mathbf{q}}^T \frac{\partial \mathbf{H}}{\partial q_i} \dot{\mathbf{q}} = \sum_{j=1}^{j=n} \sum_{k=1}^{k=n} h_{ijk} \dot{q}_j \dot{q}_k \qquad (9.13)$$

where the coefficients $h_{ijk}$ are functions only of the joint positions, and verify

$$h_{ijk} = \frac{\partial H_{ij}}{\partial q_k} - \frac{1}{2} \frac{\partial H_{jk}}{\partial q_i}$$

We can write $b_i$ as the product of a $1 \times n$ row-vector $\mathbf{c}_i(\mathbf{q}, \dot{\mathbf{q}})$ and the velocity vector $\dot{\mathbf{q}}$

$$b_i = \mathbf{c}_i \dot{\mathbf{q}}$$

Accordingly, the vector $\mathbf{b}(\mathbf{q}, \dot{\mathbf{q}})$ can be written as the product of a matrix $\mathbf{C}$ and the velocity vector $\dot{\mathbf{q}}$,

$$\mathbf{b}(\mathbf{q}, \dot{\mathbf{q}}) = \mathbf{C}(\mathbf{q}, \dot{\mathbf{q}}) \dot{\mathbf{q}} \qquad (9.14)$$

where $\mathbf{C}$ is the matrix obtained by stacking up the row vectors $\mathbf{c}_i$.

While the physical quantities $\mathbf{b}$ and $\dot{\mathbf{q}}$ are uniquely defined, there are many choices of $\mathbf{C}$ which can achieve the decomposition in (9.14). For example, a quadratic function

$$b_1 = \dot{q}_1^2 + 2\dot{q}_1 \dot{q}_2 + \dot{q}_2^2$$

can correspond to either

$$\mathbf{c}_1 = [\, \dot{q}_1 \quad \dot{q}_1 + 2\dot{q}_2 \,]$$

or

$$\mathbf{c}_1 = [\, \dot{q}_1 + \dot{q}_2 \quad \dot{q}_1 + \dot{q}_2 \,]$$

Now, by reindexing, we can rewrite $b_i$ in (9.13) as

$$b_i = \frac{1}{2} \sum_{j=1}^{n} \sum_{k=1}^{n} \frac{\partial H_{ij}}{\partial q_k} \dot{q}_j \dot{q}_k + \frac{1}{2} \sum_{k=1}^{n} \sum_{j=1}^{n} (\frac{\partial H_{ik}}{\partial q_j} - \frac{\partial H_{jk}}{\partial q_i}) \dot{q}_k \dot{q}_j$$

and therefore, a particular choice of the matrix $\mathbf{C}$ is that given by (9.11)

$$C_{ij} = \frac{1}{2} \sum_{k=1}^{n} \frac{\partial H_{ij}}{\partial q_k} \dot{q}_k + \frac{1}{2} \sum_{k=1}^{n} (\frac{\partial H_{ik}}{\partial q_j} - \frac{\partial H_{jk}}{\partial q_i}) \dot{q}_k$$

The above expression can also be written

$$C_{ij} = \frac{1}{2} \dot{H}_{ij} + \frac{1}{2} \sum_{k=1}^{n} (\frac{\partial H_{ik}}{\partial q_j} - \frac{\partial H_{jk}}{\partial q_i}) \dot{q}_k$$

which shows that, with this particular definition of $\mathbf{C}$, the matrix $\dot{\mathbf{H}} - 2\mathbf{C}$ is indeed skew-symmetric.                                                                                                                □

Note that, since $\mathbf{H}$, and therefore $\dot{\mathbf{H}}$, are symmetric matrices, the skew-symmetry of the matrix $\dot{\mathbf{H}} - 2\mathbf{C}$ can also be written

$$\dot{\mathbf{H}} = \mathbf{C} + \mathbf{C}^T$$

In practice, for manipulators with only a few degrees of freedom, this equality may be sufficient to derive the proper form of $\mathbf{C}$ without explicitly using (9.11).

With this further formalization of energy conservation, we are now ready to address the robust trajectory control problem. Let us define

$$V(t) = \frac{1}{2} [ \, \mathbf{s}^T \mathbf{H} \mathbf{s} \, ] \tag{9.15}$$

Differentiating

$$\dot{V}(t) = \mathbf{s}^T (\mathbf{H} \ddot{\mathbf{q}} - \mathbf{H} \ddot{\mathbf{q}}_r) + \frac{1}{2} \mathbf{s}^T \dot{\mathbf{H}} \mathbf{s}$$

and substituting $\mathbf{H} \ddot{\mathbf{q}}$ from the system dynamics,

$$\mathbf{H} \ddot{\mathbf{q}} \; = \; \boldsymbol{\tau} - \mathbf{C} \dot{\mathbf{q}} - \mathbf{g} = \boldsymbol{\tau} - \mathbf{C}(\mathbf{s} + \dot{\mathbf{q}}_r) - \mathbf{g}$$

yields

$$\dot{V}(t) \; = \; \mathbf{s}^T (\boldsymbol{\tau} - \mathbf{H} \ddot{\mathbf{q}}_r - \mathbf{C} \dot{\mathbf{q}}_r - \mathbf{g}) \tag{9.16}$$

where the skew-symmetry of $(\dot{\mathbf{H}} - 2\mathbf{C})$ has been used to eliminate the term $1/2 \, \mathbf{s}^T \dot{\mathbf{H}} \mathbf{s}$.

Similarly to the single-input case, let us now define the control input to be of the form

$$\boldsymbol{\tau} = \hat{\boldsymbol{\tau}} - \mathbf{k} \, sgn(\mathbf{s})$$

where $\mathbf{k} \, sgn(\mathbf{s})$ is defined as the vector of components $k_i \, sgn(s_i)$, and, similarly to the term $\hat{u}$ in chapter 7, $\hat{\boldsymbol{\tau}}$ is the control input vector which would make $\dot{V}$ equal 0 if the dynamics were exactly known

$$\hat{\boldsymbol{\tau}} = \hat{\mathbf{H}} \ddot{\mathbf{q}}_r + \hat{\mathbf{C}} \dot{\mathbf{q}}_r + \hat{\mathbf{g}}$$

We then have

$$\dot{V} \; = \; \mathbf{s}^T [\tilde{\mathbf{H}}(\mathbf{q}) \ddot{\mathbf{q}}_r + \tilde{\mathbf{C}}(\mathbf{q},\dot{\mathbf{q}}) \dot{\mathbf{q}}_r + \tilde{\mathbf{g}}(\mathbf{q})] - \sum_{1}^{n} k_i \, |s_i|$$

Given bounds on modeling errors $\tilde{\mathbf{H}}$ , $\tilde{\mathbf{C}}$ , and $\tilde{\mathbf{g}}$ , where

$$\tilde{\mathbf{H}} = \hat{\mathbf{H}} - \mathbf{H} \qquad \tilde{\mathbf{C}} = \hat{\mathbf{C}} - \mathbf{C} \qquad \tilde{\mathbf{g}} = \hat{\mathbf{g}} - \mathbf{g}$$

we see that choosing the components $k_i$ of the vector $\mathbf{k}$ such that

$$k_i \geq |\, [\tilde{\mathbf{H}}(\mathbf{q})\ddot{\mathbf{q}}_r + \tilde{\mathbf{C}}(\mathbf{q},\dot{\mathbf{q}})\dot{\mathbf{q}}_r + \tilde{\mathbf{g}}(\mathbf{q})]_i\,| \,+\, \eta_i$$

(where the constants $\eta_i$ are strictly positive) allows one to satisfy the sliding condition

$$\dot{V} \leq -\sum_{i=1}^{n} \eta_i\,|s_i|$$

As in the single-input case, the above sliding condition guarantees that the surface $\mathbf{s} = \mathbf{0}$ is reached in a finite time, and that once on the surface the trajectories remain on the surface, and therefore tend to $\mathbf{q}_d(t)$ exponentially. The control chattering implied by the switching gains $\mathbf{k}$ can be eliminated by using smooth interpolations.

**Example 9.2**    Consider again the two-link manipulator of Example 9.1. The control law is now

$$\begin{bmatrix} \tau_1 \\ \tau_2 \end{bmatrix} = \begin{bmatrix} \hat{H}_{11} & \hat{H}_{12} \\ \hat{H}_{21} & \hat{H}_{22} \end{bmatrix} \begin{bmatrix} \ddot{q}_{r1} \\ \ddot{q}_{r2} \end{bmatrix} + \begin{bmatrix} -\hat{h}\dot{q}_2 & -\hat{h}(\dot{q}_1 + \dot{q}_2) \\ \hat{h}\dot{q}_1 & 0 \end{bmatrix} \begin{bmatrix} \dot{q}_{r1} \\ \dot{q}_{r2} \end{bmatrix} - \begin{bmatrix} k_1 \, \text{sat}(s_1/0.05) \\ k_2 \, \text{sat}(s_2/0.05) \end{bmatrix}$$

The robot parameters used in the simulation are the same as in Example 9.1. The controller design assumes a maximum uncertainty of 25 % on the mass properties of the system. The actual parameter inaccuracy used in the simulations is 20 % . We let $\mathbf{\Lambda} = 20\,\mathbf{I}$ , $\eta_1 = \eta_2 = 0.1$ . The ranges $\Phi_1 = \Phi_2 = 0.05$ of the interpolations are chosen so that the maximum value of each $k_i/\Phi_i$ is approximately equal to the diagonal values of $\mathbf{K}_D$ in Example 9.1.

The robot, initially at rest at $(q_1 = 0, q_2 = 0)$ , is commanded a desired trajectory

$$q_{d1}(t) = 30^o\,(1 - \cos(2\pi t)) \qquad q_{d2}(t) = 45^o\,(1 - \cos(2\pi t))$$

The corresponding position errors and control torques are plotted in Figure 9.3.        □

# 9.2  Adaptive Robot Trajectory Control

In this section, we show that, as in the case of regulation using P.D. control, the trajectory control problem can actually be addressed *without any a priori knowledge of the system's mass properties*, by developing adaptive versions of the robust controller discussed above.

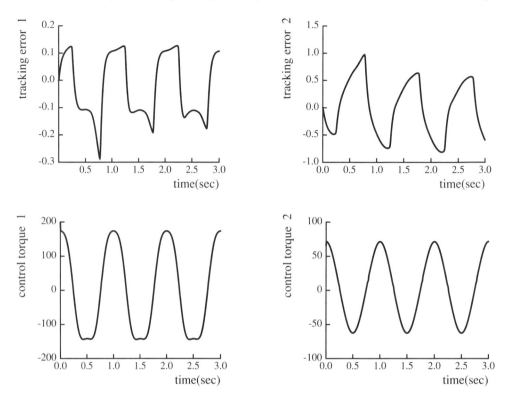

**Figure 9.3 :** Tracking errors and control torques under sliding control

## 9.2.1  The Basic Algorithm

Given the desired trajectory $\mathbf{q}_d(t)$ (we shall assume that the desired position, velocity, and acceleration are all bounded), and with some or all the manipulator parameters being unknown, the adaptive controller design problem is to derive a control law for the actuator torques, and an estimation law for the unknown parameters, such that the manipulator output $\mathbf{q}(t)$ closely tracks the desired trajectory. To this effect let us define $\tilde{\mathbf{a}} = \hat{\mathbf{a}} - \mathbf{a}$ as the parameter estimation error, with $\mathbf{a}$ being a constant vector of unknown parameters describing the manipulator's mass properties, and $\hat{\mathbf{a}}$ its estimate. We now consider, instead of (9.11), the Lyapunov function candidate

$$V(t) = \frac{1}{2}[\, \mathbf{s}^T\mathbf{H}\mathbf{s} + \tilde{\mathbf{a}}^T\mathbf{\Gamma}^{-1}\tilde{\mathbf{a}}\,] \tag{9.17}$$

where $\mathbf{\Gamma}$ is a symmetric positive definite matrix. Differentiating, and using (9.16), yields

$$\dot{V}(t) = \mathbf{s}^T(\boldsymbol{\tau} - \mathbf{H}\ddot{\mathbf{q}}_r - \mathbf{C}\dot{\mathbf{q}}_r - \mathbf{g}) + \dot{\hat{\mathbf{a}}}^T\mathbf{\Gamma}^{-1}\tilde{\mathbf{a}}$$

At this point, we exploit an additional physical property of the system, namely that, given a proper definition of the unknown parameter vector $\mathbf{a}$ describing the manipulator's mass properties, the terms $\mathbf{H}(\mathbf{q})$, $\mathbf{C}(\mathbf{q}, \dot{\mathbf{q}})$, and $\mathbf{g}(\mathbf{q})$, all depend *linearly* on $\mathbf{a}$. This property, already mentioned in chapter 8, can be immediately verified on the dynamics of the two-link manipulator in Example 9.1, and is obtained in general by referring all mass properties to the individual joint axes. Thus, we can define a *known* matrix $\mathbf{Y} = \mathbf{Y}(\mathbf{q}, \dot{\mathbf{q}}, \dot{\mathbf{q}}_r, \ddot{\mathbf{q}}_r)$ such that

$$\mathbf{H}(\mathbf{q})\ddot{\mathbf{q}}_r + \mathbf{C}(\mathbf{q},\dot{\mathbf{q}})\dot{\mathbf{q}}_r + \mathbf{g}(\mathbf{q}) = \mathbf{Y}(\mathbf{q}, \dot{\mathbf{q}}, \dot{\mathbf{q}}_r, \ddot{\mathbf{q}}_r)\,\mathbf{a}$$

Taking the control law to be

$$\boldsymbol{\tau} = \mathbf{Y}\,\hat{\mathbf{a}} \quad \mathbf{K}_D \mathbf{s} \tag{9.18}$$

which includes a "feedforward" term $\mathbf{Y}\,\hat{\mathbf{a}}$ (which is the same as the term $\hat{\boldsymbol{\tau}}$ of the robust controller) in addition to a simple P.D. term $\mathbf{K}_D\,\mathbf{s}$, leads to

$$\dot{V}(t) = \mathbf{s}^T \mathbf{Y}\tilde{\mathbf{a}} - \mathbf{s}^T \mathbf{K}_D \mathbf{s} + \dot{\hat{\mathbf{a}}}^T \boldsymbol{\Gamma}^{-1}\tilde{\mathbf{a}}$$

Updating the parameter estimates $\hat{\mathbf{a}}$ according to the correlation integrals

$$\dot{\hat{\mathbf{a}}} = -\boldsymbol{\Gamma}\,\mathbf{Y}^T \mathbf{s} \tag{9.19}$$

then yields

$$\dot{V}(t) = -\mathbf{s}^T \mathbf{K}_D\,\mathbf{s} \leq 0 \tag{9.20}$$

This implies (as intuition suggests, and as can be shown easily using Barbalat's lemma of section 4.3.2, as detailed below), that the output error converges to the surface $\mathbf{s} = \mathbf{0}$, which, given the filter-like definition (9.8) of $\mathbf{s}$, in turn shows that $\tilde{\mathbf{q}}$ and $\dot{\tilde{\mathbf{q}}}$ tend to $\mathbf{0}$ as $t$ tends to infinity. Therefore, both global stability of the system (*i.e.*, boundedness of the vectors $\mathbf{q}$, $\dot{\mathbf{q}}$, and $\hat{\mathbf{a}}$), and convergence of the tracking error, are guaranteed by the above adaptive controller.

**Proof:** Given the expression (9.20) of $\dot{V}$, it suffices to show that $\dot{V} \to 0$ as $t \to \infty$ in order to show that $\mathbf{s} \to \mathbf{0}$ as $t \to \infty$

$$\dot{V} \to 0 \quad \Rightarrow \quad \mathbf{s} \to \mathbf{0}$$

Furthermore, since $V$ is positive, Barbalat's lemma indicates that $\dot{V}$ does tend to zero if it is uniformly continuous, and in particular if $\ddot{V}$ is bounded

$$\ddot{V} \text{ bounded} \quad \Rightarrow \quad \dot{V} \to 0 \quad \Rightarrow \quad \mathbf{s} \to \mathbf{0}$$

Since, given (9.20), one has $\ddot{V} = -2\,\mathbf{s}^T \mathbf{K}_D\,\dot{\mathbf{s}}$, this shows that

$$(\mathbf{s} \text{ and } \dot{\mathbf{s}} \text{ bounded}) \quad \Rightarrow \quad \ddot{V} \text{ bounded} \quad \Rightarrow \quad \dot{V} \to 0 \quad \Rightarrow \quad \mathbf{s} \to \mathbf{0}$$

Thus, we only have to show that $\mathbf{s}$ and $\dot{\mathbf{s}}$ remain bounded.

Now since $V \geq 0$ and $\dot{V} \leq 0$, $V$ remains bounded. Given the expression (9.17) of $V$, this implies that both $\mathbf{s}$ and $\tilde{\mathbf{a}}$ are bounded. Furthermore, this in turn implies that $\mathbf{q}$, $\dot{\mathbf{q}}$, and $\hat{\mathbf{a}}$ are all bounded. Noticing that the closed-loop dynamics can be written in the simple form

$$\mathbf{H}\dot{\mathbf{s}} + (\mathbf{C} + \mathbf{K}_D)\mathbf{s} = \mathbf{Y}\tilde{\mathbf{a}} \tag{9.21}$$

this shows (since the uniform positive definiteness of $\mathbf{H}$ implies that $\mathbf{H}^{-1}$ exists and is bounded) that $\dot{\mathbf{s}}$ is also bounded.

Thus, $\mathbf{s} \to \mathbf{0}$ as $t \to \infty$, which implies that both $\tilde{\mathbf{q}}$ and $\dot{\tilde{\mathbf{q}}}$ tend to $\mathbf{0}$ as $t$ tends to infinity.   □

Note that this approach does not necessarily estimate the unknown parameters exactly, but simply generates values that allow the desired task to be achieved. Similarly to the discussion of section 8.7, "sufficient richness" conditions on the desired trajectory indicate how demanding the desired trajectory should be for tracking convergence to necessarily require parameter convergence. Also, of course, the tracking error does not merely tend "asymptotically" to zero, but for all practical purposes, converges within finite time constants determined for a given trajectory by the values of the gain matrices $\mathbf{\Lambda}$, $\mathbf{K}_D$, and $\mathbf{\Gamma}$, themselves limited by the presence of high-frequency unmodeled dynamics and measurement noise.

While it may seem somewhat superfluous to keep estimating the parameter vector $\mathbf{a}$ in order to drive the very same manipulator, the practical relevance of the above approach is that the possibly large unknown loads that the manipulator may carry can also be directly accounted for, simply by considering the (securely) grasped load as part of the last link. Actually, since the mass properties of the manipulator itself do not change, one may adapt only on the mass properties of the load (which can be described by at most 10 parameters, namely load mass, three parameters describing the position of the center of mass, and six independent parameters describing the symmetric load inertia matrix).

Furthermore, by combining this section's results with those of the previous section, the approach can be easily extended to the case where only a few significant parameters have to be adapted upon (in order to simplify the computations, for instance), while the controller is robust to residual errors resulting from inaccuracies on the *a priori* estimates of the other parameters (as well as, perhaps, to bounded time-varying additive disturbances such as stiction).

**Example 9.3**    Let us illustrate the basic properties of the adaptive controller on the two-link manipulator of Examples 9.1 and 9.2.

With $\Lambda = 20\,\mathbf{I}$ (as in Example 9.2) and $\mathbf{K}_D = 100\,\mathbf{I}$ (leading to the same P.D. gains as in Example 9.1), the control law can be written

$$\begin{bmatrix} \tau_1 \\ \tau_2 \end{bmatrix} = \begin{bmatrix} \hat{H}_{11} & \hat{H}_{12} \\ \hat{H}_{21} & \hat{H}_{22} \end{bmatrix} \begin{bmatrix} \ddot{q}_{r1} \\ \ddot{q}_{r2} \end{bmatrix} + \begin{bmatrix} -\hat{h}\dot{q}_2 & -\hat{h}(\dot{q}_1 + \dot{q}_2) \\ \hat{h}\dot{q}_1 & 0 \end{bmatrix} \begin{bmatrix} \dot{q}_{r1} \\ \dot{q}_{r2} \end{bmatrix} - \begin{bmatrix} 100\,s_1 \\ 100\,s_2 \end{bmatrix}$$

where $h$ and the $H_{ij}$ are defined as before. Therefore, the components of the matrix $\mathbf{Y}$ can be written explicitly as

$$Y_{11} = \ddot{q}_{r1} \qquad Y_{12} = \ddot{q}_{r2} \qquad Y_{21} = 0 \qquad Y_{22} = \ddot{q}_{r1} + \ddot{q}_{r2}$$

$$Y_{13} = (2\ddot{q}_{r1} + \ddot{q}_{r2})\cos q_2 - (\dot{q}_2\dot{q}_{r1} + \dot{q}_1\dot{q}_{r2} + \dot{q}_2\dot{q}_{r2})\sin q_2$$

$$Y_{14} = (2\ddot{q}_{r1} + \ddot{q}_{r2})\sin q_2 + (\dot{q}_2\dot{q}_{r1} + \dot{q}_1\dot{q}_{r2} + \dot{q}_2\dot{q}_{r2})\cos q_2$$

$$Y_{23} = \ddot{q}_{r1}\cos q_2 + \dot{q}_1\dot{q}_{r1}\sin q_2$$

$$Y_{24} = \ddot{q}_{r1}\sin q_2 - \dot{q}_1\dot{q}_{r1}\cos q_2$$

The adaptation law (9.19) uses the above expression of $\mathbf{Y}$, with $\Gamma = \text{diag}\,[.03 \quad .05 \quad .1 \quad .3]$. The actual parameters used in the simulation are the same as in Examples 9.1 and 9.2, leading to

$$\mathbf{a} = [\,a_1 \quad a_2 \quad a_3 \quad a_4\,]^T = [\,3.3 \quad 0.97 \quad 1.04 \quad 0.6\,]^T$$

Assume again, as in Example 9.2, that the desired trajectory is

$$q_{d1}(t) = 30^o\,(1 - \cos(2\pi t)) \qquad q_{d2}(t) = 45^o\,(1 - \cos(2\pi t))$$

The corresponding tracking errors, control torques, and parameter estimates are plotted in Figure 9.4, starting without any *a priori* information ($\hat{\mathbf{a}}(0) = \mathbf{0}$). The actual parameters used in the simulations are the same as in the earlier examples. We see that tracking convergence to errors of less than $0.5^o$ occurs within the first half-second.

Note that, in this example, the number of equivalent parameters is 4 while the number of original physical inertial parameters is 6 (three parameters for each link, *i.e.*, mass, center of mass, moment of inertia). This means that there are fewer equivalent parameters than physical parameters, and therefore the physical parameters cannot be determined from the equivalent parameters. This does not cause any problems in control, because in computing control inputs, we actually use the equivalent parameters or their estimates.

For comparison, Figure 9.5 shows the tracking errors and control torques using the P.D. control term $-\mathbf{K}_D\,\mathbf{s}$ alone, on the same desired trajectory. Note that since $\hat{\mathbf{a}}(0) = \mathbf{0}$, the adaptive

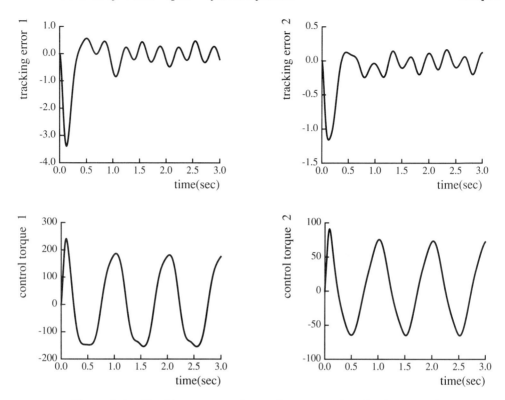

**Figure 9.4a :** Tracking errors and control torques under adaptive control

control input and the P.D. control input are initially equal.   ☐

## AN ALTERNATIVE ADAPTIVE TRAJECTORY CONTROLLER

The above tracking convergence proof is unchanged if the gain matrix $\mathbf{K}_D$ is chosen to be time-varying, provided that $\mathbf{K}_D$ remains uniformly positive definite (so that $\dot{V} \to 0$ still implies that $\mathbf{s} \to \mathbf{0}$) and of bounded time-derivative (so that $\dot{V}$ still is uniformly continuous). In the following, we show that an examination of the control problem *without uncertainty* suggests a particular time-varying choice of $\mathbf{K}_D$. We then study how this choice translates in the adaptive case, and how to incorporate it in a modified adaptive controller.

### The case of no parameter uncertainty

When there is no parameter uncertainty, there is no need for adaptation. The control law (9.18) can still be used, with the unknown parameters $\hat{\mathbf{a}}$ replaced by the known parameters $\mathbf{a}$. While the resulting controller is *different* from the computed torque controller, it can also be shown to be exponentially convergent, and thus can be viewed as an alternative to the computed torque controller.

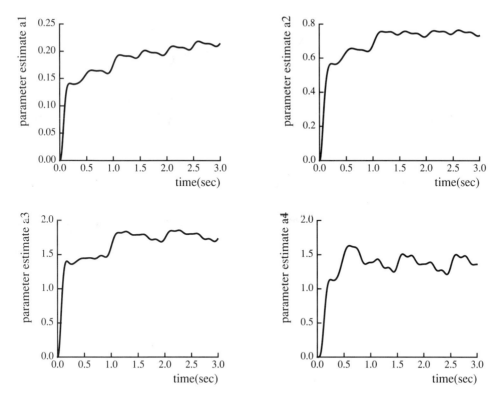

**Figure 9.4b :** Parameter estimates under adaptive control

Specifically, let us choose

$$\mathbf{K}_D = \lambda \mathbf{H}$$

where $\lambda$ is a strictly positive constant. This adjustment of $\mathbf{K}_D$ according to the (positive definite) inertia matrix $\mathbf{H}$ is intuitively appealing, since it corresponds to applying larger gains for joints of larger inertia. Let us consider the same function $V$ as before, *i.e.*,

$$V = \frac{1}{2}\mathbf{s}^T \mathbf{H}\mathbf{s}$$

Differentiation of $V$ yields

$$\dot{V} = -\lambda\, \mathbf{s}^T \mathbf{H}\mathbf{s} = -2\lambda V$$

This implies that $V(t) = V(0)e^{-2\lambda t}$. Furthermore, due to the uniform positive definiteness of $\mathbf{H}$,

$$\exists\, \alpha > 0, \ (\alpha/2)\,\|\mathbf{s}\|^2 \le V(t) = V(0)e^{-2\lambda t}$$

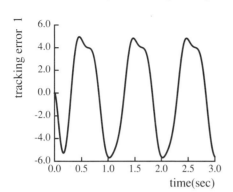
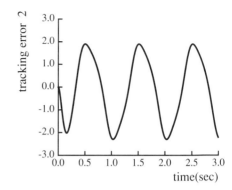

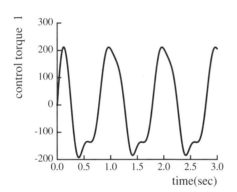
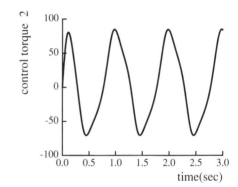

**Figure 9.5 :** Tracking errors and control torques under P.D. control

This shows that **s** is exponentially convergent to zero with a rate $\lambda$. Letting $\mathbf{\Lambda} = \lambda\,\mathbf{I}$ (where $\mathbf{I}$ is the identity matrix), this, in turn, implies the exponential convergence of tracking errors $\tilde{\mathbf{q}}$ and $\dot{\tilde{\mathbf{q}}}$ with the rate $\lambda$.

It is interesting to note that, with this choice of $\mathbf{K}_D$ and $\mathbf{\Lambda}$, the control law can be rewritten as

$$\tau = \mathbf{H}[\ddot{\mathbf{q}}_d - 2\lambda\dot{\tilde{\mathbf{q}}} - \lambda^2\tilde{\mathbf{q}}] + \mathbf{C}\,\dot{\mathbf{q}}_r + \mathbf{g}$$

Thus, the control law is almost the same as a computed torque control law, except for a minor difference in the second term. This control law is actually interesting in itself, because it does not exactly cancel the robot dynamics, but it still guarantees exponential tracking convergence at the same rate as an exact cancellation. This can be attributed to the fact that it utilizes the inherent structure of the robot dynamics.

**The adaptive case**

Motivated by the choice of $\mathbf{K}_D$ used above in the absence of uncertainty, we might be interested in using

$$\mathbf{K}_D = \lambda \hat{\mathbf{H}} \tag{9.22}$$

in the adaptive case, when uncertainty is present. However, there is a problem with this choice: the matrix $\hat{\mathbf{H}}$ may not be always positive definite in the course of parameter adaptation. We now show that this problem can be simply handled using a modification of the adaptation law.

Considering again the Lyapunov-like function candidate (9.17), and choosing $\mathbf{K}_D$ as in (9.22), we get

$$\dot{V} = \mathbf{s}^T[\tilde{\mathbf{H}}\ddot{\mathbf{q}}_r + \tilde{\mathbf{C}}\dot{\mathbf{q}}_r + \tilde{\mathbf{g}} - \lambda \hat{\mathbf{H}}\mathbf{s}] + \tilde{\mathbf{a}}^T\mathbf{\Gamma}^{-1}\dot{\tilde{\mathbf{a}}}$$

$$= \mathbf{s}^T[-\lambda \mathbf{H}\mathbf{s} + \tilde{\mathbf{H}}(\ddot{\mathbf{q}}_r - \lambda\mathbf{s}) + \tilde{\mathbf{C}}\dot{\mathbf{q}}_r + \tilde{\mathbf{g}}] + \tilde{\mathbf{a}}^T\mathbf{\Gamma}^{-1}\dot{\tilde{\mathbf{a}}}$$

Thus, if we let $\mathbf{Y}_m$ denote a modified $\mathbf{Y}$ matrix, defined by

$$\tilde{\mathbf{H}}(\ddot{\mathbf{q}}_r - \lambda\mathbf{s}) + \tilde{\mathbf{C}}\dot{\mathbf{q}}_r + \tilde{\mathbf{g}} = \mathbf{Y}_m(\mathbf{q}, \dot{\mathbf{q}}, \dot{\mathbf{q}}_r, \ddot{\mathbf{q}}_r)\,\tilde{\mathbf{a}} \tag{9.23}$$

we obtain

$$\dot{V} = -\lambda \mathbf{s}^T\mathbf{H}\mathbf{s} + \tilde{\mathbf{a}}^T[\mathbf{\Gamma}^{-1}\dot{\hat{\mathbf{a}}} + \mathbf{Y}_m\mathbf{s}]$$

The adaptation law

$$\dot{\hat{\mathbf{a}}} = -\mathbf{\Gamma}\,\mathbf{Y}_m{}^T\mathbf{s} \tag{9.24}$$

then leads to

$$\dot{V} = -\lambda \mathbf{s}^T\mathbf{H}\mathbf{s}$$

Since $\mathbf{H}$, the actual manipulator inertia matrix, is always uniformly positive definite, the above expression guarantees global stability and global tracking convergence of the adaptive control system. Note that the control law can be simply written as

$$\tau = \mathbf{Y}_m\hat{\mathbf{a}} \tag{9.25}$$

## 9.2.2 * Composite Adaptive Trajectory Control

In the previous section, the adaptation law extracts information about the parameters from the joint tracking errors. However, the tracking errors are not the only source of parameter information. As discussed in Example 8.10, prediction errors on the joint torques also contain parameter information. Here we show that, as in section 8.8, the different sources of parameter information be combined for parameter adaptation, and can further improve the performance of the adaptive controller.

## COMPOSITE ADAPTATION

The key to extracting information from both the tracking errors **s** and prediction errors **e** is to create a parameter update law which is driven by both **s** and **e**. To this purpose, while leaving the control law unchanged, we define, similarly to section 8.8, a "composite adaptation law" of the form

$$\dot{\hat{\mathbf{a}}}(t) = -\mathbf{P}(t)[\ \mathbf{Y}^T\mathbf{s} + \mathbf{W}^T\mathbf{R}(t)\mathbf{e}\ ] \tag{9.26}$$

where $\mathbf{R}(t)$ is a uniformly *p.d.* weighting matrix indicating how much attention the adaptation law should pay to the parameter information in the prediction error, and the adaptation gain $\mathbf{P}(t)$ is a uniformly *p.d.* gain matrix determined by the techniques to be described below. The quantities **Y**, **s** are defined as before, while **W** and **e** are defined in Example 8.10. In filtering the dynamics to avoid the joint acceleration (*i.e.*, in generating the matrix **W**), we can use a first-order filter, *i.e.*,

$$\mathbf{W}(\mathbf{q}, \dot{\mathbf{q}}) = \frac{\lambda_f}{p + \lambda_f}[\mathbf{Y}_1(\mathbf{q}, \dot{\mathbf{q}}, \ddot{\mathbf{q}})] \tag{9.27}$$

with $\mathbf{Y}_1$ defined by (8.83). We thus obtain a new class of adaptive controllers. Note that the earlier adaptive controller, which we shall refer to as a tracking-error-based (TEB) adaptive controller, simply corresponds to

$$\forall\, t \geq 0, \quad \mathbf{R}(t) \equiv 0 \qquad \mathbf{P}(t) \equiv \boldsymbol{\Gamma}$$

For simplicity, in the following we take $\mathbf{R}(t)$ in (9.26) to be the unity matrix. To generate the gain matrix **P** in (9.26), we may use the gain update techniques of various parameter estimators section 8.7.

Let us also point out some results on persistency of excitation relevant to our later discussion. If tracking errors $\tilde{\mathbf{q}}$ and $\dot{\tilde{\mathbf{q}}}$ converge to zero, then the persistent excitation of $\mathbf{W}_d = \mathbf{W}(\mathbf{q}_d, \dot{\mathbf{q}}_d)$ implies the *p.e.* of **W**, as shown in chapter 8. Also, as explained in section 8.7, the *p.e.* of $\mathbf{W}_d$ is itself guaranteed by the *p.e.* and uniform continuity of the matrix $\mathbf{Y}_{1d}$, which is defined by

$$\mathbf{Y}_{1d} = \mathbf{Y}_1(\mathbf{q}_d, \dot{\mathbf{q}}_d, \ddot{\mathbf{q}}_d)$$

## GLOBAL ASYMPTOTIC AND EXPONENTIAL CONVERGENCE

In the following, we show the global asymptotic and exponential convergence of the tracking errors and parameter errors for the adaptive controllers based on the composite adaptation law (9.26), using a Lyapunov-like analysis.

The convergence analysis for the composite adaptive controllers based on various gain-update techniques will all rely on the following scalar function

$$V(t) = \frac{1}{2}[\, \mathbf{s}^T \mathbf{H}\, \mathbf{s} + \tilde{\mathbf{a}}^T \mathbf{P}^{-1} \tilde{\mathbf{a}}\, ] \tag{9.28}$$

This is the same function as that used for the tracking-error-based (TEB) controller in section 9.2.1, although we now allow $\mathbf{P}$ to be possibly time-varying. Specifically, besides the case of a constant adaptation gain, we shall study the BGF exponentially forgetting least-squares algorithm of section 8.7.5.

### The Constant Gain (CG) Composite Adaptive Controller

The constant gain composite adaptive controller uses a constant (symmetric, *p.d.*) matrix $\mathbf{P}(t) = \mathbf{P}_o$ in the adaptation law (9.26). With the same Lyapunov-like function as in the TEB derivation, the composite adaptation law simply adds an extra term in the expression of $\dot{V}$

$$\dot{V}(t) = -\mathbf{s}^T \mathbf{K}_D \mathbf{s} - \tilde{\mathbf{a}}^T \mathbf{W}^T \mathbf{W} \tilde{\mathbf{a}} \tag{9.29}$$

This implies that $V(t) \le V(0)$ and, therefore, that $\mathbf{s}$ and $\tilde{\mathbf{a}}$ are upper bounded from the construction of $V$, since $\mathbf{H}$ is uniformly *p.d.* and $\mathbf{P}_o$ is *p.d.*. It is interesting to note that the scalar function here is the same as that used in (9.28) for the tracking error-based adaptive controller, while its derivative now contains an additional negative term $(-\mathbf{e}^T \mathbf{e})$, indicating that $V(t)$ will decrease as long as either the tracking error or the prediction error is not zero. More precisely,

> *If the manipulator's desired joint trajectories are bounded, then the tracking errors $\tilde{\mathbf{q}}$ and $\dot{\tilde{\mathbf{q}}}$ <u>and</u> the prediction errors $\mathbf{e}$ in the CG composite adaptive controller all globally converge to zero, and the parameter error $\tilde{\mathbf{a}}$ remains bounded. If, in addition, the trajectories are persistently exciting and uniformly continuous, the estimated parameters asymptotically converge to the true parameters.*

**Proof**: The boundedness of $\mathbf{s}$ and $\tilde{\mathbf{a}}$ has already been pointed out. The boundedness of $\tilde{\mathbf{q}}$ and $\dot{\tilde{\mathbf{q}}}$ follows from the boundedness of $\mathbf{s}$. We now prove the convergence of the tracking error measure $\mathbf{s}$ and prediction error $\mathbf{e}$ by showing the convergence of $\dot{V}$ to zero for bounded desired trajectories.

Let us first show the boundedness of $\ddot{V}(t)$

$$\ddot{V} = -2\mathbf{s}^T \mathbf{K}_D \dot{\mathbf{s}} - 2\mathbf{e}^T \dot{\mathbf{e}}$$

since this in turn guarantees the uniform continuity of $\dot{V}(t)$. The boundedness of $\tilde{\mathbf{q}}$, $\dot{\tilde{\mathbf{q}}}$, $\mathbf{q}_d$, $\dot{\mathbf{q}}_d$ and $\ddot{\mathbf{q}}_d$ implies that of $\mathbf{q}$, $\dot{\mathbf{q}}$, $\dot{\mathbf{q}}_r$ and $\ddot{\mathbf{q}}_r$. Examination of the terms in $\mathbf{Y}(\mathbf{q}, \dot{\mathbf{q}}, \dot{\mathbf{q}}_r, \ddot{\mathbf{q}}_r)$ and $\mathbf{C}(\mathbf{q}, \dot{\mathbf{q}})$ reveals that they are all bounded, reflecting the physical fact that, for a mechanical manipulator, bounded motion quantities cannot correspond to unbounded forces. Given the closed-loop dynamics (9.21),

$$\dot{\mathbf{s}} = \mathbf{H}^{-1}[\mathbf{Y}\tilde{\mathbf{a}} - (\mathbf{K}_D + \mathbf{C})\mathbf{s}] \tag{9.30}$$

and the upper boundedness of $\mathbf{H}^{-1}$, $\dot{\mathbf{s}}$ is bounded. This also implies that $\ddot{\mathbf{q}}$ is bounded.

From (9.28), we have

$$\dot{\mathbf{e}} = \dot{\mathbf{W}}\,\tilde{\mathbf{a}} + \mathbf{W}\dot{\tilde{\mathbf{a}}}$$

The second term in the right-hand side is bounded because $\mathbf{W}$, $\mathbf{Y}$, $\tilde{\mathbf{a}}$ and $\mathbf{s}$ are all bounded. Note that the boundedness of $\mathbf{q}$, $\dot{\mathbf{q}}$ and $\ddot{\mathbf{q}}$ implies the boundedness of the torque $\boldsymbol{\tau}$ and that of the matrix $\mathbf{Y}_1$. Based on the facts that the filter in (9.27) is exponentially stable and strictly proper, and that $\mathbf{Y}_1$ is bounded, one can easily show the boundedness of $\dot{\mathbf{W}}$. This in turn guarantees the boundedness of $\dot{\mathbf{e}}$.

The boundedness of $\mathbf{s}$, $\mathbf{e}$, $\dot{\mathbf{e}}$ and $\dot{\mathbf{s}}$ implies the boundedness of $\ddot{V}(t)$. Straightforward application of Corollary 2.1 then leads to $\dot{V}(t) \rightarrow 0$ as $t \rightarrow \infty$. Therefore, both the tracking error $\mathbf{s}$ and the prediction error $\mathbf{e}$ asymptotically converge to zero. The convergence of $\mathbf{s}$ to zero in turn guarantees the convergence of $\dot{\tilde{\mathbf{q}}}$ and $\tilde{\mathbf{q}}$ to zero.

If $\mathbf{W}_d$ is persistently exciting, then $\mathbf{W}$ is also *p.e.*. The convergence of the estimated parameters to the true parameters can then be shown easily by noting that the adaptation law

$$\dot{\tilde{\mathbf{a}}} = -\mathbf{P}_o \mathbf{W}^T \mathbf{W}\tilde{\mathbf{a}} - \mathbf{P}_o \mathbf{Y}^T \mathbf{s} \tag{9.31}$$

represents an exponentially stable dynamics with convergent input $\mathbf{Y}^T \mathbf{s}$.                    ∎

It is interesting to note that composite adaptive control guarantees the convergence to zero of *both* tracking error and prediction error, while direct adaptive control only guarantees that of the tracking error. This is a reflection of the fact that composite adaptation explicitly pays attention to both tracking error and prediction error.

As in section 8.8, the effect of the composite adaptation law can be given a simple intuitive interpretation.   Assume for simplicity that $\mathbf{P}(t) = \gamma\mathbf{I}$.   The TEB adaptation law (with $\boldsymbol{\Gamma} = \mathbf{P}(t)$) has a gradient-like nature, because it can be written as

$$\dot{\tilde{\mathbf{a}}} = -\gamma \frac{\partial \boldsymbol{\tau}}{\partial \tilde{\mathbf{a}}} \mathbf{s}$$

This gradient nature accounts for the poor parameter convergence when large adaptation gain is used. By contrast, using (9.28), one can rewrite the composite adaptation law as

$$\dot{\tilde{\mathbf{a}}}(t) + \gamma \mathbf{W}^T \mathbf{W} \tilde{\mathbf{a}} = -\gamma \mathbf{Y}^T \mathbf{s}$$

This means that the parameters errors are now a filtered version of the gradient direction, with the filter being time-varying. Thus, the parameter search in the composite adaptation goes along a filtered, or averaged, direction. This indicates that parameter and tracking error convergence in composite adaptive control can be smoother and faster than in TEB adaptive control.

### The Bounded-Gain-Forgetting (BGF) Composite Adaptive Controller

The composite adaptive controller with $\mathbf{P}$ in (9.26) determined by a BGF update is called the BGF composite adaptive controller. Recall from chapter 8 that the BGF update uses an exponentially forgetting least-squares gain update

$$\frac{d}{dt} \mathbf{P}^{-1}(t) = -\lambda(t) \mathbf{P}^{-1} + \mathbf{W}^T \mathbf{W} \qquad (9.32)$$

with the following variable forgetting factor

$$\lambda(t) = \lambda_o (1 - \|\mathbf{P}\|/k_o) \qquad (9.33)$$

where $k_o$ and $\lambda_o$ are two positive constants specifying the upper bound of the gain matrix norm and the maximum forgetting rate. This gain update guarantees that $\forall\, t \geq 0$, $\lambda(t) \geq 0$, and $\mathbf{P}(t) \leq k_o \mathbf{I}$ for any signal $\mathbf{W}$, and $\exists\, \lambda_1 > 0$, $\forall\, t \geq 0$, $\lambda(t) \geq \lambda_1$ if $\mathbf{W}$ is *p.e.*.

For convenience of exponential convergence analysis, we shall use the alternative adaptive trajectory controller described at the end of section 9.2.1

*The BGF composite adaptive controller has globally convergent tracking errors $\tilde{\mathbf{q}}$ and $\dot{\tilde{\mathbf{q}}}$ and prediction error $\mathbf{e}$ if the desired trajectories are bounded. Furthermore, if the desired trajectories are persistently exciting and uniformly continuous, the parameter estimation errors and tracking errors are globally exponentially convergent to zero.*

**Proof**: For the BGF adaptive controller, defined by (9.24), (9.25), (9.32) (with $\mathbf{Y}_m$ defined by (9.23) ), one has

$$\dot{V}(t) = -\lambda_s \mathbf{s}^T \mathbf{H} \mathbf{s} - (\lambda(t)/2)\, \tilde{\mathbf{a}}^T \mathbf{P}^{-1} \tilde{\mathbf{a}} - \tilde{\mathbf{a}}^T \mathbf{W}^T \mathbf{W} \tilde{\mathbf{a}} \;\leq 0 \qquad (9.34)$$

The convergence of $\tilde{\mathbf{q}}$, $\dot{\tilde{\mathbf{q}}}$ and $\mathbf{e}$ can be shown as before. Note that, additionally, the convergence of $\dot{V}(t)$ to zero leads to that of $\lambda(t) \tilde{\mathbf{a}}^T \mathbf{P}^{-1} \tilde{\mathbf{a}}$ .

As pointed out in section 8.7, $\mathbf{P}(t) \leq k_o \mathbf{I}$, and the persistent excitation of $\mathbf{W}_d$ (and consequently, that of $\mathbf{W}$) guarantees that $\lambda(t) \geq \lambda_1 > 0$. These imply

$$\lambda(t) \tilde{\mathbf{a}}^T \mathbf{P}^{-1} \tilde{\mathbf{a}} \geq \lambda_1 \tilde{\mathbf{a}}^T \tilde{\mathbf{a}}/k_o \tag{9.35}$$

Therefore the convergence of $\lambda(t) \tilde{\mathbf{a}}^T \mathbf{P}^{-1} \tilde{\mathbf{a}}$ to zero implies that of $\tilde{\mathbf{a}}$. In fact, we can more precisely show the *exponential* convergence of the tracking and estimation errors. Indeed, let $\gamma_o$ be the strictly positive constant defined by $\gamma_o = \min(2\lambda_s, \lambda_1)$. In view of (9.31), we can write

$$\dot{V}(t) + \gamma_o V(t) \leq 0$$

Therefore, $V(t) \leq V(0) \, e^{-\gamma_o t}$. This, in turn, implies the exponential convergence of $\mathbf{s}$ and $\tilde{\mathbf{a}}$ to zero. The exponential convergence of $\tilde{\mathbf{q}}$ and $\dot{\tilde{\mathbf{q}}}$ to zero follows as a result of exponential convergence of $\mathbf{s}$.     □

As in section 8.8, the composite adaptive controller can be shown to allow the adaptation gain to be increased much higher than in tracking-error-based adaptation, yielding faster parameter convergence and better tracking accuracy , and to enjoy enhanced robustness to unmodeled dynamics.

# 9.3   Putting Physics in Control

In this section, we consider extensions of the earlier discussion to larger classes of systems and control problems.

## 9.3.1 High-Frequency Unmodeled Dynamics

As we saw in chapter 7, it is usually appropriate in control design to distinguish between structured and unstructured modeling uncertainties. Structured uncertainties correspond to inaccuracies in the parameters of the model (e.g., the effects of unknown loads on the rigid-body dynamics of a manipulator), or to additive disturbances (e.g., stiction), while unstructured uncertainties reflect errors on the system order, *i.e.*, unmodeled dynamics. If modeling is adequate, the unmodeled dynamics (such as structural resonant modes, actuator dynamics, or sampling effects) are of "high frequency". One can then define control "bandwidth" as the frequency content of the control input, and design the control law so that its bandwidth is lower than the frequency range of the unmodeled dynamics. Thus, setting the control bandwidth can be thought of as a matter of consistency: while the role of feedback is to reduce the effects of structured uncertainties on performance, it is limited by the requirement that the frequency content of the control signal does not spill over into that of the unstructured uncertainties.

The discussion of the previous sections, however, which involves seeing some components of our computer-generated control law as mimicking actual physical systems, can be extended to improve performance beyond the classical bandwidth limitation. This is achieved, in a sense, by putting some implicit structure in part of the unmodeled dynamics. Consider again the earlier position control problem, but assume now that the links present some structural flexibility (as real links do). The bandwidth of the P.D. controller (*i.e.*, the corresponding upper bounds on the gains $k_{Pj}$ and $k_{Dj}$) should *not* be limited by the frequency range of the structural modes: were the links very flexible and poorly damped, equipping each joint with a *passive* spring-damper could not destabilize the system. Conversely, the control bandwidth is limited by how well the P.D. control terms do succeed at mimicking the virtual spring-damper systems, namely by the frequency ranges of actuator dynamics and sampling limitations (whichever imposes the lowest bound), *i.e.*, by the other unmodeled dynamics. It is also limited by measurement noise, yet one can easily see that large classes of measurement noise (in particular, for zero-mean noise uncorrelated with the actual system state) do not modify the analysis – inputting white noise in a passive system, for instance, does not affect stability. These effects can be easily verified experimentally.

The above discussion can be generalized. Assume that the effective limit on bandwidth (*i.e.*, the lowest bound imposed by the unmodeled dynamics) is due to a passive element of the actual system dynamics, such as e.g., structural resonant modes or aero-elastic effects. Then *components of the control law which, when combined with that passive element, mimic new passive physical dynamics, can be tuned separately from the rest of the controller by disregarding the limit on bandwidth imposed by the passive element* (see Exercise 9.6). Furthermore, the rest of the control law may then be tuned less conservatively by exploiting the presumably better behaved (e.g., better damped) behavior of the new "unmodeled dynamics" created by the previous combination.

The implications in terms of numerical implementation are obvious. To avoid that the limit on bandwidth be computational, the P.D.-like terms may be calculated at a faster sampling rate, or may be implemented using analog circuitry. Accordingly, the actual implementation may involve multiple time-scales, with the P.D.-like terms computed the most often, the feedforward terms computed at a slower rate, and the parameter update performed at a yet slower sampling rate.

## 9.3.2  Conservative and Dissipative Dynamics

In the robotic example, we dealt with a conservative physical system, subjected to external inputs such as control torques, gravity, or disturbances. Dissipative terms, such as friction, also represent additive inputs, and thus may be accounted for in a fashion similar to the adaptive compensation of the gravity torque $\mathbf{g}$ . However, it is potentially more effective to exploit the known dissipative character of these terms, rather than treat them as simple perturbations. Assume, for instance, that a viscous friction torque vector $\mathbf{D}\dot{\mathbf{q}}$ is present at the manipulator joints (with $\mathbf{D}$ a positive definite matrix). Such a term, of course, does not affect the stability of position control algorithms, since it only introduces additional damping. Let us now consider trajectory control problems. Considering the friction term as an additive disturbance $\mathbf{D}\dot{\mathbf{q}} = \mathbf{Y}_D(\dot{\mathbf{q}})\,\mathbf{a}_D$, where $\mathbf{a}_D$ is a vector of possibly unknown friction coefficients, one can use the control law

$$\tau = \mathbf{Y}\hat{\mathbf{a}} + \mathbf{Y}_D(\dot{\mathbf{q}})\,\hat{\mathbf{a}}_D - \mathbf{K}_D\mathbf{s}$$

and augment the adaptation law in $\hat{\mathbf{a}}$ with one in $\hat{\mathbf{a}}_D$

$$\dot{\hat{\mathbf{a}}}_D = -\Gamma_D\,\mathbf{Y}_D^T(\dot{\mathbf{q}})\,\mathbf{s}$$

leading to

$$\dot{V} = -\mathbf{s}^T\mathbf{K}_D\mathbf{s}$$

as before. If instead we choose to explicitly account for the dissipative character of friction, notice that $\mathbf{D}\dot{\mathbf{q}} = \mathbf{D}\dot{\mathbf{q}}_r + \mathbf{D}\mathbf{s}$ , and use the control law

$$\tau = \mathbf{Y}\hat{\mathbf{a}} + \mathbf{Y}_D(\dot{\mathbf{q}}_r)\,\hat{\mathbf{a}}_D - \mathbf{K}_D\mathbf{s}$$

and, accordingly, the modified adaptation law in $\hat{\mathbf{a}}_D$

$$\dot{\hat{\mathbf{a}}}_D = -\Gamma_D\,\mathbf{Y}_D^T(\dot{\mathbf{q}}_r)\,\mathbf{s}$$

we now obtain

$$\dot{V} = -\mathbf{s}^T(\mathbf{K}_D + \mathbf{D})\,\mathbf{s}$$

Thus, the second approach is equivalent to "increasing" $\mathbf{K}_D$ by the positive definite matrix $\mathbf{D}$. However, this is achieved using the desired velocity $\dot{\mathbf{q}}_d$ rather than the actual velocity $\dot{\mathbf{q}}$ , and therefore insensitively to the measurement noise on the velocity signals, so that one can typically tune the "effective" $\mathbf{K}_D$ at values higher than those obtained by increasing $\mathbf{K}_D$ directly. Here again, the explicit account of the dissipative nature of friction allows us in effect to increase the system's bandwidth beyond its classical limit.

A similar procedure can be used in the case of Coulomb friction. Assume that we compensate for Coulomb friction $\mathbf{D}_c \, \text{sgn}[\dot{\mathbf{q}}]$ (where $\mathbf{D}_c$ is a diagonal positive semi-definite matrix, and the sign function sgn is understood component-wise) using a term $\hat{\mathbf{D}}_c \, \text{sgn}[\dot{\mathbf{q}}_r]$ , rather than the obvious $\hat{\mathbf{D}}_c \, \text{sgn}[\dot{\mathbf{q}}]$ . Similarly to the case of viscous friction, one can easily show that this procedure adds a negative semi-definite term to $\dot{V}$. Indeed, the extra term appearing in $\dot{V}$ can be written

$$ - \mathbf{s}^T \mathbf{D}_c \, ( \, \text{sgn}(\dot{\mathbf{q}}_r ) - \text{sgn}(\dot{\mathbf{q}}) \, ) \; = \; - \sum_{i=1}^{n} s_i \, D_{ci} \, ( \, \text{sgn}(\dot{q}_i ) - \text{sgn}(\dot{q}_{ri} ) \, ) $$

where the terms $D_{ci} \geq 0$ are the diagonal elements of the Coulomb friction matrix $\mathbf{D}_c$ . Noticing that sums of the form $( \, \text{sgn}(\dot{q}_i ) - \text{sgn}(\dot{q}_{ri} ) \, )$ are either zero or of the same sign as $\text{sgn}(\dot{q}_i - \dot{q}_{ri} )$ , *i.e.*, as $\text{sgn}(s_i )$, we see that, indeed, this Coulomb friction compensation procedure adds a negative semi-definite term to $\dot{V}$. Furthermore, using $\text{sgn}[\dot{\mathbf{q}}_r]$ rather than $\text{sgn}[\dot{\mathbf{q}}]$ in the control and adaptation laws avoids the use of velocity measurements for friction compensation, and thus alleviates problems of inaccurate friction models at zero velocity.

This explicit distinction between dissipative and non-dissipative inputs is particularly relevant in cases where major components of the dynamics are involved. For instance, standard "square-law" models of hydrodynamic drag on a rigid body involve force/torque vectors of the form $\mathbf{D}(\dot{\mathbf{q}}) \, \dot{\mathbf{q}}$ , with $\mathbf{D}(\dot{\mathbf{q}})$ a symmetric positive definite matrix, and therefore are similar in structure to the previous example, with now the matrix $\mathbf{D}$ itself varying with the velocity vector $\dot{\mathbf{q}}$ .

## 9.3.3 Robotics as a Metaphor

Assuming that the full system state can be measured or estimated, conceptual extensions beyond mechanical systems can be derived: physical systems verify energy conservation equations of the form

$$ \frac{d}{dt} [ \, \text{Stored Energy} \, ] = [ \, \text{External Power Input} \, ] + [ \, \text{Internal Power Generation} \, ] $$

of which (9.9) is a particular instance. The external power input term can be written $\mathbf{y}^T \mathbf{u}$, where $\mathbf{u}$ is the input vector ("effort" or "flow"in the vocabulary of physical system modeling techniques, such as bond-graphs) and $\mathbf{y}$ is the output vector (flow or effort). By properly defining which of the physical inputs are "internal", we can always assume that the full vector $\mathbf{u}$ can be used for control purposes. Furthermore, we are not necessarily limited to dissipative or passive systems (*i.e.*, such that their internal power generation be negative or zero), as long as the internal power generated can be compensated for by the external power input, *i.e.*, by a proper choice of **u** (as in the case of gravitational forces in the robotic example). The major step is then to design a feedback controller structure (equation (9.18) in the robotic example) that translates the energy conservation equation in terms of a mapping between parametric uncertainty and a vector $\mathbf{s} = \mathbf{y} - \mathbf{y}_r$ (rather than between **u** and **y**), such that $\mathbf{s} = \mathbf{0}$ represent an exponentially stable differential equation in the tracking error – this implies, predictably, that the number of components of **s** (and, accordingly, the number of independent variables to be tracked) must be equal to (or smaller than) the dimension of the control input vector **u**. Furthermore, while an explicit *linear* parametrization of the dynamics is a key element in adaptive controller designs, bounds on the parametric uncertainty (which, in the robotic example, would correspond to bounds on the components of the vector $\mathbf{Y}\,\tilde{\mathbf{a}}$ and on disturbances) may be sufficient for simple robust designs, as seen in section 9.1.2.

**Example 9.4: Passivity Interpretation of Adaptive Manipulator Control**

The above discussion can be conveniently formalized in terms of the creation of passive mappings by means of feedback loops and adaptation loops.  Consider again the adaptive manipulator controller of section 9.1.3. Conservation of energy

$$\frac{1}{2}\frac{d}{dt}[\,\dot{\mathbf{q}}^T\,\mathbf{H}\,\dot{\mathbf{q}}\,] = \dot{\mathbf{q}}^T\,(\tau - \mathbf{g})$$

implies that the mapping $\tau - \mathbf{g} \rightarrow \dot{\mathbf{q}}$ is passive.  Taking the control law to be

$$\tau = \mathbf{Y}\,\hat{\mathbf{a}} - \mathbf{K}_D\,\mathbf{s}$$

*transforms this open-loop passive mapping*, between external input $\tau - \mathbf{g}$ and velocity $\dot{\mathbf{q}}$ , *into a closed-loop dissipative mapping* between parametric uncertainty $\mathbf{Y}\,\tilde{\mathbf{a}}$ and the "reference velocity error" vector $\mathbf{s} = \dot{\mathbf{q}} - \dot{\mathbf{q}}_r$ .  Indeed, our choice of control law yields

$$\frac{1}{2}\frac{d}{dt}[\,\mathbf{s}^T\,\mathbf{H}\,\mathbf{s}\,] = \mathbf{s}^T\,\mathbf{Y}\,\tilde{\mathbf{a}} - \mathbf{s}^T\,\mathbf{K}_D\,\mathbf{s}$$

Furthermore, using the adaptation law

$$\dot{\hat{\mathbf{a}}} = -\,\mathbf{\Gamma}\,\mathbf{Y}^T\mathbf{s}$$

then corresponds to *inserting a passive feedback block* between **s** and $-\mathbf{Y}\,\tilde{\mathbf{a}}$ (Figure 9.6), since

$$\frac{1}{2}\frac{d}{dt}[\tilde{\mathbf{a}}^T\,\mathbf{\Gamma}^{-1}\,\tilde{\mathbf{a}}] = -\,\mathbf{s}^T\,\mathbf{Y}\,\tilde{\mathbf{a}}$$

The passivity of the adaptation law can also be shown directly by noticing that the integrator

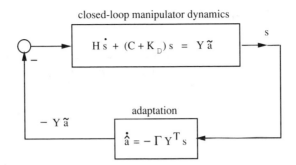

**Figure 9.6 :** Passivity interpretation of the adaptive robot controller

structure

$$\dot{\tilde{\mathbf{a}}} = \dot{\hat{\mathbf{a}}} = -\boldsymbol{\Gamma}\, \mathbf{Y}^T \mathbf{s}$$

implies that the mapping $-\mathbf{Y}^T\mathbf{s} \rightarrow \tilde{\mathbf{a}}$ is passive, and therefore that the mapping $\mathbf{s} \rightarrow -\mathbf{Y}\,\tilde{\mathbf{a}}$ is also passive. This also means that *the integrator in the adaptation law can be replaced by any operator which preserves the passivity of the mapping* $-\mathbf{Y}^T\mathbf{s} \rightarrow \tilde{\mathbf{a}}$.

For instance, the integrator, $1/p$ in the frequency domain, can be replaced by any passive transfer function of the form $h(p)/p$ (note that the factor $1/p$ has to be preserved explicitly, because the convergence proof exploits the equality $\dot{\tilde{\mathbf{a}}} = \dot{\hat{\mathbf{a}}}$). It can be replaced by any passive linear *MIMO* mapping between $-\mathbf{Y}^T\mathbf{s}$ and $\tilde{\mathbf{a}}$. It can also be replaced by any passive *nonlinear* mapping between $-\mathbf{Y}^T\mathbf{s}$ and $\tilde{\mathbf{a}}$ ; an example of performance enhancement using such a mapping is the "composite" adaptive manipulator controller of section 9.2.2.

We thus see that the passivity interpretation can allow for considerable design flexibility.  ☐

The adaptive manipulator control approach described above can be straightforwardly extended to the rigid dynamics of spacecraft, as shown in the next section. Similar concepts can apply to the nonlinear dynamics of high-performance aircraft. Furthermore, besides basic forms of energy conservation, the scalar summarizing property actually exploited for control system design may depend on the problem considered. In underwater vehicle dynamics, for instance, conservative effects such as added-mass can be directly accounted for by using the fluid's velocity potential flow. In systems with implicit interaction ports, available energy (in the case of interactions with quasi-ideal sources of temperature, pressure, or chemical potential) may be most suitable.

# 9.4 Spacecraft Control

Robotic spacecraft potentially represent a safe and economical alternative or complement to man in the construction, maintenance, and operation of the space structures to be deployed in the next decade. They present, however, specific and difficult control problems, largely due to their nonlinear dynamics. Furthermore, while spacecraft can potentially be expected to easily handle objects of masses and sizes comparable to or larger than their own (as, e.g., in releasing a payload from an orbiter, retrieving a satellite, performing docking or construction operations, or perhaps even throwing large loads accurately between robotic spacecraft), thanks to weightlessness, such tasks involve by nature large dynamic uncertainties, and therefore are likely to require effective adaptive capabilities. This section discusses the accurate attitude tracking control of rigid spacecraft handling large loads of unknown mass properties. The method is parallel to that of the robot manipulator case, and presents similar advantages over techniques based on inverse dynamics, in terms of simplicity, easier approach to robustness issues, and adaptive capabilities.

## 9.4.1 The Spacecraft Model

We consider the attitude control of a spacecraft driven by reaction wheels. In practice, the spacecraft may also be equipped with gas-jet systems (used *e.g.*, to control translational motion of the system, to compensate for non-zero translational momentum imparted by the loads, or to desaturate the reaction wheels), the control of which shall be commented upon later.

The spacecraft is treated as a rigid body whose attitude can be described by two sets of equations, namely, *kinematic* equations, which relate the time derivatives of the angular position coordinates to the angular velocity vector, and *dynamic* equations, which describe the evolution of the angular velocity vector. The development can be directly applied to the case of a spacecraft having rigidly secured a (possibly) large load of unknown mass properties. The results are also directly applicable to a spacecraft having itself inadequately known mass properties, due to, *e.g.*, reconfiguration, fuel variations in the gas-jet systems, thermal deformation, and so on.

### DYNAMIC EQUATIONS

Let us first define the reference frames in which our attitude control problem shall be described. We assume that the control torques are applied through a set of three reaction wheels along orthogonal axes (Figure 9.7). Based on these axes, we define an arbitrary orthonormal reference frame linked to the spacecraft, which we shall refer to as the *spacecraft frame*. The origin of this frame is not necessarily the center of mass of the system, nor are the axes necessarily the principal axes of the spacecraft. We also assume that an arbitrary *inertial frame* has been defined, with respect to either fixed stars or to a reference that can be considered inertial for the duration of the attitude maneuver (*e.g.*, a space station).

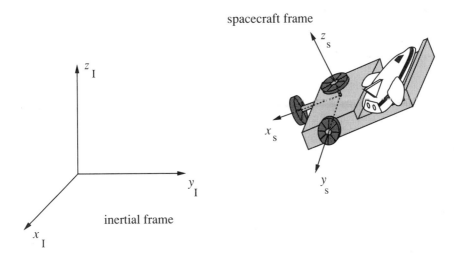

**Figure 9.7 :** Spacecraft attitude control using reaction wheels

Let $\omega$ denote the angular velocity vector of the spacecraft, expressed in the spacecraft frame. The equations describing the evolution of $\omega$ in time may be written as

$$\mathbf{H}\dot{\omega} = \mathbf{p} \times \omega + \tau$$

The "inertia" matrix $\mathbf{H}$ is symmetric positive definite, and can be written

$$\mathbf{H} = \mathbf{H}^0 - \mathbf{H}^A$$

where $\mathbf{H}^0$ is the total (spacecraft with reaction wheels) central inertia matrix, $\mathbf{H}^A$ is the (diagonal) matrix of axial wheels' inertias, and $\mathbf{p}$ is the total spacecraft angular momentum, all expressed in spacecraft coordinates. Note that since $\tau$ is the torque vector applied to the spacecraft by the reaction wheels, $-\tau$ is the vector of control torques actually applied by the reaction wheels motors. The $\times$ operator denotes the vector product operation. The notation $[\mathbf{v} \times]$ will refer to the skew-symmetric matrix defining the vector product by a vector $\mathbf{v}$

$$[\mathbf{v}\times] = \begin{bmatrix} 0 & -v_3 & v_2 \\ v_3 & 0 & -v_1 \\ -v_2 & v_1 & 0 \end{bmatrix}$$

## KINEMATIC EQUATIONS

The angular position of the body may be described in various ways. For example, one can consider the so-called Gibbs vector

$$\xi = \tan(\rho/2)\,\mathbf{e}$$

which derives from a quaternion parametrization. The vector $\xi$ represents the result of a virtual rotation of $\rho$ radians about a virtual unit axis $\mathbf{e}$, with reference to the inertial reference frame. In that case, one can write

$$\dot{\xi} = \mathbf{Z}(\xi)\,\omega$$

where

$$\mathbf{Z}[\xi] = \frac{1}{2}\,[\mathbf{I} + \xi\xi^T + \xi\times]$$

and $\mathbf{I}$ is the $3\times 3$ identity matrix. This description is valid for $-\pi < \rho < \pi$. Using this representation, the momentum $\mathbf{p}$ can be expressed as a function of $\xi$ by noting that

$$\mathbf{p} = \mathbf{C}(\xi)\,\mathbf{p}^I \tag{9.36}$$

where $\mathbf{p}^I$ is the (constant) inertial angular momentum, and the matrix $\mathbf{C}(\xi)$ represents the coordinate transformation from the inertial frame to the spacecraft frame:

$$\mathbf{C}(\xi) = 2\,(1 + \xi^T\xi)^{-1}\,[\mathbf{I} + \xi\xi^T - \xi\times] - \mathbf{I}$$

Note that $\mathbf{C}^{-1}(\xi) = \mathbf{C}^T(\xi) = \mathbf{C}(-\xi)$ .

Alternatively, the attitude of the body may be described using the classical Euler angle representation, consisting of consecutive angular clockwise rotations of angles $\phi$, $\theta$, and $\psi$ (roll, pitch, and yaw), that would bring the inertial frame in alignment with the spacecraft frame. In that case, one has

$$\dot{\gamma} = \mathbf{M}(\gamma)\,\omega$$

where $\gamma = (\phi, \theta, \psi)^T$, and

$$\mathbf{M}(\gamma) = \begin{bmatrix} 1 & \sin\phi\tan\theta & \cos\phi\tan\theta \\ 0 & \cos\phi & -\sin\phi \\ 0 & \sin\phi\sec\theta & \cos\phi\sec\theta \end{bmatrix}$$

This description is valid in the region $-\pi/2 < \theta < \pi/2$ (other representations use 4 "Euler parameters" to avoid such singularities). When using this second representation, one can write for **p** an expression similar to (9.36), namely

$$\mathbf{p} = \mathbf{R}(\gamma)\,\mathbf{p}^I$$

where $\mathbf{R}(\gamma)$ represents the coordinate transformation matrix from the inertial frame to spacecraft coordinates:

$$\mathbf{R}(\gamma) = \begin{bmatrix} \cos\psi\cos\theta & \cos\psi\sin\theta\sin\phi - \sin\psi\cos\phi & \cos\psi\sin\theta\cos\phi + \sin\psi\sin\phi \\ \sin\psi\cos\theta & \sin\psi\sin\theta\sin\phi + \cos\psi\cos\phi & \sin\psi\sin\theta\cos\phi - \cos\psi\sin\phi \\ -\sin\psi & \cos\psi\sin\theta & \cos\psi\cos\phi \end{bmatrix}$$

In summary, the equations describing the attitude of a spacecraft may be written in the synthetic form

$$\mathbf{H}\dot{\omega} = \mathbf{p} \times \omega + \tau$$

$$\dot{\mathbf{x}} = \mathbf{J}(\mathbf{x})\,\omega$$

(9.37)

where, depending on the representation used for the kinematic equations, the vector **x** is either the attitude vector $\gamma$, or the Gibbs vector $\xi$, with $\mathbf{R}(\gamma) = \mathbf{C}(\xi)$. Accordingly, the matrix **J** represents either the matrix **M** or the matrix **Z**.

Note that control-moment-gyroscopes (CMGs) may be used in place of reaction wheels while keeping a similar formalism. Instead of controlling angular momentum by varying the angular speeds of rotors of constant orientation, as in the case of reaction wheels, CMGs obtain the same effect by using simple or double gimbal mechanisms to vary the orientation of constant-speed rotors. In this case, however, the potential singularities of the gimbals have to be accounted for explicitly.

## 9.4.2  Attitude Control

**CHOICE OF STATES**

Consider the system described by (9.37), and choose as state-space coordinates the components of the vectors **x** and $\dot{\mathbf{x}}$. This choice is well-defined, since the matrix **J** remains invertible in the domain of validity of the kinematic representation.   By differentiating the expression of $\dot{\mathbf{x}}$, the equations of motion can be written as

$$\mathbf{H}^*(\mathbf{x})\ddot{\mathbf{x}} + \mathbf{C}^*(\mathbf{x},\dot{\mathbf{x}})\dot{\mathbf{x}} = \mathbf{F}$$

with

$$\boldsymbol{\tau} = \mathbf{J}^T\mathbf{F}$$

$$\mathbf{H}^*(\mathbf{x}) = \mathbf{J}^{-T}\mathbf{H}(\mathbf{x})\mathbf{J}^{-1} \tag{9.38}$$

$$\mathbf{C}^*(\mathbf{x},\dot{\mathbf{x}}) = -\mathbf{J}^{-T}\mathbf{H}\mathbf{J}^{-1}\dot{\mathbf{J}}\mathbf{J}^{-1} - \mathbf{J}^{-T}[\mathbf{p}\times]\mathbf{J}^{-1} \tag{9.39}$$

Note that it can be easily shown that one has the simple expression

$$\mathbf{J}^{-1}(\mathbf{x}) = 2\,(1 + \mathbf{x}^T\mathbf{x})^{-1}\,[\mathbf{I} - \mathbf{x}\times\,] \tag{9.40}$$

As in the robotic case, two properties of the above dynamics are exploited in the adaptive controller design.   First, the matrix $(\dot{\mathbf{H}}^* - 2\mathbf{C}^*)$ is *skew-symmetric* (a property which can be interpreted as reflecting conservation of energy).   Indeed, using expressions (9.38) and (9.39), we have

$$\dot{\mathbf{H}}^* - 2\,\mathbf{C}^* = \frac{d}{dt}(\mathbf{J}^{-T})\,\mathbf{H}\,\mathbf{J}^{-1} - \mathbf{J}^{-T}\,\mathbf{H}\,\frac{d}{dt}(\mathbf{J}^{-1}) + 2\,\mathbf{J}^{-T}[\mathbf{p}\times]\mathbf{J}^{-1}$$

which, given the skew-symmetry of the matrix $[\mathbf{p}\times]$ , implies the skew-symmetry of $(\dot{\mathbf{H}}^* - 2\,\mathbf{C}^*)$ .  Second, the dynamics is *linear* in terms of a properly defined constant parameter vector $\mathbf{a}$. While many such parametrizations are possible, we choose $\mathbf{a}$ to consist of the 6 independent components of the symmetric central inertia matrix $\mathbf{H}$, and of the 3 components of the constant inertial angular momentum $\mathbf{p}^I$.   Given expressions (9.38) and (9.39) and the relation (9.37), the matrices $\mathbf{H}^*$ and $\mathbf{C}^*$ are indeed linear in the constant parameter vector $\mathbf{a}$.

Based on the above dynamic formulation, it is now straightforward to derive, for spacecraft control, results similar to those obtained in the case of robot manipulators.   This is not surprising, since we are dealing in both cases with trajectory-controllable Hamiltonian systems.   Specifically, we first show the effectiveness of a simple proportional-derivative (P.D.) control for regulation applications. We then derive a globally convergent adaptive controller for fast attitude tracking applications, similar to the adaptive robot manipulator controller. We assume that the system's state vector, namely $\mathbf{x}$ and $\dot{\mathbf{x}}$, is available (or computable) from measurements.

## ATTITUDE REGULATION

We first show that a simple P.D.-like control law yields stable *position* control (i.e., given a *constant* $\mathbf{x}_d$, has $\mathbf{x}$ converge to $\mathbf{x}_d$), using essentially the same proof as in the robot manipulator case.

Consider a simple P.D. control law in $\mathbf{F}$, of the form

$$\boldsymbol{\tau} = \mathbf{J}^T \mathbf{F} = -\mathbf{J}^T [ \mathbf{K}_P \tilde{\mathbf{x}} + \mathbf{K}_D \dot{\tilde{\mathbf{x}}} ] \qquad (9.41)$$

Physically, this control law is actually mimicking the effect of a *passive* mechanical system composed of springs and dampers. This remark can be formalized by taking as a Lyapunov function candidate

$$V_1(t) = \frac{1}{2} [ \dot{\mathbf{x}}^T \mathbf{H}^* \dot{\mathbf{x}} + \tilde{\mathbf{x}}^T \mathbf{K}_P \tilde{\mathbf{x}} ]$$

Intuitively, $V_1(t)$ can be interpreted as the sum of the pseudo "kinetic energy" $1/2\, \dot{\mathbf{x}}^T \mathbf{H}^* \dot{\mathbf{x}}$ and of the "potential energy" $1/2\, \tilde{\mathbf{x}}^T \mathbf{K}_P \tilde{\mathbf{x}}$ associated with the virtual spring torques. Given the skew-symmetry of the matrix $(\dot{\mathbf{H}}^* - 2\,\mathbf{C}^*)$, the time-derivative of $V_1(t)$ can be written as

$$\dot{V}_1(t) = \dot{\mathbf{x}}^T ( \mathbf{F} + \mathbf{K}_P \tilde{\mathbf{x}} )$$

which, using the P.D. control law (9.41), simplifies as

$$\dot{V}_1(t) = -\dot{\mathbf{x}}^T \mathbf{K}_D \dot{\mathbf{x}} \le 0$$

Not surprisingly, we recognize $\dot{V}_1(t)$ as the power dissipated by the virtual dampers. We now only need to check that the system cannot get "stuck" at a stage where $\dot{V}_1$ equals 0 while $\mathbf{x}$ does not equal $\mathbf{x}_d$, or, to put it more technically, invoke the invariant-set theorem. Since $\dot{V}_1 = 0$ implies that $\dot{\mathbf{x}} = \mathbf{0}$, which in turn implies that $\ddot{\mathbf{x}} = (\mathbf{H}^*)^{-1} \mathbf{K}_P \tilde{\mathbf{x}}$, one has that $\dot{V}_1 \equiv 0$ only if $\tilde{\mathbf{x}} = \mathbf{0}$ (note that, given the Lyapunov derivation, $\mathbf{x}$ is bounded as long as $\mathbf{x}_d$ is bounded, so that the singularity of the Gibbs-vector kinematic representation is not reached, and therefore $\mathbf{H}^*$ and $\mathbf{H}^{*-1}$ exist). Thus, the system does converge to the desired state, as the physical reasoning suggested.

Note that the position convergence of this very simple controller does not require *a priori* knowledge of the system's mass properties, although, of course, for a given set of gains, its transient performance does depend on the actual mass properties of the spacecraft. Similarly, the proof extends to the case when the spacecraft is equipped with flexible appendages (such as solar panels or large antennas) because such structures are themselves mechanically passive systems, although the transient performance depends on the natural damping of the appendages.

## ADAPTIVE ATTITUDE TRACKING

The adaptive control problem applies to more demanding, fast attitude tracking applications. It consists in generating a *control law* to be applied by the reaction wheels, and a parameter vector *estimation law*, such that the tracking error asymptotically converge to zero, *i.e.*, such that the actual attitude vector $\mathbf{x}(t)$ "catch up" with the desired attitude trajectory $\mathbf{x}_d(t)$. Although it yields a much higher performance controller than the simple P.D. approach described above, the adaptive controller also avoids any required *a priori* knowledge of the system's mass properties.

We assume that the desired $\mathbf{x}_d$, $\dot{\mathbf{x}}_d$, and $\ddot{\mathbf{x}}_d$ are all bounded. Let $\hat{\mathbf{a}}$ be the (time-varying) parameter vector estimate, $\tilde{\mathbf{a}} = \hat{\mathbf{a}} - \mathbf{a}$ be the parameter estimation error, and $\tilde{\mathbf{x}} = \mathbf{x} - \mathbf{x}_d$ be the tracking error. Consider the Lyapunov-like function

$$V(t) = \frac{1}{2} [ \, \mathbf{s}^T \mathbf{H}^* \mathbf{s} + \tilde{\mathbf{a}}^T \mathbf{\Gamma}^{-1} \tilde{\mathbf{a}} \, ]$$

with the vector $\mathbf{s}$, a measure of tracking error, defined as

$$\mathbf{s} = \dot{\tilde{\mathbf{x}}} + \lambda \tilde{\mathbf{x}} \tag{9.42}$$

where $\lambda$ is a strictly positive constant. We shall write (9.42) as

$$\mathbf{s} = \dot{\mathbf{x}} - \dot{\mathbf{x}}_r$$

where

$$\dot{\mathbf{x}}_r = \dot{\mathbf{x}}_d - \lambda \tilde{\mathbf{x}} \tag{9.43}$$

As in the robotic case, the "reference velocity" vector $\dot{\mathbf{x}}_r$ is formed by shifting the desired velocity $\dot{\mathbf{x}}_d$ according to the position error $\tilde{\mathbf{x}}$. Intuitively, the first term in the Lyapunov function candidate $V(t)$ is obtained by replacing the actual velocity $\dot{\mathbf{x}}$ by the "reference velocity error" $\mathbf{s}$ in the expression of the pseudo "kinetic energy" $1/2 \, \dot{\mathbf{x}}^T \mathbf{H}^* \dot{\mathbf{x}}$ ; the second term in $V(t)$ represents a quadratic measure of parameter estimation error.

Differentiating $V(t)$, and using the skew-symmetry of the matrix $(\dot{\mathbf{H}}^* - 2\mathbf{C}^*)$, leads to

$$\dot{V}(t) = \mathbf{s}^T (\mathbf{F} - \mathbf{H}^* \ddot{\mathbf{x}}_r - \mathbf{C}^* \dot{\mathbf{x}}_r ) + \tilde{\mathbf{a}}^T \mathbf{\Gamma}^{-1} \dot{\tilde{\mathbf{a}}}$$

Taking the control law to be

$$\mathbf{F} = \hat{\mathbf{H}}^*(\mathbf{x})\ddot{\mathbf{x}}_r + \hat{\mathbf{C}}^*(\mathbf{x},\dot{\mathbf{x}})\dot{\mathbf{x}}_r - \mathbf{K}_D\mathbf{s} \tag{9.44}$$

where $\mathbf{K}_D$ is a symmetric positive definite matrix, yields

$$\dot{V}(t) = \mathbf{s}^T(\tilde{\mathbf{H}}^*\ddot{\mathbf{x}}_r + \tilde{\mathbf{C}}^*\dot{\mathbf{x}}_r - \mathbf{K}_D\mathbf{s}) + \tilde{\mathbf{a}}^T\mathbf{\Gamma}^{-1}\dot{\tilde{\mathbf{a}}}$$

where

$$\tilde{\mathbf{H}}^* = \hat{\mathbf{H}}^* - \mathbf{H}^* \qquad \tilde{\mathbf{C}}^* = \hat{\mathbf{C}}^* - \mathbf{C}^*$$

The linear parametrization of the dynamics allows us to define a *known* matrix $\mathbf{Y}^*(\mathbf{x},\dot{\mathbf{x}},\dot{\mathbf{x}}_r,\ddot{\mathbf{x}}_r)$ such that

$$\tilde{\mathbf{H}}^*(\mathbf{x})\ddot{\mathbf{x}}_r + \tilde{\mathbf{C}}^*(\mathbf{x},\dot{\mathbf{x}})\dot{\mathbf{x}}_r = \mathbf{Y}^*(\mathbf{x},\dot{\mathbf{x}},\dot{\mathbf{x}}_r,\ddot{\mathbf{x}}_r)\,\tilde{\mathbf{a}}$$

so that

$$\dot{V}(t) = -\mathbf{s}^T\mathbf{K}_D\mathbf{s} + \tilde{\mathbf{a}}^T[\mathbf{\Gamma}^{-1}\dot{\tilde{\mathbf{a}}} + \mathbf{Y}^{*T}\mathbf{s}]$$

Note that, given definition (9.24) of the reference velocity vector $\dot{\mathbf{x}}_r$, the computation of the matrix $\mathbf{Y}^*(\mathbf{x},\dot{\mathbf{x}},\dot{\mathbf{x}}_r,\ddot{\mathbf{x}}_r)$ only requires $\mathbf{x}$ and $\dot{\mathbf{x}}$ to be measured. Choosing the adaptation law as

$$\dot{\hat{\mathbf{a}}} = -\mathbf{\Gamma}(\mathbf{Y}^*)^T\mathbf{s}$$

yields

$$\dot{V}(t) = -\mathbf{s}^T\mathbf{K}_D\mathbf{s} \leq 0 \tag{9.45}$$

This shows, using the uniform continuity of $\dot{V}$ as in the robotic case, that $\dot{V} \rightarrow 0$ as $t \rightarrow \infty$, which in turn implies that $\mathbf{s} \rightarrow \mathbf{0}$, which shows that $\tilde{\mathbf{x}} \rightarrow \mathbf{0}$ and $\dot{\tilde{\mathbf{x}}} \rightarrow \mathbf{0}$. Therefore, both global stability of the system and convergence of the tracking are guaranteed by the above adaptive controller.

Note that one can show easily that adding a term ($\lambda\,\tilde{\mathbf{x}}^T\mathbf{K}_D\,\tilde{\mathbf{x}}$) to the function $V$ still maintains $\dot{V} \leq 0$, which shows directly that $\tilde{\mathbf{x}}$ is bounded, and therefore that the singularity of the quaternion-based (Gibbs vector) kinematic representation is not reached as long as the rotation corresponding to the desired maneuver does not exceed 180 degrees. In the case that the desired maneuver does not correspond to this requirement, one can simply decompose the control problem in subtasks, and change reference frames between tasks.

The control law, expressed in terms of the applied reaction wheels control torques, can be written from (9.44) as

$$\boldsymbol{\tau} = \mathbf{J}^T(\hat{\mathbf{H}}^*\ddot{\mathbf{x}}_r + \hat{\mathbf{C}}^*\dot{\mathbf{x}}_r - \mathbf{K}_D\mathbf{s}) = \mathbf{J}^T(\mathbf{Y}^*\hat{\mathbf{a}} - \mathbf{K}_D\mathbf{s})$$

Note that while we are estimating the *central* inertia matrix **H** as part of the adaptation process, the actual position of the center of mass of the system is not assumed to be known. Also, note that the scheme does not necessarily estimate the unknown parameters exactly, but simply generates values that allow the tracking error to converge to zero (i.e., the actual trajectory to "catch up" with the desired trajectory). As in the robotic case, "sufficient richness" conditions indicate how demanding the desired trajectory should be for tracking convergence to necessarily require parameter convergence. For a given trajectory, the speed of tracking convergence is determined by the values of the gains $\lambda$, $\mathbf{K}_D$, and $\Gamma$, themselves limited by the presence of high-frequency unmodeled dynamics and measurement noise. Composite adaptive versions of the algorithm can also be developed.

As in the robotic case, the methodology can be easily extended to the case where only a few significant parameters are adapted upon (in order to simplify the computations, for instance), while the controller is robust to residual errors resulting from inaccuracies on the *a priori* estimates of the other parameters, as well as to bounded time-varying additive disturbances (*e.g.*, such as those that may be created by vibrations of flexible appendages). Also, as in section 9.3.1, the P.D. term $-\mathbf{J}^T\mathbf{K}_D\mathbf{s}$ in **F** can be replaced by any term of the form $-\mathbf{J}^T[\mathbf{K}_D\dot{\tilde{\mathbf{x}}} + \mathbf{K}_P\tilde{\mathbf{x}}]$ (where the constant matrices $\mathbf{K}_D$ and $\mathbf{K}_P$ are symmetric positive definite), and thus be tuned independently of $\lambda$ so as to be more robust to unmodeled structural vibrations.

Finally, note that we only assumed that the system's state (*i.e.*, **x** and $\dot{\mathbf{x}}$) is available from measurements. If the initial angular velocity $\Omega(0)$ of the reaction wheels about their axes can also be reliably measured, then, using the fact that

$$\mathbf{p}^I = \mathbf{R}(-\mathbf{x}(0))\,[\mathbf{H}^0\boldsymbol{\omega}(0) + \mathbf{H}^A\boldsymbol{\Omega}(0)] = \mathbf{R}(-\mathbf{x}(0))\,\mathbf{H}^A\,[\boldsymbol{\omega}(0) + \boldsymbol{\Omega}(0)] + \mathbf{R}(-\mathbf{x}(0))\,\mathbf{H}\,\boldsymbol{\omega}(0)$$

the above adaptive controller can be expressed (under the mild assumption that the matrix $\mathbf{H}^A$ of axial wheels inertias is known) in terms of only the 6 independent components of the inertia matrix **H**. Of course, if $\mathbf{p}^I$ itself is known (*e.g.*, to be zero), then again the adaptive controller need only estimate the 6 independent components of the inertia matrix **H** .

**Example 9.5**   Let us illustrate the development with simple simulations.   Assume that the spacecraft, initially at $\mathbf{x}(0) = \mathbf{0}$, is required to follow a hypothetical desired attitude trajectory consisting of five parts, namely, a 1-second constant acceleration phase to $\mathbf{x}_d(1) = [0.5\ \ 0.5\ \ 0.5]^T$, a 1-second constant deceleration phase to $\mathbf{x}_d(2) = [1.\ \ 1.\ \ 1.]^T$, a 3-second constant attitude phase, and the reverse trajectory back to the original attitude. Assume that the actual central inertia matrix **H** is

$$\mathbf{H} = \begin{bmatrix} 15 & 5 & 5 \\ 5 & 10 & 7 \\ 5 & 7 & 20 \end{bmatrix}$$

and that the spacecraft has an inertial angular momentum $\mathbf{p}^I = [1. -1. 0.]^T$ (imparted *e.g.*, by the original spinning of the load). The sampling rate used in the simulations is 100 Hz.

For comparison purposes, we first consider the performance of a simple P.D. controller in $\mathbf{F}$, with $\mathbf{K}_P = 200\,\mathbf{I}$, $\mathbf{K}_D = 20\,\mathbf{I}$. The corresponding tracking errors are given in Figure 9.8.

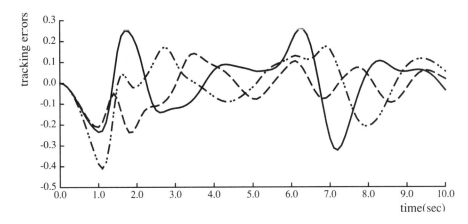

**Figure 9.8 :** Tracking errors, P.D. controller

Although the practical purpose of the adaptive controller is to achieve consistent performance in the presence of large load variations, the strongest demonstration of the algorithm consists in adapting to the mass properties of the whole spacecraft. That is, by setting $\hat{\mathbf{a}}(0) = \mathbf{0}$, no initial information is assumed about the inertia matrix or the angular momentum, and therefore, in this example, the adaptive controller starts exactly as the P.D. The gains $\mathbf{K}_D$ and $\lambda$ are also chosen such that the term $\mathbf{K}_D\mathbf{s}$ ( $= \mathbf{K}_D\dot{\tilde{\mathbf{x}}} + \lambda\mathbf{K}_D\tilde{\mathbf{x}}$ ) be exactly equal to the P.D. control law (namely, with $\mathbf{K}_D$ as in the P.D. controller, and $\lambda = 10$). The adaptation gain matrix is $\Gamma = 30\,\mathbf{I}$ (a conservative choice, as is indicated by the smoothness of the parameter estimates). The corresponding tracking errors (Figure 9.9) show a five-time performance improvement over the P.D. in the first leg, and a ten-time improvement in the second leg (where more initial information is available from the first-leg adaptation process). The adaptive controller torques are similar in magnitude and activity to those of the P.D. controller. Also, this controller performance is obtained along a smooth and rather "unexciting" desired trajectory, which represents a challenge to an adaptive controller.                                                                                           □

Advanced control algorithms present exceptional potential in the dynamically clean, weightless environment of space. The above development can in principle be

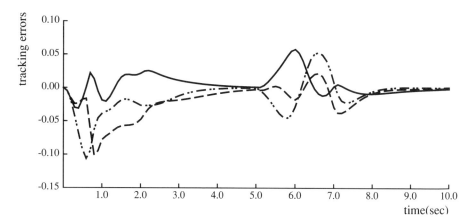

**Figure 9.9 :** Tracking errors, adaptive controller

extended easily to include translational control of the spacecraft using gas-jets, by adding the system's mass and position of the center of mass to the list of components of the unknown parameter vector **a**. Again, if we assume the reaction wheels' axial angular velocities to be measured, and their axial inertias to be known, then only a total of 10 parameters need to be estimated as part of the adaptation process. However, the necessity of averaging on-off gas-jet action presents problems of its own.

In addition, the lightness requirements in space components may also present difficulties linked to the presence of low-frequency structural modes. In particular, while the previous discussion can be extended easily to control the rigid dynamics of manipulators mounted on the spacecraft, practical implementation may require flexibility issues to be explicitly addressed. In many space robotics applications, however, distributed flexibility effects can be adequately modeled using simple lumped approximations, which can in turn be easily handled using *e.g.*, singularly perturbed versions of the original rigid-model adaptive control results.

## 9.5  Summary

Simplicity and enhanced performance may be achieved by astutely exploiting the known physical properties of the controlled systems. In the case of robot manipulators, explicitly accounting for energy conservation allows us to study easily the stability of basic control laws such as P.D.'s, and to derive simple robust controllers. The additional property of linear parametrization, common to all mechanical systems, then also allows us to construct globally tracking-convergent adaptive robot controllers. Similar approaches can be taken in the contexts of, *e.g.*, spacecraft control and process control. Physical insights may also allow us to exploit dissipation terms (such as friction in a mechanical system) rather than fight them, and they may also guide the study of robustness to unmodeled dynamics.

## 9.6  Notes and References

The P.D. stability result of section 9.1.1 was first proved in a robotics context by [Takegaki and Arimoto, 1981], and independently rederived by [De Jonckheere, 1981; Koditschek, 1984; Van der Schaft, 1986]. The slightly simpler proof used here is from [Slotine, 1988]. [Arimoto and Miyazaki, 1983] extended the result to P.I.D. control in the presence of gravitational torques. [Hogan, 1985] provides a discussion of the closely related topic of impedance control. Details on robot kinematics and dynamics can be found, *e.g.*, in [Asada and Slotine, 1986].

The development of sections 9.1.2 and 9.2 is adapted from [Slotine and Li, 1986-1989]. Related results can be found, *e.g*, in [Craig, *et al.*, 1986; Bastin and Campion, 1986; Middleton and Goodwin, 1986; Bayard and Wen, 1987; Hsu, *et al.*, 1987; Sadegh and Horowitz, 1987; Koditschek, 1987; Paden and Panja, 1988; Middleton, 1988; Ortega and Spong, 1988; Li and Slotine, 1989]. Passivity interpretations are discussed in [Kelly and Carelli, 1988; Landau and Horowitz, 1988; Ortega and Spong, 1988]. Section 9.3 is adapted from [Slotine, 1988]. Computational aspects are studied in [Walker, 1988; Niemeyer and Slotine, 1988, 1989]. The development is extensively demonstrated experimentally in [Slotine and Li, 1987a, 1988] and in [Niemeyer and Slotine, 1988]. Extensions of the approach to process control are discussed in [Slotine and Yidstie, 1989, 1990]. Issues of persistency of excitation are discussed, *e.g.*, in [Slotine and Li, 1987c].

Section 9.4 is based on [Slotine and Di Benedetto, 1988, 1990]. Related recent results can be found, *e.g.*, in [Baillieul and Levi, 1987; Wen, 1988; Koditschek, 1988; Dwyer and Kim, 1989; Singh, 1989] and references therein.

## 9.7  Exercises

**9.1**   Consider a robot control law of the form

$$\tau_i = -\tau_i^{max} \, \text{sgn} \, s_i$$

where $\tau_i^{max}$ is the maximum torque that the $i^{\text{th}}$ actuator can produce (and $s_i = \dot{\tilde{q}}_i + \lambda \tilde{q}_i$). Assuming that control chattering is not an issue, what are the trajectories that can be tracked by this simple controller?

Check your results in simulations on the manipulator of Example 9.2.

Assume now that the manipulator is in a vertical plane. How are the above results modified?

**9.2**   Show that the term $-\mathbf{k}\ \text{sgn}[\mathbf{s}]$ in the sliding controller of section 9.1.2 can be replaced by a term of the form $-\mathbf{k}\ \mathbf{s}/\|\mathbf{s}\|$ while preserving global tracking convergence. What associated inequality must $\mathbf{k}$ satisfy?

**9.3**   A mathematical result in input-output stability states that a (strictly) lowpass filtered version of a square-integrable signal necessarily tends to zero. Show that this result allows to conclude immediately from (9.17) and (9.20) that $\tilde{\mathbf{q}} \to \mathbf{0}$ as $t \to \infty$.

**9.4**   For the two-link manipulator of Example 9.3, check explicitly the uniform positive-definiteness of the matrix $\mathbf{H}$, and the skew-symmetry of the matrix $(\dot{\mathbf{H}} - 2\mathbf{C})$.

**9.5**   Simulate the adaptive controller of Example 9.3 on various "typical" desired trajectories. Discuss parameter convergence as a function of the trajectory excitation. Using (8.91), discuss intuitively why

$$\xi = \frac{1}{T} \int_{o}^{T} \mathbf{Y}_d^T \mathbf{Y}_d \, dr$$

is reasonable measure of a trajectory's excitation (where $\mathbf{Y}_d = \mathbf{Y}(\mathbf{q}_d, \dot{\mathbf{q}}_d, \dot{\mathbf{q}}_d, \ddot{\mathbf{q}}_d)$ and $T$ is the time length of the trajectory). Show that it is reasonable to pick $\boldsymbol{\Gamma}$ approximately proportional to the average value of $\mathbf{Y}_d^T \mathbf{Y}_d$ on typical trajectories.

**9.6**   Consider again the adaptive control of a manipulator, and assume for simplicity that $\boldsymbol{\Lambda} = \lambda\,\mathbf{I}$. Show that tracking convergence is preserved if, in the control law, an extra term of the form $\mathbf{K}_P\,\tilde{\mathbf{q}}$ is added to the term $\mathbf{K}_D\,\mathbf{s}$, where $\mathbf{K}_P$ is an arbitrary symmetric positive definite matrix. Specifically, show that adding $1/2\,\tilde{\mathbf{q}}^T\,\mathbf{K}_P\,\tilde{\mathbf{q}}$ to the original Lyapunov-like function $V$ then adds a negative (semi-definite) term $-\lambda\,\tilde{\mathbf{q}}^T\,\mathbf{K}_P\,\tilde{\mathbf{q}}$ to its derivative $\dot{V}$.

   From a practical point of view, the above result means that the P.D. control gains can be tuned independently of $\lambda$. Assuming that the presence of structural resonant modes is the active limit on the magnitude of $\lambda$, illustrate in simulations the advantages that such design flexibility provides, on the two-link manipulator of Example 9.3.

**9.7**   Show that if the BGF update in the composite adaptive robot controller of section 9.2.2 is replaced simply by an exponentially forgetting least-squares update with a *constant* forgetting factor, then $V$ converges to zero exponentially, regardless of excitation. Does this implies exponential convergence of the tracking and of the parameters?

**9.8**   Given the results of Exercise 9.6, and using the BGF adaptive controller of section 9.2.2., suggest reasonable *a priori* choices of $\lambda$, $\mathbf{K}_P$, $\mathbf{R}$, and the filter used to obtain $\mathbf{W}$ (including the filter's d.c. gain), given a characterization of unmodeled dynamics as in section 7.3. Check your results in simulations.

**9.9** Show that a manipulator can be viewed as a passive mapping between joint torques and joint velocities, even in the presence of gravitational torques. Does the manipulator necessarily have to be rigid?

**9.10** Discuss sliding and adaptive controllers for systems of the form

$$\mathbf{H}\, \mathbf{q}^{(n)} = \mathbf{f} + \mathbf{u}$$

where the notation $\mathbf{q}^{(n)}$ refers to a vector of components $q_1^{(n_1)}, q_2^{(n_2)}, \dots$, with the state vector being

$$\mathbf{x} = [\, q_1, \dot{q}_1, \dots, q_1^{(n_1-1)}, q_2, \dot{q}_2, \dots, q_2^{(n_2-1)}, \dots\,]^T$$

$\mathbf{f}$ being a nonlinear function of the state $\mathbf{x}$, and $\mathbf{H}$ being a *symmetric positive definite matrix* function of the state $\mathbf{x}$ and not depending explicitly on any of the $q_i^{(n_i-1)}$.

**9.11** In the single-input case, extend the discussion of section 9.3.2 to the compensation of monotonous increasing functions of $\dot{q}$ (or, more generally, of $q^{(n-1)}$, for $n^{\text{th}}$-order dynamics rather than second-order dynamics), such as *e.g*, a "square drag" $\dot{q}\,|\dot{q}|$ in an underwater vehicle. Generalize in the multi-input case to the compensation of functions with positive semi-definite gradients.

**9.11** The manipulator control algorithms described in this chapter can be implemented *directly* in cartesian space, *i.e.*, using the end-point position (and orientation) $\mathbf{x}$ and velocity $\dot{\mathbf{x}}$ as the new state, in place of the joint positions and velocities. Two approaches are possible:

(a) Write the kinematic relation between $\mathbf{q}$ and $\mathbf{x}$

$$\dot{\mathbf{x}} = \mathbf{J}\,\dot{\mathbf{q}}$$

where $\mathbf{J} = \mathbf{J}(\mathbf{q})$ is the so-called manipulator Jacobian. Use that kinematic relation to compute $\ddot{\mathbf{x}}$ from the robot dynamics. Show that this leads to writing the dynamics in the form [Khatib, 1983]

$$\mathbf{H}^*\ddot{\mathbf{x}} + \mathbf{C}^*\dot{\mathbf{x}} + \mathbf{g}^* = \mathbf{F} \qquad \text{with} \qquad \boldsymbol{\tau} = \mathbf{J}^T\mathbf{F}$$

Complete the adaptive controller design.

(b) With obvious notations, compute $\dot{\mathbf{q}}_r$ according to

$$\dot{\mathbf{q}}_r = \mathbf{J}^{-1}\dot{\mathbf{x}}_r$$

where we assume that the Jacobian remains invertible over the domain of operation, and complete the design by using the usual joint-space adaptive controller.

Show that the two approaches actually lead to *identical* equations, except for the form of $\mathbf{K}_D$ . Which is easier to implement computationally?

How do these results translate to spacecraft control? (Adapted from [Slotine and Li, 1986; Niemeyer and Slotine, 1989].)

# BIBLIOGRAPHY

**Alexander, H. L., and Cannon, R. H.**, Experiments on the Control of a Satellite Manipulator, *Stanford University* (1987).

**Anderson, B. D. O.**, Exponential Stability of Linear Systems Arising From Adaptive Identification. *I.E.E.E. Trans. on Auto. Contr.*, 22-2 (1977).

**Anderson, B. D. O., Bitmead, R. R., Johnson, C. R., Jr, Kokotovic, P. V., Kosut, R. L., Mareels, I. M. Y., Praly, L., Riedle, B. D.**, Stability of Adaptive Systems: Passivity and Averaging Analysis, *M.I.T. Press* (1986).

**Anderson, B. D. O. and Johnson, C. R. Jr.**, Exponential Convergence of Adaptive Identification and Control Algorithms, *Automatica*, 19, 1 (1982).

**Arimoto, S.**, Learning Control Theory for Dynamic Systems - A Survey, *I.E.E.E. Conf. Decision and Control*, Fort Lauderdale (1985).

**Arimoto, S. and Miyazaki, F.**, Stability and Robustness of P.I.D. Feedback Control for Robot Manipulators of Sensory Capability, *1st Int. Symp. Robotics Res.* (1983).

**Arnold, V. I.**, Ordinary Differential Equations, *MIT Press* (1973).

**Asada, H.**, Studies In Prehension and Handling By Robot Hands With Elastic Fingers, *Doctoral Dissertation, Kyoto University* (1979).

**Asada, H. and Kanade, T.**, Design of Direct-Drive Arms, *A.S.M.E. Journal of Vibration, Stress, and Reliability In Design*, **105**, 3, July (1983).

**Asada, H. and Slotine, J.-J. E.**, Robot Analysis and Control, *John Wiley and Sons* (1986).

**Asakawa, K., Akiya, F. and Tabata, F.**, A Variable Compliance Device and Its Application for Automatic Assembly, *Fujitsu Laboratories*, Kawasaki, Japan (1982).

**Astrom, K. J.**, Analysis of Rohrs Counterexamples to Adaptive Control, *Proc. 22nd I.E.E.E. Conf. on Dec. and Contr.*, San Antonio, TX, p. 982 (1983a).

**Astrom, K. J.**, Theory and Applications of Adaptive Control: A Survey, *Automatica*, 19, (5) (1983b).

**Astrom, K. J.**, Interactions Between Excitation and Unmodeled Dynamics in Adaptive Control,

*Proc. 23rd I.E.E.E. Conf. on Dec. and Contr.*, Las Vegas, NV, pp. 1276-1281 (1984).

**Astrom, K. J. and Wittenmark, B.**, Computer Controlled Systems: Theory and Design, *Prentice-Hall* (1984).

**Astrom, K. J. and Wittenmark, B.**, Adaptive Control, *Addison-Wesley* (1989).

**Atkeson, C. G., An, C. M. and Hollerbach, J. M.**, Estimation of Inertial Parameters of Manipulator Loads and Links, *Int. Symp. Robotics Res.*, Gouvieux (1985).

**Aizerman, M. A. and Gantmakher, F. R.**, On Some Features of Switchings in Nonlinear Control with a Piecewise-Smooth Response of the Nonlinear Element, *Avtomatika i Telemekhanika*, 18, 11 (1957).

**Baillieul, J.**, Controllability and Observability of Polynomial Systems, *Nonlin. Anal.* 5, pp. 543-552 (1981).

**Baillieul, J. and Levi, M.**, Rotational Elastic Dynamics, *Physica, Norht-Holland Physics Publishing* (1987).

**Banks, S. P.**, A Note on Nonlinear Observers, *Int. J. Contr.*, 34, 1, pp. 185-190 (1981).

**Banks, S. P.**, Control Systems Engineering: Modelling and Simulation, Control Theory and Microprocessor Implementation, *Prentice-Hall* (1986).

**Barmish, B. R., Corless, M. J. and Leitmann, G.**, A New Class of Stabilizing Controllers for Uncertain Dynamical Systems, **S.I.A.M. J. Contr. Optimization**, 21, 2, pp. 264-255 (1983).

**Bastin, G. and Campion, G.**, Indirect Adaptive Control of Linearly Parameterized Nonlinear Systems, *3rd I.F.A.C. ACASP*, preprint (1989).

**Bayard, D. S. and Wen, J. T.**, Simple Adaptive Control Laws for Robotic Manipulators, *Proc. of Fifth Yale Workshop on the Applications of Adaptive Systems Theory* (1987).

**Bergen, A. R. and Franks, R. L.**, Justification of the Describing Function Method, *S.I.A.M. J. Control*, 9, pp. 568-589 (1971).

**Bergmann, E. V., Walker, B. K., and Levy, D. R.**, Mass Property Estimation for Control of Asymmetrical Satellites, *A.I.A.A. Journal of Guidance, Control, and Dynamics*, 10, 5 (1987).

**Bobrow, J. E., Dubowsky, S., and Gibson, J. S.**, Time-Optimal Control of Robotic Manipulators Along Specified Paths, *Int. J. Robotics Research*, **4**, 3 (1985).

**Bodson, M.**, Stability, Convergence, and Robustness of Adaptive Systems, *Doctoral thesis, College of Engineering, University of California, Berkeley* (1986).

**Bondarev, A. G., Bondarev, S. A., Kostylyova, N. Y., and Utkin, V. I.**, Sliding Modes in Systems with Asymptotic State Observers, *Autom. Remote Contr.*, **6** (1985).

**Book W. J.**, Recursive Lagrangian Dynamics of Flexible Manipulator Arms, *2nd Int. Sym. of Robotics Research*, Kyoto (1984).

**Bottema, O. and Roth, B.**, Theoretical Kinematics, *North Holland, Amsterdam* (1979).

**Brady, M. T., Hollerbach, M., Johnson, T. L., Lozano-Perez, T. and Mason, M. T.**, Robot Motion, *M.I.T. Press* (1982).

**Brockett, R. W.**, Volterra Series and Geometric Control Theory, *Automatica*, 12, pp. 167-176 (1976).

**Brockett, R. W.**, Feedback Invariants for Non-linear Systems, *I.F.A.C. Congress*, 6, pp. 1115-1120 (1978).

**Brockett, R. W.**, Asymptoticstability and Feedback Stabilization, *Differential Geometric Control Theory*, R. W. Brockett, R. S. Millman and H. Sussmann eds., *Birkhauser*, pp. 181-191 (1983).

**Brunovsky, P.**, A Classification of Linear Controllable Systems, *Kybernetika*, 6, pp. 173-188 (1970).

**Bryson, A. E.**, Control of Spacecraft and Aircraft, *Department of Aeronautics and Astronautics Report*, Stanford University (1985).

**Bryson, A. and Ho, Y.**, Applied Optimal Control, *Hemisphere* (1975).

**Byrnes, C. I. and Isidori, A.**, A Frequency Domain Philosophy for Nonlinear Systems, *I.E.E.E. Conf. Dec. Contr.*, 23, pp. 1569-1573 (1984)

**Byrnes, C. I., and Isidori, A.**, On the Attitude Stabilization of Rigid Spacecraft, *preprint* (1986).

**Byrnes, C. I. and Isidori, A.**, Local Stabilization of Minimum-Phase Nonlinear Systems, *Syst. Contr. Lett.*, 11, pp. 9-17 (1988a).

**Byrnes, C. I. and Isidori, A**, Feedback Design from the Zero Dynamics Point of View, *Bozeman Work. Comp. Contr.* (1988b).

**Byrnes, C. I. and Isidori, A.**, Analysis and Design of Nonlinear Feedback Systems, *I.E.E.E. Trans. Aut. Contr.* (1989a).

**Byrnes, C. I. and Isidori, A.**, Output Regulation of Nonlinear Systems, preprint (1989b).

**Cannon, R. H. and Schmitz, E.**, Initial Experiments on the End-Point Control of a Flexible One-

Link Robot, *Int. J. Robotics Research*, 3, 3 (1984).

**Canudas, C., Astrom, K. J. and Braun, K.**, Adaptive Friction Compensation in D.C. Motor Drives, *I.E.E.E. Int. Conf. Robotics and Automation*, San Francisco (1986).

**Cetaev, N. G.**, A Theorum on Instability (Russian), *Dok. Akad. Nauk. SSSR*, 1, pp. 529-531 (1934).

**Cetaev, N. G.**, On the Instability of Equilibrium in Some Cases Where the Force Function is Not Maximum (Russian), *Prikl. Mat. Meh.*, 16, pp. 89-93 (1952).

**Cetaev, N. G.**, The Stability of Motion, *Pergamon Press*, New York (1961); translation of the Russian edition, Moskow (1955).

**Cho, D. and Hedrick, J. K.**, A Nonlinear Controller Design Method for Fuel-Injected Automotive Engines, *American Control Conf.* (1988).

**Chua, L. O. and Lin, G. N.**, Nonlinear Programming Without Computation, *I.E.E.E. Trans. Circuits and Systems*, 31, 2 (1984).

**Corless, M. J. and Leitmann, G.**, Continuous State Feedback Guaranteeing Uniform Ultimate Boundedness for Uncertain Dynamic Systems, *I.E.E.E. Trans. Aut. Contr.*, 26, 5, pp. 850-861 (1982).

**Craig, J.**, Adaptive Control of Manipulators through Repeated Trials, *American Control Conf.*, San Diego (1984).

**Craig, J. J., Hsu, P. and Sastry, S.**, Adaptive Control of Mechanical Manipulators, *I.E.E.E. Int. Conf. Robotics and Automation*, San Francisco (1986).

**Crouch, P.**, Spacecraft Attitude Control and Stabilization: Application of Geometric Control Theory to Rigid Body Models, *I.E.E.E. Trans. Autom. Control*, 29, 4 (1984).

**De Kleer, J.**, Qualitative and Quantitative Knowledge in Classical Mechanics, *Artificial Intelligence Laboratory, Massachusetts Institute of Technology*, 352 (1975).

**Denavit, J. and Hartenberg, R. S.**, A Kinematic Notation for Lower Pair Mechanisms Bases On Matrices, *J. Applied Mechanics 22* (1955).

**Desa, S. and Roth, B.**, Synthesis of Control Systems for Manipulators Using Multivariable Robust Servomechanism Theory, *Int. J. Robotics Research*, 4, 3 (1985).

**Desoer, C.A., Liu, R. and Auth, Jr., L.V.**, Linearity vs. Nonlinearity and Asymptotic Stability in the Large. *I.R.E. Transactions on Circuit Theory*, vol. 12:117-118 (1965).

**Desoer, C. A. and Vidyasagar, M.**, Feedback Systems: Input Output Properties, *Academic Press*

(1975).

**Di Benedetto, M. D., Grizzle, J. W., and Moog, C. H.**, Rank Invariants of Nonlinear Systems, *S.I.A.M. J. Contr. Optimiz.*, to appear.

**Di Benedetto, M. D. and Isidori, A.**, The Matching of Nonlinear Models via Dynamic State Feedback, *S.I.A.M. J. Contr. Optimiz.* 24, pp. 1063-1075 (1986).

**Drake, S. H.**, Using Compliance In Lieu of Sensory Feedback for Automatic Assembly, *Doctoral Dissertation, Department of Mechanical Engineering, Massachusetts Institute of Technology* (1977).

**Drakunov, S. V.**, An Adaptive Quasi-Optimal Filter with Discontinuous Parameters, *Avtomatika i Telemekhanika*, 9, pp. 76-86 (1983).

**Duarte, M. A. and Narendra, K. S.**, Combined Direct and Indirect Adaptive Control of Plants with Relative Degree Greater than One, Technical Report No. 8715, *Center for Systems Science*, Yale University, New Haven, CT (1987).

**Duarte, M. A. and Narendra, K. S.**, Combined Direct and Indirect Adaptive Control of Plants with Relative Degree Greater than One, *Automatica*, 25, 3 (1989).

**Dubowsky, S. and DesForges, D. T.**, The Application of Model-Referenced Adaptive Control to Robotic Manipulators, *J. Dynamic Systems, Measurement, Control*, 101 (1979).

**Dwyer, T. A. W.**, Exact Nonlinear Control of Large Angle Rotational Maneuvers, *I.E.E.E. Trans. Autom. Control*, 29 (1984)

**Dwyer, T. A. W. and Batten, A. L.**, Exact Spacecraft Detumbling and Reorientation Maneuvers with Gimbaled Thrusters and Reaction Wheels, *A.A.S. J. of the Astronautical Sciences*, 3 (1985).

**Dwyer, T. A .W., Fadali, M. S. and Ning Chen**, Single Step Optimization of Feedback - Decoupled Spacecraft Attitude Maneuvers, *I.E.E.E. 24th Conf. Dec. Control*, Fort Lauderdale, FL (1985).

**Dwyer, T. A. W. and Kim, J.**, Bandwidth-Limited Robust Nonlinear Sliding Control of Pointing and Tracking Maneuvers, *American Control Conf.* (1989).

**Emelyanov, S. V.**, A Technique to Develop Complex Control Equations by Using Only the Error Signal of Control Variable and Its First Derivative, *Avtomatika i Telemekhanika*, 18, 10, pp. 873-885 (1957).

**Erdmann, M. A.**, Using Backprojections for Fine Motion Planning with Uncertainty, *I.E.E.E. Int. Conf. Robotics and Automation*, St. Louis (1985).

**Featherstone, R.**, The Calculation of Robot Dynamics Using Articulated-Body Inertias, *Int.*

*J. Robotics Research*, 2, 1 (1983a).

**Featherstone, R.**, Position and Velocity Transformations Between Robot End Effector Coordinates and Joint Angles, *Int. J. Robotics Research*, 2, 1 (1983b).

**Fernandez, B. R. and Hedrick, J. K.**, Control of Multivariable Nonlinear Systems by the Sliding Mode Method, *Int. J. Contr.*, 46, 3, pp. 1019-1040 (1987).

**Feynman, R. P., Leighton R. B. and Sands, M.**, The Feynman Lectures On Physics, *Addison-Wesley* (1963).

**Filippov, A. F.**, Differential equations with discontinuous right-hand sides, Mathematicheskii Sbornik, **51**, 1 (1960), in Russian. Translated in English, *Am. Math. Soc. Trans.*, **62**, 199 (1964).

**Fliess, M.**, A Note on the Invertibility of Nonlinear Input-Output Differential Systems, *Syst. Contr. Lett.* 8, pp. 147-151 (1986a).

**Flugge-Lotz, I., Taylor, C. E. and Lindberg, H. E.**, Investigation of Nonlinear Control Systems, *NACA*, ref. 1391 (1958).

**Freund, E.**, Decoupling and Pole Assignment In Nonlinear Systems, *Electronics Letter*, 9, 16 (1973).

**Freund, E.**, The Structure of Decoupled Nonlinear Systems, *Int. J. Contr.*, 21, pp. 443-450 (1975).

**Gelb, A. and Vander Velde, W. E.**, Multiple-Input Describing Functions and Nonlinear System Design, *McGraw-Hill* (1968).

**Gilbert, E. G. and Ha, I. J.**, An Approach to Nonlinear Feedback Control With Applications to Robotics, *I.E.E.E. Conf. Decision and Control*, San Antonio (1983).

**Goldstein, H.** Classical Mechanics, *Addison-Wesley* (1981).

**Golla, D. F., Garg, S. C. and Hughes, P. C.**, Linear State-Feedback Control of Manipulators, *Mech. Machine Theory*, 16 (1981).

**Goodwin, G. C., and Sin, K. S.**, Adaptive Filtering, Prediction, and Control, *Prentice Hall* (1984).

**Graham, D. and D. McRuer**, Analysis of Nonlinear Control Systems, *John Wiley & Sons* (1961).

**Grizzle, J. W. and Isidori, A.**, Block Noninteracting Control with Stability via Static State Feedback, *Math. Contr. Sign. Syst.* (1989).

**Guckenheimer, J. and Holmes, P.**, Nonlinear Oscillations, Dynamical Systems, and Bifurcations of Vector Fields, *Springer-Verlag* (1983).

**Gutman, S.**, Uncertain Dynamical Systems - A Lyapunov Min-Max Approach, *I.E.E.E. Trans. Autom. Control*, 24, pp. 437-443 (1979).

**Gutman, S. and Palmor, Z.** Properties of Min-Max Controller in Uncertain Dyanical Systems, *S.I.A.M. J. Contr. Optimization*, 20, 6, pp. 850-861 (1982).

**Ha, I. J. and Gilbert, E. G.**, Robust Tracking in Nonlinear Systems and its Applications to Robotics, *I.E.E.E. Conf. Decision and Control*, Fort Lauderdale (1985).

**Ha, I. J. and Gilbert, E. G.**, A Complete Characterization of Decoupling Control Laws for a General Class of Nonlinear Systems, *I.E.E.E. Trans. Aut. Contr.*, 31, pp. 823-830 (1986).

**Hahn, W.**, Theory and Application of Lyapunov's Direct Method, *Prentice Hall* (1963).

**Hahn, W.**, Stability of Motion, *Springer Verlag* (1967).

**Hale, J. K.**, Ordinary Differential Equations, *Kreiger, Molaban*, FL (1980); originally published, *Wiley (Interscience)*, New York (1969).

**Hanafusa, H. and Asada, H.**, Mechanics of Gripping Form by Artificial Fingers, *Trans. Society of Instrument and Control Engineers*, 12, 5 (1976).

**Hanafusa, H. and Asada, H.**, A Robotic Hand With Elastic Fingers and its Application to Assembly Process, *I.F.A.C. Symp. Information and Control Problems in Manufacturing Technology*, Tokyo (1977a).

**Hanafusa, H. and Asada, H.**, Stable Prehension of Objects by the Robot Hand with Elastic Fingers, *7th Int. Symp. Industrial Robots*, Tokyo (1977b).

**Hanafusa, H., Yoshikawa, T., and Nakamura, Y.**, Analysis and Control of Articulated Robot Arms with Redundancy, *Prep. 8th I.F.A.C. World Congress* (1981).

**Hauser, J. E.**, Approximate Tracking for Nonlinear Systems with Application to Flight Control, *Doctoral thesis, College of Engineering, University of California, Berkeley* (1989).

**Hedrick, J. K. and Gopalswamy, S.**, Nonlinear Flight Control Design via Sliding Methods, *Dept. of Mechanical Engineering, Univ. of California, Berkeley* (1989).

**Hermann, R. and Krener, A. J.**, Nonlinear Controllability and Observability, *I.E.E.E. Trans. Aut. Contr.*, 22, pp. 728-740 (1977).

**Hilbert, D., and Cohn-Vossen, S.**, Geometry and The Imagination, *Chelsea* (1952).

**Hirschorn, R. M.**, Invertibility of Multivariable Control Systems, *I.E.E.E. Trans. Autom. Control*, 24, pp. 855-865 (1979).

**Hirschorn, R. M.**, Output Tracking in Multivariable Nonlinear Systems, *I.E.E.E. Trans. Autom. Control*, 26, pp. 593-595 (1981).

**Hirschorn, R. M.**, Output Tracking for Nonlinear Systems with Singular Points, *S.I.A.M. J. Control & Optimization*, 25, pp. 547-557 (1987).

**Ho, J. Y.**, Direct Path Method for Flexible Multibody Spacecraft Dynamics, *A.I.A.A. J. Spacecraft and Rockets*, 14, 2 (1977).

**Hogan, N.**, Impedance Control of a Robotic Manipulator, *Winter Ann. Meeting of the A.S.M.E.*, Washington, D.C. (1981).

**Hogan, N.**, Impedance Control: An Approach to Manipulation, *J. Dyn. Syst., Meas., and Control*, 107, 1, pp. 1-7 (1985a).

**Hogan, N.**, Control Strategies for Computer Movements Derived from Physical Systems Theory, *Int. Symp. Synergetics*, Bavaria (1985b).

**Hollerbach, J. M.**, A Recursive Formulation of Lagrangian Manipulator Dynamics, *I.E.E.E. Trans. Systems, Man, Cybernetics*, **10**, 11 (1980).

**Hollerbach, J. M.**, Wrist-Partitioned Inverse Kinematic Accelerations and Manipulator Dynamics, *Int. J. Robotics Res.*, 2, 4 (1983).

**Horn, B. K. P.**, Robot Vision, *McGraw Hill* (1986).

**Hsia, T. C.**, Adaptive Control of Robot Manipulators - A Review, *I.E.E.E. Int. Conf. Robotics and Automation*, San Francisco (1986).

**Hsu, J. C. and Meyer, A. U.**, Modern Control Principles and Applications, *McGraw-Hill* (1968).

**Hsu, P., Sastry, S., Bodson, M. and Paden, B.**, Adaptive Identification and Control of Manipulators Without Joint Acceleration Measurements, *I.E.E.E. Int. Conf. Robotics and Automation*, Raleigh, NC (1987).

**Hughes, P. C.**, Dynamics of a Flexible Manipulator Arm for the Space Shuttle, *Amer. Astronautical Soc. and Amer. Inst. of Aeronautics and Astrodynamics Specialist Conf.*, Jackson Hole, Wyoming (1977).

**Hughes, P.C.**, Dynamics of a Chain of Flexible Bodies, *J. Astronautical Sciences*, 27, 4 (1979).

**Hunt, L. R., Su, R., and Meyer, G.**, Global Transformations of Nonlinear Systems, *I.E.E.E. Trans. Aut. Contr.* 28, pp. 24-31 (1983b).

**Inoue, H.**, Force Feedback in Precise Assembly Tasks, *Artificial Intelligence Laboratory,*

*Massachusetts Institute of Technology*, 308 (1977).

**Ioannou, P. A. and Kokotovic, P. V.**, Instability Analysis and Improvement of Robustness of Adaptive Control, *Automatica*, 20, 5, pp. 583-594 (1984).

**Ioannou, P. A. and Kokotovic, P. V.**, Adaptive Systems with Reduced Models, *Springer-Verlag, 1983*

**Isidori, A.**, Nonlinear Control Systems:  An Introduction, *Springer Verlag* (1989).

**Isidori, A., De Luca, A. and Nicolo, F.**, Control of Robot Arms with Elastic Joints via Nonlinear Dynamic Feedback, *I.E.E.E. Conf. Decision and Control*, Fort Lauderdale (1985).

**Isidori, A. and Grizzle, J. W.**, Fixed Modes and Nonlinear Noninteracting Control with Stability, *I.E.E.E. Trans. Aut. Contr.*, 33, pp. 907-914 (1988).

**Isidori, A., Krener, A. J., Giorgi, C. G. and Monaco, S.**, Nonlinear Decoupling via Feedback:  A Differential Geometric Approach, *I.E.E.E. Trans. Aut. Contr.* ,26, pp. 331-345 (1981).

**Isidori, A., and Moog, C.**, On the Nonlinear Equivalent of the Notion of Transmission Zeros, *I.I.A.S.A. Conf. Modeling and Adaptive Control*, Hungary (1986).

**Isidori, A. and Moog, C.**, On the Nonlinear Equivalent of the Notion of Transmission Zeros, *Modelling and Adaptive Control*, C. I. Byrnes and A. Kurzhanski eds., *Springer Verlag:  Lec. notes Contr. Info. Scie.*, 105, pp. 445-471 (1988).

**Isidori, A., and Ruberti, A.**, On the Synthesis of Linear Input-Output Responses for Nonlinear Systems, *Syst. Contr. Lett.* 4, pp. 17-22 (1984).

**Jakubczyk, B. and Respondek, W.**, On Linearization of Control Systems, *Bull. Acad. Polonaise Sci. Ser. Sci. Math.*, 28, pp. 517-522 (1980).

**Jonckheere, E. A.**, Lagrangian Theory of Large Scale Systems, *European Conf. Circuit Theory and Design, The Hague, Netherlands* (1981).

**Junkins, J. L. and Turner, J. D.**, Optimal Spacecraft Rotational Maneuvers, *Elsevier* (1986).

**Kane, T. R., and Levinson, D. A.**, The Use of Kane's Dynamic Equations in Robotics, *Int. J. Robotics Research*, 2, 3 (1983).

**Kanellakopoulos, I., Kokotovic, P. V. and Marino, R.**, Robustness of Adaptive Nonlinear Control Under an Extended Matching Condition, *I.F.A.C. NOLCOS* preprint, pp. 192-197 (1989).

**Kahn, M. E.**, The Near-Minimum-Time Control of Open-loop Articulated Kinematic Chains, *Artificial Intelligence Laboratory, Stanford University*, 106 (1969).

**Kahn, M. E. and Roth, B.**, The Near Minimum-Time Control of Open-Loop Articulated Kinematic Chains, *J. Dynamic Systems, Measurement, Control*, 93 (1971).

**Kalman, R. E. and Bertram, J. E.**, Control System Analysis and Design via the "Second Method" of Lyapunov I: Continuous-Time Systems, *Journal of Basic Engineering*, pp. 394-400 (1960).

**Kanade, T., Khosla, P. K. and Tanaka, N.**, Real-Time Control of CMU Direct-Drive Arm II Using Customized Inverse Dynamics, *I.E.E.E. Conf. On Decision and Control*, Las Vegas (1984).

**Kane, T. R.**, Dynamics of Nonholonomic Systems, *J. Applied Mechanics*, 28 (1961).

**Kazerooni, H.**, A Robust Design Method for Impedance Control of Constrained Dynamic Systems, *Doctoral Dissertation, Massachusetts Institute of Technology* (1985).

**Kelly, R., and Carelli, R.**, Unified Approach to Adaptive Control of Robotic Manipulators, *Proc. 27th Conf. on Dec. and Contr.*, pp. 1598-1603 (1988).

**Khatib, O.**, Commande Dynamique dans L'Espace Operationnel des Robots Manipulateurs en Presence D'obstacles, *Docteur Ingenieur Thesis, Ecole Nationale Superieure de L'Aeronautique et de L'Espace*, Toulouse, France (1980).

**Khatib, O.**, Dynamic Control of Manipulators in Operational Space, *Sixth IFTOMM Congress on Theory of Machines and Mechanisms, New-Delhi* (1983).

**Khorasani, K. and Spong, M. W.**, Invariant Manifolds and Their Application to Robot Manipulators With Flexible Joints, *I.E.E.E. Int. Conf. Robotics and Automation*, St. Louis (1985).

**Khosla, P. K. and Kanade, T.**, Parameter Identification of Robot Dynamics, *I.E.E.E. Conf. Decision and Control*, Fort Lauderdale (1985).

**Khosla, P. K., and Kanade, T.**, An Algorithm to Determine the Identifiable Parameters in the Dynamic Robot Model, *I.E.E.E. Int. Conf. Robotics and Automation*, San Francisco (1986).

**Koditschek, D. E.**, Natural Motion of Robot Arms. *I.E.E.E. Conf. on Dec. and Contr.*, Las Vegas, NV (1984).

**Koditschek, D. E.**, Adaptive Techniques for Mechanical Systems, *Proc. of Fifth Yale Workshop on the Applications of Adaptive Systems Theory* (1987).

**Koivo, A. J. and Guo, T. K.**, Control of Robotic Manipulator With Adaptive Control, *I.E.E.E. Conf. Decision and Control* (1981).

**Kokotovic, P. V.**, Applications of Singular Perturbation Techniques to Control Problems, *S.I.A.M. Rev.*, 26, pp. 501-550 (1984).

**Kokotovic, P. V.**, Control Theory in the 80's: Trends in Feedback Design, *Automatica*, 21, pp. 225-236 (1985).

**Kokotovic, P. V., Khalil, H. K. and O'Reilly, J.**, Singular Perturbation Methods in Control: Analysis and Design, *Academic Press* (1986).

**Kokotovic, P. V., Riedle, B. D. and Praly, L.**, On a Stability Criterion for Slow Adaptation, *Systems and Contr. Letters*, 6, pp. 7-14 (1985).

**Kokotovic, P. V. and Sussmann, H. J.**, Positive Real Lemma and the Global Stabilization of Nonlinear Systems, *Syst. and Contr. Lett.*, 13, pp. 125-133 (1989).

**Krasovskii, N. N.**, The Inversion of the Theorems of Liapunov's Second Method and the Question of Stability of Motion Using the First Approximation, *Prikl. Mat. Meh.*, 20, pp. 255-265 (1956).

**Krasovskii, N. N.**, Problems of the Theory of Stability of Motion, *Stanford Univ. Press*, Stanford, CA (1963); translation of the Russian edition, Moskow (1959).

**Krener, A. J. and Isidori, A.**, Linearization by Output Injection and Nonlinear Observers, *Syst. Contr. Lett.*, 3, pp. 47-52 (1983).

**Krener, A. J., Isidori, A. and Respondek, W.**, Partial and Robust Linearization by Feedback, *I.E.E.E. Conf. Dec. Contr.*, 22, pp. 126-130 (1983).

**Krener, A. J. and Respondek, W.**, Nonlinear Observers with Linearizable Error Dynamics *S.I.A.M. J. Contr. Optimiz.*, 23, pp. 197-216 (1985).

**Landau, Y. D.**, Adaptive Control, *Dekker* (1979).

**Landau, I. D.**, Adaptive Control Techniques for Robot Manipulators – The Status of the Art, *I.F.A.C. Symp. Robot Control*, Barcelona (1985).

**Landau, I., and Horowitz, R.**, Synthesis of Adaptive Controllers for Robot Manipulators Using a Passive Feedback Systems Approach, *I.E.E.E. Int. Conf. Robotics and Automation* (1988).

**La Salle, J. and Lefschetz, S.**, Stability by Liapunov's Direct Method, *Academic Press* (1961).

**Lathrop, R.M.**, Parallelism In Manipulator Dynamics, *I.E.E.E. Conf. On Robotics and Automation*, St. Louis (1985).

**Lefschetz, S.**, Stability of Nonlinear Control Systems, *Academic Press* (1962).

**Li, W., and Slotine, J.-J. E.**, Parameter Estimation Strategies for Robotic Applications., *A.S.M.E. Winter Annual Meeting* (1987).

**Li, W. and Slotine, J.-J. E.**, Indirect Adaptive Robot Control, *System and Control Letters* (1989).

**Lions, J. L.**, Some Methods in the Mathematical Analysis of Systems and Their Control, *Gordon and Breach*, New-York (1981).

**Ljung, L.**, Systems Identification: Theory for the User, *Prentice-Hall* (1987).

**Lozano-Perez, T.**, Robot Programming, *I.E.E.E. Proceedings* (1983).

**Luenberger, D. G.**, Introduction to Dynamic Systems, *J. Wiley & Sons*, New York (1979).

**Luh, J. Y. S., Walker, M. W. and Paul, R. P. C.**, On-Line Computational Scheme for Mechanical Manipulators *A.S.M.E. J. Dyn. Syst. Meas. Contr.*, 102 (1980a).

**Luh, J. Y. S., Walker, M. H. and Paul, R. P. C.**, Resolved Acceleration Control of Mechanical Manipulators, *I.E.E.E. Trans. Automatic Control*, 25, 3 (1980b).

**Lyapunov, A. M.**, The General Problem of Motion Stability, (1892), in Russian. Translated in French, *Ann. Fac. Sci. Toulouse* 9, pp. 203-474 (1907). Reprinted in Ann. Math. Study No. 17, *Princeton Univ. Press* (1949).

**Marino, R.**, High-gain Feedback in Nonlinear Control Systems, *Int. J. Control*, 42, pp. 1369-1385 (1985).

**Marino, R.**, On the Largest Feedback Linearizable Subsystem, *Syst. Contr. Lett.*, 7, pp. 345-351 (1986).

**Marino, R. and Kokotovic, P. V.**, A Geometric Approach to Nonlinear Singularly Perturbed Control Systems, *Automatica*, 24, pp. 31-41 (1988).

**Marino, R. and Nicosia, S.**, On the Feedback Control of Industrial Robots with Elastic Joints: A Singular Perturbation Approach, *University of Rome*, R-84.01 (1984).

**Marino, R., and Spong, M. W.**, Nonlinear Control Techniques for Flexible Joint Manipulators: A Single Link case Study, *I.E.E.E. Conf. Rob. Autom.* (1986).

**Mason, M. T.**, Compliance and Force Control for Computer Controlled Manipulators, *I.E.E.E. Trans. Systems, Man and Cybernetics*, 11, 6 (1981).

**Mason, M. T.**, Manipulator Grasping and Pushing Operations, *Doctoral Dissertation, Massachusetts Institute of Technology* (1982).

**Mason, M. T. and Salisbury, J. K.**, Robot Hands and the Mechanics of Manipulation, *M.I.T. Press* (1985).

**Massera, J. L.**, On Liapunov's Conditions of Stability, *Ann. of Math.*, 50, pp. 705-721 (1949).

**Massera, J. L.**, Contributions to Stability Theory, *Ann. of Math.*, 64, pp. 182-206 (1956), Erratum *Ann. of Math.*, 68, p. 202 (1958).

**Meerkov, S. M.**, Principle of Vibrational Control: Theory and Applications, *I.E.E.E. Trans. on Aut. Contr.*, 25, pp. 755-762 (1980).

**Meyer, G. and Cicolani, L.**, Application of Nonlinear System Inverses to Automatic Flight Control Design, *Theory and Application of Optimal Control in Aerospace Systems*, P. Kant ed., *NATO AGARD - AG251*, pp. 10.1-10.29 (1980).

**Middleton, R. H. and Goodwin, G. C.**, Adaptive Computed Torque Control for Rigid Link Manipulators, *I.E.E.E. Conf. on Dec. and Contr.*, Athens, Greece (1986).

**Middleton, R. H.**, Hybrid Adaptive Control of Robot Manipulators, *I.E.E.E. Conf. on Dec. and Contr.*, Austin, TX (1988).

**Misawa, E. A. and Hedrick, J. K.**, Nonlinear Observers: A State-of-the-art Survey, *A.S.M.E. J. Dyn. Syst. and Meas.*, 111, 3, pp. 344-352 (1989).

**Monaco, S. and Stornelli, S.**, A Nonlinear Attitude Control Law for a Satellite with Flexible Appendages, *I.E.E.E. 24th Conf. Dec. Control*, Fort Lauderdale, FL (1985).

**Moog, C. H.**, Nonlinear Decoupling and Structure at Infinity, *Mathematics of Control, Signals and Systems*, 1 pp. 257-268 (1988).

**Morgan, A. P. and Narendra, K. S.**, On the Uniform Asymptotic Stability of Certain Linear Nonautonomous Differential Equations, *S.I.A.M. J. of Control and Optimization*, 15 (1977).

**Nakamura, Y. and Hanafusa, H.**, Task Priority Based Redundancy Control of Robot Manipulators, *Int. Symp. of Robotics Research*, Kyoto (1984).

**Nam, K. and Arapostathis, A.**, A Model-Reference Adaptive Control Scheme fo r Pure-Feedback Nonlinear Systems, *I.E.E.E. Trans. Aut. Contr.*, 33, pp. 1123-1131 (1989).

**Narendra, K. S. and Annasswamy, A. M.**, Stable Adaptive Systems, *Prentice Hall* (1989).

**Narendra, K. S. and Goldwyn, R. M.**, A Geometrical Criterion for the Stability of Certain Nonlinear Nonautonomous Systems, *I.E.E.E. Trans. of the Circuit Theory Group*, CT-11 (3) (1964).

**Narendra, Kumpati S. and Annaswamy, Anuradha M.**, Stable Adaptive Systems, *Prentice Hall* (1989).

**Narendra, K. S. and Taylor, J. H.**, Frequency Domain Criteria for Absolute Stability, *Academic*

*Press* (1973).

**Nevins, J. L. and Whitney, D. E.** Computer Controlled Assembly, *Scientific American*, 238, 2 (1978).

**Nicosia, S., Nicolo, F. and Lentinit, D.**, Dynamical Control of Industrial Robots With Elastic and Dissipative Joints, *8th Triennial I.F.A.C. World Congress*, Kyoto, Japan (1981).

**Niemeyer, G., and Slotine, J.-J. E.**, Performance in Adaptive Manipulator Control, *Proc. 27th I.E.E.E. Conf. Decision and Control*, Austin, TX (1988).

**Niemeyer, G., and Slotine, J.-J. E.**, Computational Algorithms for Adaptive Compliant Motion, *I.E.E.E. Int. Conf. Robotics and Automation* (1989).

**Nijmeijer, H.**, Controlled Invariance for Affine Control Systems, *Intr. J. Contr.*, 2, pp. 824-833 (1981).

**Nijmeijer, H.**, Controllability Distributions for Nonlinear Systems, *Syst. Contr. Lett.*, 2, pp. 122-129 (1982).

**Nijmeijer, H. and Schumacher, J. M.**, Zeros at Infinity for Affine Nonlinear Systems, *I.E.E.E. Trans. Aut. Contr.*, 30, pp. 566-573 (1986).

**Nijmeijer, H. and Van der Schaft, A. J.**, Controlled Invariance for Nonlinear Systems, *I.E.E.E. Trans. Aut. Contr.*, 27, pp. 904-914 (1982).

**Nijmeijer, H. and Van der Schaft, A.J.**, Nonlinear Dynamical Control Systems, *Springer-Verlag* (1990).

**Nilsson, N.**, Principles of Artificial Intelligence, *Tioga Publishing*, California (1980).

**Ohwovoriole, M. S. and Roth, B.**, A Theory of Parts Mating for Assembly Automation, *Ro. Man. Sy. -81*, Warsaw, Poland (1981).

**Ogata, K.**, Modern Control Engineering, *Prentice-Hall* (1970).

**Okada, T.**, Computer Control of Multi-Jointed Finger System, *6th Int. Joint Conf. on Artificial Intelligence*, Tokyo (1979).

**Orin, D. E. and Schrader, W. W.**, Efficient Computation of the Jacobian for Robot Manipulators, *Int. J. Robotics Research* (1984).

**Ortega, R., and Spong, M.**, Adaptive Motion Control of Rigid Robots: A Tutorial, *I.E.E.E. Int. Conf. Decision and Control* (1988).

**Paden, B. and Riedle, B.**, A Positive-Real Modification of a Class of Nonlinear Controllers for

Robot Manipulators, *Proc. American Control Conf.* (1988).

**Paden, B. and Panja, R.**, A Globally Asymptotically Stable 'PD+' Controller for Robot Manipulators, *Int. J. Contr.*, 47, 6 (1988).

**Parks, P. D.**, Lyapunov Redesign of Model Reference Adaptive Control Systems, *I.E.E.E. Trans. on Aut. Contr.*, 11, p. 362 (1966).

**Paul, R. P. and Shimano, B.**, Compliance and Control, *1976 Joint Automatic Control Conf.*, San Francisco (1976).

**Phillips, C. L. and Harbor, R. D.**, Feedback Control Systems, *Prentice-Hall* (1988).

**Pieper, D. L.**, The Kinematics of Manipulators Under Computer Control, *Doctoral Dissertation, Stanford University* (1968).

**Pieper, D. L. and Roth, B.**, The Kinematics of Manipulators Under Computer Control, *2nd Int. Conf. Theory of Machines and Mechanisms, Warsaw* (1969).

**Pomet, J. B. and Praly, L.**, Adaptive Nonlinear Regulation: Equation Error from the Lyapunov Equation, *Proc. 28th I.E.E.E. CDC*, pp. 1008-1013 (1989).

**Popov, V. M.**, Absolute Stability of Nonlinear Control Systems of Automatic Control, *Automation and Remote Control*, 22 (1962).

**Popov, V. M.**, Hyperstability of Automatic Control Systems. *Springer Verlag*, New York (1973).

**Praly, L.**, A Geometric Approach for the Local Analysis of a One Step Ahead Adaptive Controller, *Proc. 4th Yale Workshop on Applications of Adaptive Systems Theory*, Yale University, CT (1985).

**Praly, L., d'Andrea-Novel, B. and Coron, J. M.**, Lyapunov Design of Stabilizing Controllers, *Proc. I.E.E.E. Conf. Dec. and Contr.* (1989).

**Prigogine, I.**, Introduction to Thermodynamics of Irreversible Processes, 3rd Ed., *Wiley-InterScience*, New York (1967).

**Raibert, M. H.**, Analytical Equations vs. Table Look-Up for Manipulation: A Unifying Concept, *I.E.E.E. Conf. Decision and Control*, New Orleans (1977).

**Raibert, M. H.**, A Model for Sensory Motor Control and Learning, *Biological Cybernetics*, 29 (1978).

**Raibert, M. H.**, Legged Robots that Balance, *M.I.T. Press* (1986).

**Raibert, M. H. and Craig J. J.**, Hybrid Position/Force Control of Manipulators, *J. Dynamic*

*Systems, Measurement, Control*, 102 (1981).

**Raibert, M. H. and Horn, B. K. P.**, Manipulator Control Using the Configuration Space Method, *Industrial Robot*, 5, 2 (1978).

**Reboulet, C. and Champetier, C.**, A New Method for Linearizing Non-Linear Control Systems: The Pseudolinearization, *Int. J. Contr.*, 40, 4, pp. 631-638 (1984).

**Redding, D.C., and Adams, N.J.**, *A.I.A.A. Journal of Guidance, Control, and Dynamics*, 10, 1 (1987).

**Riedle, B. D. and Kokotovic**, Stability Analysis of an Adaptive System with Unmodeled Dynamics, *Int. J. Contr.*, 41, pp. 389-402 (1985a).

**Riedle, B. D. and Kokotovic, P. V.**, A Stability-Instability Boundary for Disturbance-Free Slow Adaptation and Unmodeled Dynamics, *I.E.E.E. Trans. on Aut. Contr.*, 30, pp. 1027-1030 (1985b).

**Rohrs, C. E., Valavani, L. S., Athans, M. and Stein, G.**, Robustness of Continuous-Time Adaptive Control Algorithms in the Presence of Unmodeled Dynamics, *I.E.E.E. Trans. on Aut. Contr.*, 30, pp. 881-889 (1985).

**Roth, B.**, Screws and Wrenches That Cannot Be Bought at a Hardware Store, *Int. Symp. of Robotics Research*, Bretton Woods (1983).

**Rouche, N., Habets, P. and Laloy, M.**, Stability Theory by Liapunov's Direct Method, *Springer Verlag* (1977).

**Rugh, W.J.**, Analytical Framework for Gain Scheduling, *I.E.E.E. Control Systems Magazine*, 11, 1 (1991).

**Sadegh, N., and Horowitz, R.**, Stability Analysis of an Adaptive Controller for Robotic Manipulators, *I.E.E.E. Int. Conf. Robotics and Automation*, Raleigh, NC (1987).

**Safonov, M. G. and Athans, M.**, Robustness and Computational Aspects of Nonlinear Stochastic Estimators and Regulators, *I.E.E.E. Trans. Aut. Contr.*, 26, 1, pp. 717-725 (1978).

**Sahar, J. and Hollerbach, J.**, Planning of Minimum-Time Trajectories for Robot Arms, *Artificial Intelligence Laboratory, Massachusetts Institute of Technology*, 804 (1984).

**Salisbury, J. K.**, Active Stiffness Control of a Manipulator in Cartesian Coordinates, *I.E.E.E. Conf. Decision and Control*, Albuquerque, New Mexico (1980).

**Salisbury, J. K.**, Kinematic and Force Analysis of Articulated Hands, *Doctoral Dissertation, Stanford University* (1982).

**Salisbury, J. K.**, Design and Control of an Articulated Hand, *Int. Symp. On Design and Systems*, Tokyo (1984).

**Salisbury, J. K., and Craig, J. J.**, Articulated Hands: Force Control and Kinematic Issues, *Int. J. Robotics Research*, 1, 1 (1982).

**Sandberg, I. W.**, A Frequency-Domain Condition for the Stability of Feedback Systems Containing a Single Time-Varying Nonlinear Element, *Bell Syst. Tech. J.*, 43, pt. 2, p. 1601 (1964).

**Sastry, S. and Bodson, M.**, Adaptive Control: Stability, Convergence, and Robustness, *Prentice-Hall* (1989).

**Sastry, S. S. and Isidori, A.**, Adaptive Control for Linearizable Systems, *I.E.E.E. Trans. Aut. Contr.* (1989).

**Scheinman V. C.**, Design of a Computer Controlled Manipulator, *Stanford Artificial Intelligence Laboratory*, 92 (1969).

**Schwartz, L.**, Cours d'Analyse, *Hermann*, Paris (1981).

**Shimano, B.**, The Kinematic Design and Force Control of Computer Controlled Manipulators, *Artificial Intelligence Laboratory, Stanford University* (1978).

**Shinners, M. W.**, Modern Control System Theory and Application, *Addison-Wesley* (1978).

**Siljak, D. D.**, Large Scale Dynamic Systems, *North-Holland* (1978).

**Silverman, L. M.**, Inversion of Multivariable Linear Systems, *I.E.E.E. Trans. Aut. Contr.*, 14, pp. 270-276 (1969).

**Simon, H. A.**, The Sciences of the Artificial, 2nd Ed., *M.I.T. Press* (1981).

**Singh, S. N.**, Decoupling of Invertible Nonlinear Systems with State Feedback and Precompensation, *I.E.E.E. Trans. Aut. Contr.* 25, pp. 1237-1239 (1980).

**Singh, S. N.**, A Modified Algorithm for Invertibility in Nonlinear Systems, *I.E.E.E. Trans. Autom. Control*, 26, pp.595 - 598 (1981).

**Singh, S. N.**, Generalized Functional Reproducibility Condition for Nonlinear Systems, *I.E.E.E. Trans. Autom. Control*, 27, pp. 958-960 (1982).

**Singh, S. N. and W. J. Rugh**, Decoupling in a Class of Nonlinear Systems by State Variable Feedback, *J. Dynamic Systems, Measurement, Control* (1972).

**Slotine, J.-J. E.**, Tracking Control of Nonlinear Systems using Sliding Surfaces, *Doctoral*

*Dissertation, Massachusetts Institute of Technology* (1983).

**Slotine, J.-J. E.** Sliding Controller Design for Nonlinear Systems, *Int. J. Control.*, 40, 2 (1984).

**Slotine, J.-J. E.** The Robust Control of Robot Manipulators, *Int. J. Robotics Research*, 4, 2 (1985).

**Slotine, J.-J. E. and Coetsee, J. A.**, Adaptive Sliding Controller Synthesis for Nonlinear Systems, *Int. J. Control*, 43, 6 (1986).

**Slotine, J.-J. E. and Di Benedetto, M. D.**, Hamiltonian Adaptive Control of Spacecraft, M.I.T.-N.S.L. Report 880105, Jan. (1988) .

**Slotine, J.-J. E. and Di Benedetto, M. D.**, Hamiltonian Adaptive Control of Spacecraft, *I.E.E.E. Trans. on Auto. Control*, 35-7 (1990) .

**Slotine, J.-J. E., Hedrick, J. K. and Misawa, E. A.**, On Sliding Observers, *A.S.M.E. J. Dyn. Syst. and Meas.*, 109, 3, pp. 245-252 (1986).

**Slotine, J.-J. E., and Li W.**, On The Adaptive Control of Robot Manipulators, in Robotics: Theory and Applications (ed. F.W. Paul and K. Youcef-Toumi) from *A.S.M.E. Winter Annual Meeting,* Anaheim, CA (1986).

**Slotine, J.-J. E., and Li W.**, Adaptive Robot Control, A Case Study, *I.E.E.E. Int. Conf. Robotics and Automation,* Raleigh, NC (1987a).

**Slotine, J.-J. E., and Li W.**, Adaptive Strategies in Constrained Manipulation, *I.E.E.E. Int. Conf. Robotics and Automation,* Raleigh, NC (1987b).

**Slotine, J. J. E. and Li W.**, Theoretical Issues in Adaptive Manipulator Control, *Proc. of Fifth Yale Workshop on Applications of Adaptive Systems Theory* (1987c).

**Slotine, J.-J. E. and Li W.**, Adaptive Robot Control - A New Perspective, *I.E.E.E. Conf. Decision and Control* L.A., CA (1987d).

**Slotine, J.-J. E., and Li W.**, On the Adaptive Control of Robot Manipulators, *Int. J. Robotics Res.*, 6, 3 (1987e).

**Slotine, J.-J. E., and Li W.**, Adaptive Manipulator Control: A Case Study, *I.E.E.E. Trans. on Auto. Control*, 33-11 (1988).

**Slotine, J.-J. E. and Li W.**, Composite Adaptive Control of Robot Manipulators, *Automatica*, 25, 4 (1989).

**Slotine, J.-J. E. and Sastry S. S.**, Tracking Control of Nonlinear Systems Using Sliding Surfaces With Applications to Robot Manipulators, *Int. J. Control*, 39, 2 (1983).

**Slotine, J.-J. E. and Spong, M. W.**, Robust Robot Control with Bounded Input Torques, *Int. J. Robotics Systems*, 2, 4 (1985).

**Slotine, J.-J. E., and Yoerger, D. R.**, Inverse Kinematics Algorithm for Redundant Manipulators, *Int. J. Robotics and Automation*, 1, 2 (1986).

**Sontag, E. and Sussmann, H.**, Remarks on Continuous Feedback, *I.E.E.E. Conf. Dec. Contr.*, 19, pp.916-921 (1980).

**Spong, M. W., and Vidyasagar M.**, Robust Nonlinear Control of Robot Manipulators, *I.E.E.E. Conf. Decision and Control*, Fort Lauderdale (1985).

**Spong, M. W., and Vidyasagar M.**, Robot Dynamics and Control, *John Wiley and Sons* (1989).

**Stepanenko, Y. and Vukobratovic, M.**, Dynamics of Articulated Open-Chained Active Mechanisms, *Mathematical Biosciences*, 28 (1976).

**Su, R.**, On the Linear Equivalents of Nonlinear Systems, *Syst. Contr. Lett.*, 2, pp. 48-52 (1982).

**Sussmann, H.**, Single Input Observability of Continuous Time Systems, *Math. Syst. Theory*, 12, pp. 371-393 (1979).

**Sussmann, H.**, Lie Brackets and Local Controllability, *S.I.A.M. J. Contr. Optimiz.*, 21, pp. 686-713 (1983).

**Sussmann, H.**, On a General Theorem on Local Controllability, *S.I.A.M. J. Contr. Optimiz.*, 25, pp. 158-194 (1987).

**Sussmann, H. J.**, Limitations to the Stabilizability of Globally Minimum Phase Systems, *I.E.E.E. Trans. Aut. Contr.*, 35, 1 (1990).

**Sweet, L. M. and Good, M. C.**, Redefinition of the Robot Control Problem: Effects of Plant Dynamics, Drive System Constraints, and User Requirements, *I.E.E.E. Conf. Decision and Control*, Las Vegas (1984).

**Takegaki, M. and Arimoto, S.**, A New Feedback Method for Dynamic Control of Manipulators, *J. Dynamic Systems, Measurement, Control*, 102 (1981).

**Taylor, D. G., Kokotovic, P. V., Marino, R. and Kanellakopoulos, I.**, Adaptive Regulation of Nonlinear Systems with Unmodeled Dynamics, *I.E.E.E. Trans. Aut. Contr.*, 34, pp. 405-412 (1989).

**Thau, F. E.**, Observing the State of Nonlinear Dynamic Systems, *Int. J. Contr.*, 17, 3, pp. 471-479 (1973).

**Thomson, W. and Tait, P.**, Treatise on Natural Philosophy, *University of Cambridge Press* (1886).

**Tsypkin, J.**, Theory of Relay Systems in Automatic Control, Moscow (1955).

**Tsypkin, J.**, *Aut. Remote Contr.*, 17 (1956).

**Uicker, J. J.**, On the Dynamic Analysis of Spatial Linkages Using 4 by 4 Matrices, *Doctoral Dissertation, Department of Mechanical Engineering and Astronautical Sciences*, Northwestern University (1965).

**Uicker, J. J., Denavit, J. and Hartenberg, R. S.**, An Iterative Method for the Displacement Analysis of Spatial Mechanisms, *J. Applied Math.*, 31 (1964).

**Utkin, V. I**, Equations of Sliding Mode in Discontinuous Systems I, II, *Automation and Remote Control* (1972).

**Utkin, V. I.**, Variable Structure Systems With Sliding Mode: A Survey, *I.E.E.E. Trans. Automatic Control*, 2 (1977).

**Utkin, V.I.**, Sliding Modes and Their Application to Variable Structure Systems, *MIR Publishers*, Moscow (1978).

**Vafa, Z. and Dubowsky, S.**, On the Dynamics of the Manipulators in Space Using the Virtual Manipulator Approach, *Proc. of the 1987 I.E.E.E. Int. Conf. on Robotics and Automation*, Raleigh, NC (1987).

**Van der Schaft, A. J.**, Observability and Controllability for Smooth Nonlinear Systems, *S.I.A.M. J. Contr. Optimiz.*, 20, pp. 338-354 (1982).

**Van der Schaft, A. J.**, Stabilization of Hamiltonian Systems , *Nonl. An. Theor. Meth. Appl.*, 10 (1986).

**Van der Schaft, A. J.**, On Clamped Dynamics of Nonlinear Systems, *Mem. n. 634*, University of Twente (1987).

**Van der Schaft, A. J.**, On Clamped Dynamics of Nonlinear Systems, *Analysis and Control of Nonlinear Systems*, C. I. Byrnes, C. F. Martin and R. E. Saeks eds., *North Holland*, pp. 499-506 (1988).

**Verghese, G., Fernandez, B. R. and Hedrick, J. K.**, Stable, Robust Tracking by Sliding Mode Control, *Systems and Contr. Lett.* (1988).

**Vidyasagar, M.**, Nonlinear System Analysis, *Prentice-Hall*, Englewood Cliffs, NJ (1978).

**Vidyasagar, M.**, Decomposition Techniques for Large-Scale Systems with Nonadditive

Interactions: Stability and Stabilizability, *I.E.E.E. Trans. Aut. Contr.*, 25, pp. 773-779 (1980a).

**Vidyasagar, M.**, On the Stabilization of Nonlinear Systems Using State Detection, *I.E.E.E. Trans. Aut. Contr.*, 25, 3, pp. 504-509 (1980b).

**Walcott, B. and Zak, S. H.**, State Observation of Nonlinear Uncertain Dynamical Systems, *I.E.E.E. Trans. Aut. Contr.*, 32, 2, pp. 166-169 (1987).

**Walcott, B. and Zak, S. H.**, Combined Observer-Controller Synthesis for Uncertain Dynamical Systems with Applications, *I.E.E.E. Trans. on System, Man, and Cyber.*, 18, 1, pp. 88-104 (1988).

**Walker, M. W. and Orin, D. E.**, Efficient Dynamic Computer Simulation of Robot Mechanisms, *A.S.M.E. J. Dynamic Systems, Measurement, Control*, 104 (1982).

**Walker, M.W.**, An Efficient Algorithm for the Adaptive Control of a Manipulator, *Proc. 5th Int. Conf. Robotics and Automation* (1988).

**Whitney, D. E.**, Resolved Motion Rate Control of Manipulators and Human Prostheses, *I.E.E.E. Trans. Man-Machine Systems*, 10 (1969).

**Whitney, D. E.**, Use of Resolved Rate to Generate Torque Histories for Arm Control, *C. S. Draper Laboratory*, MAT-54 (1972a).

**Whitney, D. E.**, The Mathematics of Coordinated Control of Prostheses and Manipulators, *J. Dynamics Systems, Measurement, Control* (1972b).

**Whitney, D. E.**, Historical Perspectives and State of the Art in Robot Force Control, *I.E.E.E. Int. Conf. Robotics and Automation, St. Louis* (1985).

**Wiener, N.**, Cybernetics, or Control and Communication in the Animal and the Machine, *M.I.T. Press* (1961).

**Winston, P. H.**, Artificial Intelligence, *Addison-Wesley* (1984).

**Wolovich, W. A.**, Linear Multivariable Systems, *Springer-Verlag*, New York (1974).

**Wonham, W. M.**, Linear Multivariable Control: A Geometric Approach, *Springer Verlag* (1979).

**Yoerger, D. R., Newman, J. B. and Slotine, J.-J. E.**, Supervisory Control System for the JASON ROV, *I.E.E.E. J. Oceanic Eng.*, OE-11, 3 (1986).

**Yoerger, D. R. and Slotine, J.-J. E.**, Robust Trajectory Control of Underwater Vehicles, *I.E.E.E. J. Oceanic Eng.*, 10, 4 (1985).

**Yoerger, D. R. and Slotine, J.-J. E.**, Task-Resolved Motion Control of Vehicle Manipulator

Systems, *Int. J. Robotics and Automation*, 1, 2 (1986).

**Yoshizawa, T.**, Stability Theory by Liapunov's Second Method, *The Mathematical Society of Japan* (1966).

**Yoshizawa, T.**, Stability Theory and the Existence of Periodic Solutions and Almost Periodic Solutions, *Springer Verlag* (1975).

**Young, K.-K. D.**, Controller Design for A Manipulator Using Theory of Variable Structure Systems, *I.E.E.E. Trans. Systems, Man, Cybernetics*, 8, 2 (1978).

**Young, K.-K. D., Kokotovic, P. K. and Utkin, V. I.**, A Singular Perturbation Analysis of High Gain Feedback Systems, *I.E.E.E. Trans. Automatic Control*, 22 (1977).

**Zames, G.**, On the Input-Output Stability of Time-Varying Nonlinear Feedback Systems, *I.E.E.E. Trans. on Aut. Contr.*, Part I: 11, 2, pp. 228-238, Part II: 11, 3, pp. 465-476 (1966).

# INDEX